Finite Element Techniques for Fluid Flow

Finite Element Techniques for Fluid Flow

J. J. CONNOR, SC.D.
Professor of Civil Engineering
Massachusetts Institute of Technology, USA

C. A. BREBBIA, PH.D.
Senior Lecturer in Civil Engineering
University of Southampton, UK

NEWNES–BUTTERWORTHS
LONDON BOSTON
Sydney Wellington Durban Toronto

THE BUTTERWORTH GROUP

UNITED KINGDOM
Butterworth & Co (Publishers) Ltd
London: 88 Kingsway, WC2B 6AB

AUSTRALIA
Butterworths Pty Ltd
Sydney: 586 Pacific Highway, NSW 2067
Also at Melbourne, Brisbane, Adelaide and Perth

CANADA
Butterworth & Co (Canada) Ltd
Toronto: 2265 Midland Avenue, Scarborough, Ontario M1P 4S1

NEW ZEALAND
Butterworths of New Zealand Ltd
Wellington: 26–28 Waring Taylor Street, 1

SOUTH AFRICA
Butterworth & Co (South Africa) (Pty) Ltd
Durban: 152–154 Gale Street

USA
Butterworth (Publishers) Inc
Boston: 19 Cummings Park, Woburn, Mass. 01801

First published in 1976

ISBN 0 408 00176 3

Filmset and printed in England by J. W. Arrowsmith Ltd, Bristol

Bound in England by The Pitman Press, Bath

Preface

Until recently, finite element techniques were almost exclusively used in structural engineering problems but now there is a growing awareness of their potential in other engineering fields, especially in fluid mechanics.

This book presents these recent advances in a simple way. The authors have been mainly concerned with producing a text for teaching which can be easily followed by the self-taught student. The last part will undoubtedly also be of use to research workers.

The reader will be led from the basic principles of Chapter 1, and the simple finite element concepts and models given in Chapters 2 and 3, step by step to more complex applications. A chapter (4) on the governing equations of fluid flow has been included to provide a more complete progression, though this is not intended for those already well versed in fluid dynamics. Chapter 5 is concerned with the solution of potential type problems and Chapter 6 sets out viscous flow problems in porous media; both are topics well suited to finite element solutions and of general interest to the engineer, applied mathematician and physicist.

In the remaining chapters the solutions of more specialised problems are presented. Chapter 7 describes how circulation problems can be tackled using finite elements, Chapter 8 deals with the solution of the mass transfer equation and Chapter 9 discusses ways of solving general transient incompressible flows.

Since this book contains more material than could be used in a standard course, the authors have also indicated in the contents some sections which could be omitted without affecting the general structure. Those students not interested in coastal engineering and transport type problems could in addition leave out Chapters 7 and 9.

Finally, the authors wish to thank all those who made this book possible, especially their research associates Dr. R. Adey, Mr. J. Rodenhuis, Dr. S. Smith and Dr. J. Wang.

The Authors
Southampton 1976

Contents

* Topics which may be omitted in an introductory course without affecting continuity.

1 Weighted Residual and Variational Methods

1.1 BASIC DEFINITIONS

We start by introducing some basic definitions and properties for a sequence of functions such as

$$\phi_1(x), \phi_2(x), \phi_3(x) \ldots \phi_n(x) \tag{1.1}$$

The functions are assumed to satisfy certain given conditions, called admissibility conditions, relating to the boundary conditions and the degree of continuity. We will study them in more detail in the following.

If the functions can be linearly combined, for instance,

$$\phi = \alpha\phi_1 + \beta\phi_2 \tag{1.2}$$

where α and β are numbers, they are called *elements* of a linear space R, and the following properties hold:

$$\begin{aligned} \phi_1 + \phi_2 &= \phi_2 + \phi_1 \\ (\alpha + \beta)\phi &= \alpha\phi + \beta\phi \\ \alpha(\phi_1 + \phi_2) &= \alpha\phi_1 + \alpha\phi_2 \end{aligned} \tag{1.3}$$

The *inner product* of two functions ϕ_1 and ϕ_2 is denoted by

$$\langle \phi_1, \phi_2 \rangle \tag{1.4}$$

and it represents an operation on ϕ_1 and ϕ_2, such as

$$\langle \phi_1, \phi_2 \rangle = \int_{x_1}^{x_2} \phi_1(x)\phi_2(x)\,\mathrm{d}x \tag{1.5a}$$

or

$$\langle \phi_1, \phi_2 \rangle_c = \int_0^t \phi_1(t - \tau)\phi_2(\tau)\,\mathrm{d}\tau \tag{1.5b}$$

The second definition is called the convolution. We will consider only the first type of product here.

For real functions, the inner product has the following properties:

$$\begin{aligned}
\langle \phi_1, \phi_2 \rangle &= \langle \phi_2, \phi_1 \rangle \\
\alpha\langle \phi_1, \phi_2 \rangle &= \langle \alpha\phi_1, \phi_2 \rangle \\
\langle \phi_1, \phi_2 + \phi_3 \rangle &= \langle \phi_1, \phi_2 \rangle + \langle \phi_1, \phi_3 \rangle \\
\langle \phi_1, \phi_1 \rangle &> 0 \quad \text{if } \phi_1 \neq 0 \\
&= 0 \quad \text{if } \phi_1 = 0
\end{aligned} \tag{1.6}$$

where $\phi_1 = 0$ is a 'null' function which exists in the space R.

A measure (*norm*) of the function ϕ can be taken as the square root of the inner product of ϕ by itself and is denoted by $\|\phi\|$.

$$\|\phi\| = \sqrt{\langle \phi, \phi \rangle} \tag{1.7}$$

A sequence of functions such as (1.1) is said to be *linearly independent* if

$$\alpha_1\phi_1 + \alpha_2\phi_2 + \ldots + \alpha_n\phi_n = 0 \tag{1.8}$$

only when all α_i are zero.

A sequence of linearly independent functions is said to be *complete* if a number N and a set of constants α_i can be found such that, given an admissible but otherwise arbitrary function u, we have

$$\left\| u - \sum_{i=1}^{N} \alpha_i\phi_i \right\| < \varepsilon \tag{1.9}$$

where ε is any small quantity.

The functions ϕ_i are called *basis functions* and the coefficients α_i are the *Fourier coefficients*.

If the normalised basis functions are mutually orthogonal,

$$\begin{aligned} \langle \phi_i, \phi_j \rangle &= 0 \quad \text{if } i \neq j \\ \langle \phi_i, \phi_i \rangle &= 1 \end{aligned} \tag{1.10}$$

Each additional term we take in the *linearly independent* and *complete* sequence ϕ_i will introduce a further α_i. For the Nth approximation, we have

$$u^{(N)} = \sum_{1}^{N} \alpha_i \phi_i$$

Thus

$$\|u^{(N)}\| \rightarrow \|u\| \quad \text{as} \quad N \rightarrow \infty \tag{1.11}$$

The norm of $u^{(N)}$ for a mutually orthogonal complete sequence (if the sequence is not orthogonal we will accept that we can always reduce it to an orthogonal one) is

$$\begin{aligned} \|u^{(N)}\| &= \sqrt{\left\{ \left\langle \sum_{i=1}^{N} \alpha_i \phi_i, \sum_{j=1}^{N} \alpha_j \phi_j \right\rangle \right\}} \\ &= \sqrt{\left\{ \sum_{i=1}^{N} \sum_{j=1}^{N} \alpha_i \alpha_j \langle \phi_i, \phi_j \rangle \right\}} \end{aligned} \tag{1.12}$$

and since $\langle \phi_i, \phi_j \rangle = 0$ if $i \neq j$, we have

$$\|u^{(N)}\| = \sqrt{\left\{ \sum_{i=1}^{N} \alpha_i^2 \langle \phi_i, \phi_i \rangle \right\}} \tag{1.13}$$

Each term in the summation is positive, thus $\|u^{(N)}\|$ approaches $\|u\|$ from below as N increases.

$$\|u^{(N)}\| \leqslant \|u^{(M)}\| \leqslant \|u\| \quad \text{with } N < M \tag{1.14}$$

An *operator* $\mathscr{L}(\ )$ is defined as a process which, when applied to a given function u, produces another function p:

$$\mathscr{L}(u) = p \tag{1.15}$$

An operator is *linear* if

$$\mathscr{L}(\alpha u_1 + \beta u_2) = \alpha \mathscr{L}(u_1) + \beta \mathscr{L}(u_2) \tag{1.16}$$

This definition is general but we will consider here only differential operators.

Properties analogous to symmetry and positive definiteness for a matrix can also be defined for an operator. Consider a square matrix, $\mathbf{a} = [a_{ij}]$. We say $\mathbf{a}$ is symmetrical when $\mathbf{a}^T = \mathbf{a}$, where $\mathbf{a}^T$ (the transpose of $\mathbf{a}$) is formed by interchanging the rows and columns of $\mathbf{a}$. Symmetry requires $a_{ij} \equiv a_{ji}$. Another way of defining symmetry is to require

$$\langle \mathbf{y}, \mathbf{ax} \rangle \equiv \langle \mathbf{x}, \mathbf{ay} \rangle \tag{a}$$

for arbitrary vectors $\mathbf{x}$ and $\mathbf{y}$. Expanding (a) and noting that $(\mathbf{bc})^T = \mathbf{c}^T\mathbf{b}^T$,

$$\langle \mathbf{y}, \mathbf{ax} \rangle = \mathbf{y}^T\mathbf{ax} \tag{b}$$

$$\langle \mathbf{x}, \mathbf{ay} \rangle = \mathbf{x}^T\mathbf{ay} = \mathbf{y}^T\mathbf{a}^T\mathbf{x} \tag{c}$$

shows that (a) is equivalent to $\mathbf{a}^T = \mathbf{a}$. The latter definition is more convenient for extension to operators. Positive definiteness is defined by

$$\langle \mathbf{x}, \mathbf{ax} \rangle \geqslant 0 \tag{d}$$

for all $\mathbf{x}$ and equals 0 only when $\mathbf{x}$ is a null vector. This property is extremely valuable in establishing solution schemes and also in constructing variational statements.

With this as background, let us consider the problem represented by a set of homogeneous equations in the interior of a domain, V

$$\mathscr{L}(u) = 0, \qquad x \in V \tag{1.17}$$

We form the inner product of $\mathscr{L}(u)$ with another function, say v. The matrix transpose operations in (b), (c) are equivalent here to integration by parts of $\langle \mathscr{L}(u), v \rangle$ until the derivatives of u are eliminated. This leads to the 'transposed' form of the inner product and also to boundary terms. We write the results as

$$\langle \mathscr{L}(u), v \rangle = \langle u, \mathscr{L}^*(v) \rangle + \int_S (F(v)G(u) - F(u)G^*(v))\,\mathrm{d}S \tag{1.18}$$

where S is the exterior surface and F, G are differential operators whose forms follow naturally from the integration by parts. By definition, $F(v)$ contains the v terms resulting from the first phase of the partial integration and $G(u)$ contains the corresponding u terms. Some examples which illustrate this operation are included below.

The operator $\mathscr{L}^*$ is called the *adjoint* of $\mathscr{L}$. If $\mathscr{L}^* = \mathscr{L}$, $\mathscr{L}$ is said to be self-adjoint. In this case, $G^* = G$ also. Self-adjointness of an operator is analogous to symmetry of a matrix. In addition to determining whether the operator is self-adjoint, the partial integration also generates two different categories of boundary conditions. The set $F(u)$ prescribed are called the *essential* boundary conditions and $G(u)$ prescribed are the *nonessential* or *natural* boundary conditions. One can specify either type of boundary condition on the surface of a domain. However, the essential boundary conditions must be enforced at some point in order for the solution to be unique. Letting S_1 and S_2 represent complementary portions of the total surface, S, we can state the boundary conditions for the self-adjoint problem ($\mathscr{L}^* = \mathscr{L}$) as:

$$\begin{aligned} &F(u) \text{ prescribed on } S_1 \\ &G(u) \text{ prescribed on } S_2 \\ &S_1 + S_2 = S \end{aligned} \tag{1.19}$$

The positive definite property of a self-adjoint operator is defined by the requirement that

$$\langle \mathscr{L}(u), u \rangle > 0 \tag{1.20}$$

for all nontrivial u which satisfy the homogeneous form of the boundary conditions. One determines whether $\mathscr{L}$ is positive definite by integrating the inner product until it contains only products of derivatives of the same order. This operation is the mid-point in the transformation of $\mathscr{L}$ into $\mathscr{L}^*$ (i.e. equation (1.18)).

Example 1.1

(i) Consider

$$\mathscr{L}(u) = \frac{d^2u}{dx^2} \qquad 0 < x < 1 \tag{a}$$

Forming the inner product and integrating yields

$$\begin{aligned} \int_0^1 v\mathscr{L}(u)\,dx &= \int_0^1 v\frac{d^2u}{dx^2}\,dx \\ &= \left| v\frac{du}{dx} \right|_0^1 - \int_0^1 \frac{dy}{dx}\frac{du}{dx}\,dx \\ &= \left| v\frac{du}{dx} - u\frac{dv}{dx} \right|_0^1 + \int_0^1 u\frac{d^2v}{dx^2}\,dx \end{aligned} \tag{b}$$

Using the notation of (1.18), we have

$$F(v) = v$$

$$G(u) = \mathrm{d}u/\mathrm{d}x \qquad G^*(v) = \mathrm{d}v/\mathrm{d}x \tag{c}$$

$$\mathscr{L}^* = \mathscr{L}$$

and the operator is self-adjoint. The essential boundary condition is u prescribed and the natural boundary condition is $\mathrm{d}u/\mathrm{d}x$ prescribed.

Referring back to (b), if we take $v = u$ and homogeneous boundary conditions, the first partial integration yields

$$\int_0^1 u\mathscr{L}(u)\,\mathrm{d}x = -\int_0^1 \left(\frac{\mathrm{d}u}{\mathrm{d}x}\right)^2 \mathrm{d}x \tag{d}$$

Then $\mathscr{L}(u) = \mathrm{d}^2u/\mathrm{d}x^2$ is *negative definite*.

(ii) We examine next a more general operator,

$$\mathscr{L}(u) = \frac{\mathrm{d}^2}{\mathrm{d}x^2}\left(a_1(x)\frac{\mathrm{d}^2u}{\mathrm{d}x^2}\right) + \frac{\mathrm{d}}{\mathrm{d}x}\left(a_2(x)\frac{\mathrm{d}u}{\mathrm{d}x}\right) + a_3(x)u \qquad 0 < x < 1 \tag{a}$$

The first partial integration operation results in

$$\int_0^1 v\mathscr{L}(u)\,\mathrm{d}x = \int_0^1 \left\{a_1\frac{\mathrm{d}^2u}{\mathrm{d}x^2}\frac{\mathrm{d}^2v}{\mathrm{d}x^2} - a_2\frac{\mathrm{d}u}{\mathrm{d}x}\frac{\mathrm{d}v}{\mathrm{d}x} + a_3uv\right\}\mathrm{d}x + \left| v\left\{\frac{\mathrm{d}}{\mathrm{d}x}\left(a_1\frac{\mathrm{d}^2u}{\mathrm{d}x^2}\right) + a_2\frac{\mathrm{d}u}{\mathrm{d}x}\right\} + \frac{\mathrm{d}v}{\mathrm{d}x}\left\{-a_1\frac{\mathrm{d}^2u}{\mathrm{d}x^2}\right\}\right|_0^1 \tag{b}$$

If we continue, we would find that $\mathscr{L}^* = \mathscr{L}$. The boundary terms follow from (b) and are summarised below.

Essential boundary conditions

$$\left.\begin{aligned} F_1(u) &= u \\ F_2(u) &= \frac{\mathrm{d}u}{\mathrm{d}x} \end{aligned}\right\}\text{prescribed} \tag{c}$$

Natural boundary conditions

$$\left.\begin{aligned} G_1(u) &= \frac{\mathrm{d}}{\mathrm{d}x}\left(a_1\frac{\mathrm{d}^2u}{\mathrm{d}x^2}\right) + a_2\frac{\mathrm{d}u}{\mathrm{d}x} \\ G_2(u) &= -a_1\frac{\mathrm{d}^2u}{\mathrm{d}x^2} \end{aligned}\right\}\text{prescribed} \tag{d}$$

Considering (b), we can write

$$\langle u\mathscr{L}(u)\rangle = \int_0^1 \left\{ a_1\left(\frac{d^2u}{dx^2}\right)^2 - a_2\left(\frac{du}{dx}\right)^2 + a_3u^2 \right\} dx \tag{e}$$

for u satisfying homogeneous boundary conditions. If a_1, a_3 are > 0 and $a_3 < 0$ in the interval $0 < x < 1$, the operator is obviously *positive definite.*

(iii) Operating on

$$\mathscr{L}(u) = \frac{d^2u}{dx^2} + \frac{du}{dx} + u \tag{a}$$

yields

$$\int_0^1 \left(\frac{d^2u}{dx^2} + \frac{du}{dx} + u\right) v\, dx = \int_0^1 \left(\frac{d^2v}{dx^2} - \frac{dv}{dx} + v\right) u\, dx + \left| v\left(\frac{du}{dx} + u\right) - u\left(\frac{dv}{dx}\right)\right|_0^1 \tag{b}$$

The operator is not self-adjoint due to the presence of the first derivative term. Odd-order derivatives will lead to skew-symmetric terms in $\mathscr{L}^*$ and G^*. The essential boundary condition is u prescribed. In this case, we take the nonessential (natural) boundary condition as du/dx prescribed.

1.2 WEIGHTED RESIDUAL METHODS

Weighted residual methods are numerical procedures for approximating the solution of a set of differential (or integral) equations of the form

$$\mathscr{L}(u_0) = p \qquad x \in V \tag{1.21}$$

with boundary conditions

$$\mathscr{S}(u_0) = g \qquad x \in S \tag{1.22}$$

where x represents the spatial coordinates x_1, x_2 and x_3; S is the external surface of the continuum; and u_0 is the exact solution. The function u_0 is approximated by a set of functions $\phi_k(x)$,

$$u = \sum_{k=1}^{N} \alpha_k \phi_k \tag{1.23}$$

where α_k are undetermined parameters and ϕ_k are linearly independent functions taken from a complete sequence.

We will initially require that these functions satisfy *all* the boundary conditions of the problem [equation (1.22)] and have the necessary degree of continuity as to make the left-hand side of (1.21) different from zero. A procedure for relaxing the boundary condition requirements is discussed in the next section.

Substitution of (1.23) into (1.21) produces an error function ε, which is called the *residual*:

$$\varepsilon = \mathscr{L}(u) - p \neq 0 \tag{1.24}$$

Note that ε is equal to zero for the exact solution. This error is forced to be zero, in an average sense, by setting weighted integrals of the residual equal to zero:

$$\langle \varepsilon, w_i \rangle = 0, \qquad i = 1, 2 \ldots N \tag{1.25}$$

where w_i is a set of weighting functions. In what follows, we first review a few of the weighted residual methods and then discuss the Galerkin method in greater detail.

(a) THE COLLOCATION METHOD

In this method, we satisfy the differential equations only at a set of chosen points. For a given approximating function

$$u = \sum_{k=1}^{N} \alpha_k \phi_k \tag{1.26}$$

we have

$$\varepsilon = \mathscr{L}(u) - p = \sum_{k=1}^{N} \alpha_k \mathscr{L}(\phi_k) - p \tag{1.27}$$

The parameters are determined by enforcing the condition $\varepsilon = 0$ at N points in the domain.

We can express these conditions in the same form as (1.25) by introducing a Dirac function $\Delta(x_i)$ such that $\Delta(x_i) = 0$ for x outside the interval $x_i \pm c$ and

$$\int_{x_i - c}^{x_i + c} \Delta(x_i)\,\mathrm{d}x = \int_{x_i - c}^{x_i + c} \Delta_i\,\mathrm{d}x = 1 \tag{1.28}$$

where c is a small value (for point collocation $c \to 0$). Then collocation is equivalent to

$$\langle \varepsilon, \Delta_i \rangle = \langle \mathscr{L}(u) - p, \Delta_i \rangle = 0, \qquad i = 1, 2 \ldots N \tag{1.29}$$

Example 1.2

Consider the following second-order equation, which applies in the domain $0 < x < 1$:

$$\mathscr{L}(u) = \frac{\mathrm{d}^2 u}{\mathrm{d}x^2} + u + x = 0 \tag{a}$$

with boundary conditions

$$u = 0 \quad \text{at } x = 0$$
$$u = 0 \quad \text{at } x = 1 \tag{b}$$

We propose as an approximating function

$$u = x(1 - x)(\alpha_1 + \alpha_2 x + \ldots) \tag{c}$$

which satisfies the boundary conditions for arbitrary α_i.

If only two terms in the approximation are taken,

$$u = x(1 - x)(\alpha_1 + \alpha_2 x) \tag{d}$$

the error is

$$\varepsilon = \mathscr{L}(u) - p = x + (-2 + x - x^2)\alpha_1 + (2 - 6x + x^2 - x^3)\alpha_2 \tag{e}$$

We choose $x = \frac{1}{4}$, $x = \frac{1}{2}$ as collocation points. This choice requires

$$\begin{bmatrix} \frac{29}{16} & -\frac{35}{64} \\ \frac{7}{4} & \frac{7}{8} \end{bmatrix} \begin{Bmatrix} \alpha_1 \\ \alpha_2 \end{Bmatrix} = \begin{Bmatrix} \frac{1}{4} \\ \frac{1}{2} \end{Bmatrix} \tag{f}$$

Solving (f) yields

$$\alpha_1 = \tfrac{6}{31}, \qquad \alpha_2 = \tfrac{40}{217}$$
$$u = \frac{x(x-1)}{217}(42 + 40x) \tag{g}$$

Comparing this result with the exact solution, the following table can be drawn:

x	u_{app}	u_{exact}
0·25	0·045	0·044014
0·50	0·071	0·069747
0·75	0·062	0·060056

$$u_{exact} = \frac{\sin x}{\sin 1} - x \tag{h}$$

(b) THE LEAST-SQUARE METHOD

In this method we take the inner product of the error by itself, and the quantity thus obtained is required to be a minimum. Starting with

$$\varepsilon = \mathscr{L}(u) - p \tag{1.30}$$

we define F as

$$F = \langle \varepsilon, \varepsilon \rangle = \langle \mathscr{L}(u) - p, \mathscr{L}(u) - p \rangle \tag{1.31}$$

If the approximating function is

$$u = \sum_{k=1}^{N} \alpha_k \phi_k \tag{1.32}$$

we minimise F by differentiating with respect to α_i

$$\frac{\partial F}{\partial \alpha_i} = 0 \qquad i = 1, 2 \ldots N \tag{1.33}$$

This yields

$$\frac{\partial F}{\partial \alpha_i} = \frac{\partial}{\partial \alpha_i}\langle \varepsilon, \varepsilon \rangle = \frac{\partial}{\partial \alpha_i}\left\{\left\langle \mathscr{L}\left(\sum \alpha_k \phi_k\right), \mathscr{L}\left(\sum \alpha_k \phi_k\right)\right\rangle - 2\left\langle \mathscr{L}\left(\sum \alpha_k \phi_k\right), p\right\rangle + \langle p, p \rangle\right\} \tag{1.34}$$

When $\mathscr{L}$ is a linear operator, the equations simplify to

$$2\left\langle \mathscr{L}\left(\sum \alpha_k \phi_k\right), \mathscr{L}(\phi_i)\right\rangle - 2\langle \mathscr{L}(\phi_i), p \rangle = 0 \tag{1.35}$$

which can be written as

$$\left\langle \mathscr{L}\left(\sum \alpha_k \phi_k\right) - p, \mathscr{L}(\phi_i)\right\rangle = 0 \tag{1.36}$$

Example 1.3

Consider the equation treated in Example 1.2. We take the second-order approximation here also.

$$u = x(1 - x)\alpha_1 + x^2(1 - x)\alpha_2 \tag{a}$$

$$\varepsilon = x + (-2 + x - x^2)\alpha_1 + (2 - 6x + x^2 - x^3)\alpha_2 \tag{b}$$

Squaring ε and minimising it with respect to α_1 and α_2 we obtain

$$\begin{aligned} &\int_0^1 \varepsilon(-2 + x - x^2)\,dx = 0 \\ &\int_0^1 \varepsilon(2 - 6x + x^2 - x^3)\,dx = 0 \end{aligned} \tag{c}$$

Expanding (c),

$$\begin{bmatrix} 202 & 101 \\ 101 & 1532 \end{bmatrix} \begin{Bmatrix} \alpha_1 \\ \alpha_2 \end{Bmatrix} = \begin{Bmatrix} 55 \\ 393 \end{Bmatrix} \tag{d}$$

and solving results in

$$\alpha_1 \simeq 0{\cdot}192, \qquad \alpha_2 \simeq 0{\cdot}165 \tag{e}$$

In the following table this solution is compared to the exact solution:

x	u_{app}	u_{exact}
0·25	0·043	0·044014
0·50	0·068	0·069747
0·75	0·059	0·060056

In the conventional collocation method, the number of points is equal to the number of unknown parameters. One can extend the method to treat more points than unknowns, in which case α_i are obtained by minimising in a least-square sense.

The error

$$\varepsilon = \mathscr{L}(u) - p \tag{1.37}$$

is evaluated at M points, where $M > N$, and a function F, which is the sum of ε^2 at the M different points, is formed:

$$F = \langle \{\mathscr{L}(u) - p\}^2, \Delta_m \rangle \tag{1.38}$$

where Δ_m is the Dirac function as defined previously ($m = 1, \ldots, M$). Minimising (1.38), one obtains for the ith equation ($i = 1, 2, \ldots, N$)

$$\left\langle \{\mathscr{L}(u) - p\} \left\{ \frac{\partial \mathscr{L}(u)}{\partial \alpha_i} \right\}, \Delta_m \right\rangle = 0 \tag{1.39}$$

When $\mathscr{L}$ is a linear operator, the expanded form is

$$\left\langle \left\{ \mathscr{L}\left(\sum \alpha_k \phi_k \right) - p \right\} \mathscr{L}(\phi_i), \Delta_m \right\rangle = 0, \qquad i = 1, 2 \ldots N \tag{1.40}$$

Equation (1.40) gives a square symmetric system of equations.

Example 1.4

Let us consider again Example 1.2 with the corresponding boundary conditions and the approximation

$$u = x(1 - x)(\alpha_1 + \alpha_2 x) \tag{a}$$

The residual is

$$\begin{aligned} \varepsilon &= x + (-2 + x - x^2)\alpha_1 + (2 - 6x + x^2 - x^3)\alpha_2 \\ \varepsilon &= x + \mathscr{L}(\phi_1)\alpha_1 + \mathscr{L}(\phi_2)\alpha_2 \end{aligned} \tag{b}$$

We now apply collocation at three points, $x_1 = \frac{1}{4}, x_2 = \frac{1}{2}, x_3 = \frac{3}{4}$, which gives

$$\begin{gathered} \begin{Bmatrix} \varepsilon_1 \\ \varepsilon_2 \\ \varepsilon_3 \end{Bmatrix} = \begin{bmatrix} \mathscr{L}(\phi_1)_1 & \mathscr{L}(\phi_2)_1 \\ \mathscr{L}(\phi_1)_2 & \mathscr{L}(\phi_2)_2 \\ \mathscr{L}(\phi_1)_3 & \mathscr{L}(\phi_2)_3 \end{bmatrix} \begin{Bmatrix} \alpha_1 \\ \alpha_2 \end{Bmatrix} - \begin{Bmatrix} -x_1 \\ -x_2 \\ -x_3 \end{Bmatrix} \\ \Downarrow \\ \begin{Bmatrix} \varepsilon_1 \\ \varepsilon_2 \\ \varepsilon_3 \end{Bmatrix} = \begin{bmatrix} -\frac{29}{16} & \frac{35}{64} \\ -\frac{7}{4} & -\frac{7}{8} \\ -\frac{29}{16} & -\frac{151}{64} \end{bmatrix} \begin{Bmatrix} \alpha_1 \\ \alpha_2 \end{Bmatrix} - \begin{Bmatrix} -\frac{1}{4} \\ -\frac{1}{2} \\ -\frac{3}{4} \end{Bmatrix} \end{gathered} \tag{c}$$

where $\mathscr{L}(\ )_i$ indicates the $\mathscr{L}(\ )$ function at x_i $(i = 1, 2, 3)$.

The squaring and minimisation implied in (1.40) are equivalent to the following matrix multiplication:

$$\begin{gathered} \begin{bmatrix} \mathscr{L}(\phi_1)_1 & \mathscr{L}(\phi_1)_2 & \mathscr{L}(\phi_1)_3 \\ \mathscr{L}(\phi_2)_1 & \mathscr{L}(\phi_2)_2 & \mathscr{L}(\phi_2)_3 \end{bmatrix} \begin{bmatrix} \mathscr{L}(\phi_1)_1 & \mathscr{L}(\phi_2)_1 \\ \mathscr{L}(\phi_1)_2 & \mathscr{L}(\phi_2)_2 \\ \mathscr{L}(\phi_1)_3 & \mathscr{L}(\phi_2)_3 \end{bmatrix} \begin{Bmatrix} \alpha_1 \\ \alpha_2 \end{Bmatrix} \\ - \begin{bmatrix} \mathscr{L}(\phi_1)_1 & \mathscr{L}(\phi_1)_2 & \mathscr{L}(\phi_1)_3 \\ \mathscr{L}(\phi_2)_1 & \mathscr{L}(\phi_2)_2 & \mathscr{L}(\phi_2)_3 \end{bmatrix} \begin{Bmatrix} -x_1 \\ -x_2 \\ -x_3 \end{Bmatrix} = \begin{Bmatrix} 0 \\ 0 \end{Bmatrix} \\ \Downarrow \\ \begin{bmatrix} -\frac{29}{16} & -\frac{7}{4} & -\frac{29}{16} \\ \frac{35}{64} & -\frac{7}{8} & -\frac{151}{64} \end{bmatrix} \begin{bmatrix} -\frac{29}{16} & \frac{35}{64} \\ -\frac{7}{4} & -\frac{7}{8} \\ -\frac{29}{16} & -\frac{151}{64} \end{bmatrix} \begin{Bmatrix} \alpha_1 \\ \alpha_2 \end{Bmatrix} \\ - \begin{bmatrix} -\frac{29}{16} & -\frac{7}{4} & -\frac{29}{16} \\ \frac{35}{64} & -\frac{7}{8} & -\frac{151}{64} \end{bmatrix} \begin{Bmatrix} -\frac{1}{4} \\ -\frac{1}{2} \\ -\frac{3}{4} \end{Bmatrix} \end{gathered} \tag{d}$$

Equation (d) reduces to

$$\begin{bmatrix} 9{\cdot}66 & 4{\cdot}82 \\ 4{\cdot}82 & 6{\cdot}61 \end{bmatrix} \begin{Bmatrix} \alpha_1 \\ \alpha_2 \end{Bmatrix} = \begin{Bmatrix} 2{\cdot}69 \\ 2{\cdot}06 \end{Bmatrix} \tag{e}$$

$$\alpha_1 = 0{\cdot}192; \qquad \alpha_2 = 0{\cdot}172$$

The results can be tabulated as follows:

x	u_{app}	u_{exact}
0·25	0·044	0·044014
0·50	0·069	0·069747
0·75	0·060	0·060056

(c) THE METHOD OF MOMENTS

Given the equation

$$\langle \varepsilon, w_i \rangle = 0, \qquad i = 1, 2 \ldots N \tag{1.41}$$

we could use any set of linearly independent and complete functions as weighting functions w_i. The simplest choice for a one-dimensional problem is the series

$$1, x, x^2, x^3 \ldots \tag{1.42}$$

In this way successive higher 'moments' of the residual are forced to be zero.

$$\langle \varepsilon, x^i \rangle = 0 \qquad i = 0, 1, 2 \ldots \tag{1.43}$$

This scheme is called the method of moments.

Example 1.5

The error function for Example 1.2 is orthogonalised with respect to 1 and x.

$$\int_0^1 \varepsilon \,.\, 1 \, \mathrm{d}x = 0 \qquad \int_0^1 \varepsilon \,.\, x \, \mathrm{d}x = 0 \tag{a}$$

Substituting for ε,

$$\varepsilon = \alpha_1(-2 + x - x^2) + \alpha_2(2 - 6x + x^2 - x^3) + x \tag{b}$$

and integrating gives

$$\begin{bmatrix} \frac{11}{6} & \frac{11}{6} \\ \frac{11}{22} & \frac{19}{20} \end{bmatrix} \begin{Bmatrix} \alpha_1 \\ \alpha_2 \end{Bmatrix} = \begin{Bmatrix} \frac{1}{2} \\ \frac{1}{3} \end{Bmatrix} \tag{c}$$

The solution and comparison with the exact values are

$$\alpha_1 = \tfrac{122}{649}, \qquad \alpha_2 = \tfrac{110}{649} \tag{d}$$

x	u_{app}	u_{exact}
0·25	0·043191	0·044014
0·50	0·068181	0·069747
0·75	0·059084	0·060056

(d) GALERKIN'S METHOD

Galerkin's method is a particular weighted residual method in which the weighting functions are the same as the trial functions. Given the system of equations

$$\begin{aligned} \mathscr{L}(u) - p &= 0 \qquad x \in V \\ \mathscr{S}(u) - q &= 0 \qquad x \in S \end{aligned} \tag{1.44}$$

and an approximating function

$$u = \sum_{k=1}^{N} \alpha_k \phi_k \tag{1.45}$$

which satisfies the boundary conditions $\mathscr{S}(u) = q(S)$, $x \in S$. The residual

$$\varepsilon = \mathscr{L}\left(\sum \alpha_k \phi_k\right) - p \tag{1.46}$$

is orthogonalised with respect to the trial functions ϕ_i:

$$\langle \varepsilon, \phi_i \rangle = 0 \tag{1.47}$$

$$\Downarrow$$

$$\int \left\{ \mathscr{L}\left(\sum \alpha_k \phi_k\right) - p \right\} \phi_i \, \mathrm{d}V = 0 \qquad i = 1, 2, \ldots, N$$

If $\mathscr{L}$ is a linear operator, (1.47) produces a system of linear equations from which the α_k coefficients can be obtained. Note, however, that Galerkin's method is applicable to nonlinear problems as well.

Example 1.6

Applying the method for Example 1.2 leads to the following integrals:

$$\int_0^1 \varepsilon x(1-x)\,\mathrm{d}x = 0$$

$$\int_0^1 \varepsilon x^2(1-x)\,\mathrm{d}x = 0 \tag{a}$$

Integrating (a), and solving for α_1, α_2, we obtain

$$\begin{bmatrix} \frac{3}{10} & \frac{3}{20} \\ \frac{3}{20} & \frac{13}{105} \end{bmatrix} \begin{Bmatrix} \alpha_1 \\ \alpha_2 \end{Bmatrix} = \begin{Bmatrix} \frac{1}{12} \\ \frac{1}{20} \end{Bmatrix} \tag{b}$$

$$\alpha_1 = \tfrac{71}{369}, \qquad \alpha_2 = \tfrac{7}{41}$$

The approximate solution is

$$u = x(1-x)\{\tfrac{71}{369} + \tfrac{7}{41}x\} \tag{c}$$

which compares with the exact solution as follows:

x	u_{app}	u_{exact}
0·25	0·0440	0·044014
0·50	0·0698	0·069747
0·75	0·0600	0·060056

Contrary to other weighted residual methods in which the error is orthogonalised with respect to a set of functions different from the trial functions, in Galerkin's procedure the weighting functions are the same as the trial ones. This choice gives a physical significance to the Galerkin method in many engineering problems, as we shall show later.

Consider the orthogonality conditions:

$$\int \{\mathscr{L}(u) - p\}\phi_i\,\mathrm{d}V = 0; \qquad i = 1, 2 \ldots N \tag{1.48}$$

If we define

$$\delta u = \delta\alpha_1\,\phi_1 + \delta\alpha_2\,\phi_2 + \ldots + \delta\alpha_N\,\phi_N \tag{1.49}$$

where the $\delta\alpha_i$ terms are arbitrary increments, (1.48) can be written

$$\int \{\mathscr{L}(u) - p\}\, \delta u \, \mathrm{d}V = 0 \tag{1.50}$$

for arbitrary δu. It is understood that arbitrary δu is equivalent to $\delta\alpha_i\, \phi_i$ for $k = 1, 2, \ldots, N$. We use this equivalent notation, i.e. (1.50), only with the Galerkin method.

Example 1.7

Consider Poisson's equation:

$$\frac{\partial^2 u}{\partial x^2} + \frac{\partial^2 u}{\partial y^2} = c \tag{a}$$

with $u = 0$ at $x = 0, a$ and $y = 0, b$ (Figure 1.1). We can take as a

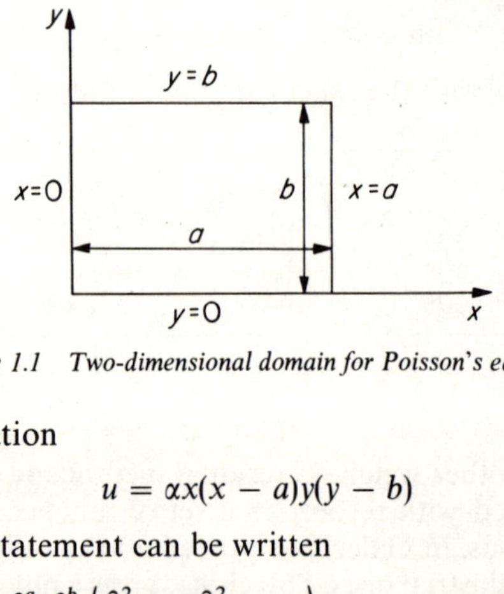

Figure 1.1 Two-dimensional domain for Poisson's equation

first approximation

$$u = \alpha x(x - a)y(y - b) \tag{b}$$

The Galerkin statement can be written

$$\int_0^a \int_0^b \left(\frac{\partial^2 u}{\partial x^2} + \frac{\partial^2 u}{\partial y^2} - c\right) \delta u \, \mathrm{d}x \, \mathrm{d}y = 0 \tag{c}$$

where $\delta u = \delta\alpha\, x(x - a)y(y - b)$. After substituting (b) we find that

$$\frac{\alpha}{90}[a^3 b^3 (a^2 + b^2)] - \frac{c a^3 b^3}{36} = 0 \tag{d}$$

$$\alpha = \frac{5}{2}\frac{c}{a^2 + b^2} \tag{e}$$

Hence

$$u = \frac{5c}{a^2 + b^2}(x^2 - ax)(y^2 - by) \tag{f}$$

Evaluating (f) at the centre point $x = a/2$, $y = b/2$ gives

$$u_c = \frac{5}{32} c \frac{a^2 b^2}{a^2 + b^2} \tag{g}$$

Let us now solve the same example but using a trigonometric series to approximate u. We assume, taking symmetry into consideration, that

$$u = \sum_k \sum_l \alpha_{kl} \sin \frac{k\pi x}{a} \sin \frac{l\pi y}{b} \tag{h}$$

This expression satisfies the boundary conditions and the sinusoidal functions are orthogonal, i.e.

$$\begin{aligned} \int_0^a \sin \frac{m\pi x}{a} \sin \frac{n\pi x}{a} \, \mathrm{d}x &= 0 \quad \text{for } m \neq n \\ &= \frac{a}{2} \quad \text{for } m = n \end{aligned} \tag{i}$$

After substituting (h) into (c) and taking orthogonality into account, we have

$$\alpha_{kl} \int_0^a \int_0^b \left(\frac{\pi^2 k^2}{a^2} + \frac{\pi^2 l^2}{b^2} \right) \sin^2 \frac{k\pi}{a} x \sin^2 \frac{l\pi}{b} y \, \mathrm{d}x \, \mathrm{d}y - \int_0^a \int_0^b c \sin \frac{k\pi}{a} x \sin \frac{l\pi}{b} y \, \mathrm{d}x \, \mathrm{d}y = 0 \tag{j}$$

This equation gives, after integration,

$$\alpha_{kl} = \frac{16 a^2 b^2}{\pi^4 kl(b^2 k^2 + a^2 l^2)} c \tag{k}$$

Note that the equations are uncoupled due to the orthogonality of the basis functions. The approximate solution is

$$u = \sum_k \sum_l c \frac{16 a^2 b^2}{\pi^4 kl(b^2 k^2 + a^2 l^2)} \sin \frac{k\pi x}{a} \sin \frac{l\pi y}{b} \tag{l}$$

When the number of terms is infinite, we obtain the exact solution. For the case $a = b$, the centre point ($x = a/2$, $y = b/2$) value is

$$u_c = (8 + \tfrac{16}{15} + \tfrac{8}{81} + \ldots)\frac{a^2}{\pi^4}c = u_{\text{exact}} = \frac{36{\cdot}64}{\pi^4}\left(\frac{a}{2}\right)^2 c \qquad \text{(m)}$$

1.3 WEAK FORMULATIONS

The examples treated in the previous section were restricted to self-adjoint operators and boundary conditions coinciding with the essential boundary conditions. Weighted residual methods are applicable for arbitrary operators and boundary conditions, and we will discuss in this section a general procedure for formulating weighted residual statements which allow only partial satisfaction of the boundary conditions, and, of more significance, the use of basis functions having relaxed continuity requirements.

We need first to introduce a classification for the degree of continuity of a function. Consider a function $u(x)$ defined over a region V and having a shape illustrated in Figure 1.2. The function

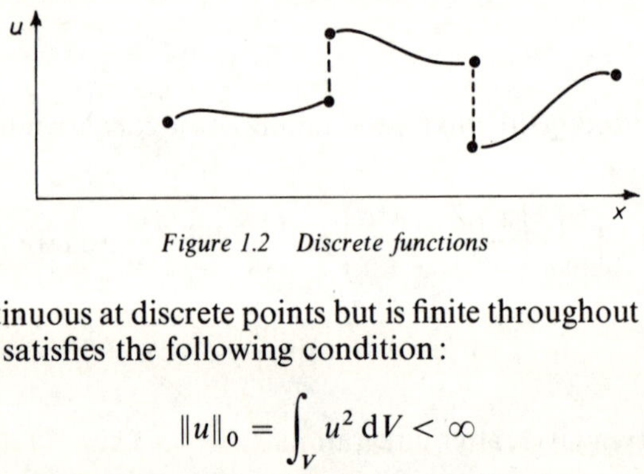

Figure 1.2 Discrete functions

is discontinuous at discrete points but is finite throughout the region. Its norm satisfies the following condition:

$$\|u\|_0 = \int_V u^2 \, \mathrm{d}V < \infty \qquad (1.51)$$

All functions satisfying (1.51), i.e. those that are square integrable, are said to belong to the L_2 function space. This space includes most of the functions that we shall deal with.

Imposing restrictions on the continuity of the derivatives leads to subsets of spaces, called Sobolev spaces. The space, $W_2^{(1)}$, contains all functions whose first derivative is square integrable. Its definition equation is (for a one dimensional function)

$$\|u\|_1 = \int \left(u^2 + \left(\frac{\mathrm{d}u}{\mathrm{d}x}\right)^2\right) \mathrm{d}x < \infty \qquad (1.52)$$

The superscript on w refers to the *order* of the highest finite derivative, the subscript to the *square* norm measure. Higher-order spaces are defined in an analogous way. For example, the $W_2^{(2)}$ space contains all functions which satisfy

$$\|u\|_2 = \int \left(u^2 + \left(\frac{\mathrm{d}u}{\mathrm{d}x}\right)^2 + \left(\frac{\mathrm{d}^2u}{\mathrm{d}x^2}\right)^2 \right) \mathrm{d}x < \infty \tag{1.53}$$

Examples of $W_2^{(1)}$ and $W_2^{(2)}$ functions are shown in Figure 1.3. Note that differentiation lowers the order of the space. If u belongs to $W_2^{(2)}$, then $\mathrm{d}u/\mathrm{d}x$ belongs to $W_2^{(1)}$. The above definitions can be

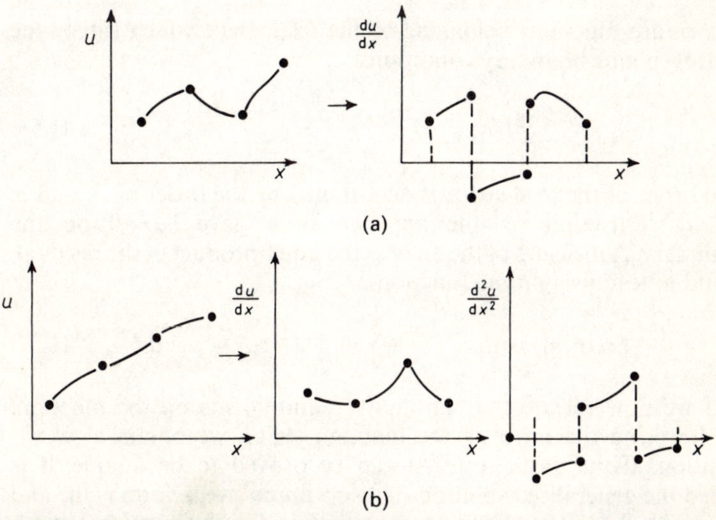

Figure 1.3 Types of function: (a) first-derivative square integrable, $W_2^{(1)}$; (b) second-derivative square integrable, $W_2^{(2)}$

extended to two- and three-dimensional problems by replacing the scalar operators with vector products.

Let us now return to the problem of solving the differential equation

$$\mathscr{L}(u) - p = 0 \qquad x \in V \tag{a}$$

subject to the boundary conditions

$$\mathscr{S}(u) - g = 0 \qquad x \in S \tag{b}$$

We suppose $\mathscr{L}$ is an nth order operator and $p, g \in W_2^{(1)}$. Also, we do not distinguish between essential and natural boundary conditions. Thus $\mathscr{S}(u) - g$ may represent a combination of essential and natural boundary conditions over the exterior domain. We will illustrate this point later. The classical solution is a function belonging to the $W_2^{(n+1)}$ space and satisfies $\varepsilon = \mathscr{L}(u) - p = 0$ for all $x \in V$ and $\varepsilon_B = \mathscr{S}(u) - g = 0$ for all $x \in S$.

In the conventional weighted-residual methods described in the previous section, we approximate the solution with

$$u = u_B + \sum_{j=1}^{N} \alpha_j \phi_j \tag{1.54}$$

where u_B satisfies the prescribed boundary condition,

$$\mathscr{S}(u_B) - g = 0 \qquad x \in S \tag{1.55}$$

and ϕ_j are functions belonging to the *trial* space which satisfy the homogeneous boundary conditions,

$$\mathscr{S}(\phi_j) = 0 \qquad \begin{matrix} j = 1, 2, \ldots, N \\ x \in S \end{matrix} \tag{1.56}$$

The order of the trial space is determined by the order of $\mathscr{L}$ and p.

Our restriction on the approximation gave $\varepsilon_B = 0$ on the boundary. A measure of the error is the inner product of the residual, ε, and a 'test' (weighting) function, w.

$$\text{Error measure} = \langle \varepsilon, w \rangle = \int_V (\mathscr{L}(u) - p) w \, \mathrm{d}x \tag{1.57}$$

If we now release the continuity requirements on the function, i.e. lowering the order of the function space, we obtain a 'weak' solution. If the weak solution can be proved to be unique, it is called the generalised solution. The 'optimum' weak form is the one for which the trial and test spaces coincide. Optimum refers here to the balance between uniqueness and existence. We interpret weighted residual methods as a particular numerical scheme for generating a weak solution.

In order to illustrate how continuity requirements can be released let us study the following second-order equation:

$$\mathscr{L}(u) = \frac{\mathrm{d}^2 u}{\mathrm{d}x^2} + u + x = 0 \tag{a}$$

with an essential boundary condition

$$u = f \quad \text{at } x = 0 \tag{b}$$

and a natural condition

$$\frac{\mathrm{d}u}{\mathrm{d}x} = g \quad \text{at } x = 1 \tag{c}$$

We require the trial solution to satisfy the essential boundary condition. The test function satisfies the homogeneous form of the essential condition, i.e.

$$\begin{aligned} u &= f \quad \text{at } x = 0 \\ w &= 0 \quad \text{at } x = 0 \end{aligned} \tag{d}$$

The following statement can now be written:

$$\int \left(\frac{d^2u}{dx^2} + u - x \right) w \, dx + \left| \left(g - \frac{du}{dx} \right) w \right|_{x=1} = 0 \tag{e}$$

where $w \in L_2$ and $u \in W_2^{(2)}$, i.e. function and first derivatives are continuous. Integrating equation (e) by parts one obtains:

$$\int \left\{ (u - x)w - \frac{du}{dx}\frac{dw}{dx} \right\} dx + |gw|_{x=1} = 0 \tag{f}$$

where u and $w \in W_2^{(1)}$, i.e. only the function needs to be continuous. Integrating again we find:

$$\int \left\{ (u - x)w + u \frac{d^2w}{dx^2} \right\} dx + \left| gw - u \frac{dw}{dx} \right|_0^1 = 0 \tag{g}$$

where $u \in L_2$ and $w \in W_2^{(2)}$.

Scheme (f) is the most widely applied. Note that in Galerkin's method the trial and test functions are the same and one can replace w with δu (see equation (1.50)).

Example 1.8

Consider the second-order equation:

$$\mathscr{L}(u) = \frac{d^2u}{dx^2} + u + x = 0 \tag{a}$$

with boundary conditions

$$\begin{aligned} u(0) &= 0 \quad \text{at } x = 0 \\ \frac{du(1)}{dx} &= 0 \quad \text{at } x = 1 \end{aligned} \tag{b}$$

We first propose an approximate solution satisfying both boundary conditions, i.e.

$$u^{(1)} = \alpha^{(1)} x \left(1 - \frac{x}{2} \right) = \alpha^{(1)} \phi_1 \tag{c}$$

The residual is

$$\varepsilon^{(1)} = -\alpha^{(1)} + \alpha^{(1)}\left(x - \frac{x^2}{2}\right) + x \tag{d}$$

which is orthogonalised now with respect to ϕ_1, using Galerkin's

$$\int_0^1 \varepsilon^{(1)} x\left(1 - \frac{x}{2}\right) \mathrm{d}x = 0 \tag{e}$$

After integration, we obtain

$$\alpha^{(1)} = \tfrac{25}{24} \tag{f}$$

Let us now approximate u by identically satisfying the $u(0) = 0$ condition while $(\mathrm{d}u/\mathrm{d}x)$ at $x = 1$ will be satisfied only in an average way. One can write the corresponding integral as

$$\int_0^1 \left\{\frac{\mathrm{d}^2 u}{\mathrm{d}x^2} + u + x\right\} \delta u \, \mathrm{d}x = \left[\frac{\mathrm{d}u}{\mathrm{d}x} - g\right]_{x=1} \delta u \tag{g}$$

Note that in this case $g = 0$, that is we want to impose $\mathrm{d}u/\mathrm{d}x = 0$.

We can propose

$$u^{(2)} = \alpha^{(2)}x + \beta^{(2)}x^2 = \alpha^{(2)}\phi_1 + \beta^{(2)}\phi_2 \tag{h}$$

Hence

$$\frac{\mathrm{d}u^{(2)}}{\mathrm{d}x} = \alpha^{(2)} + 2\beta^{(2)}x$$

We now find a residual in the 0–1 domain

$$\varepsilon^{(2)} = 2\beta^{(2)} + \alpha^{(2)}x + \beta^{(2)}x^2 + x \tag{i}$$

plus the boundary residual

$$\varepsilon_{\mathrm{B}} = \left[\frac{\mathrm{d}u}{\mathrm{d}x}\right]_{x=1} = [\alpha^{(2)} + 2\beta^{(2)}x]_{x=1} = \alpha^{(2)} + 2\beta^{(2)} \tag{j}$$

and orthogonalise both with respect to ϕ_1 and ϕ_2. This gives

$$\begin{aligned} \int_0^1 \varepsilon^{(2)} x \, \mathrm{d}x &= [\varepsilon_{\mathrm{B}} x]_{x=1} = \varepsilon_{\mathrm{B}} \\ \int_0^1 \varepsilon^{(2)} x^2 \, \mathrm{d}x &= [\varepsilon_{\mathrm{B}} x^2]_{x=1} = \varepsilon_{\mathrm{B}} \end{aligned} \tag{k}$$

These equations can be written, after integration,

$$\begin{bmatrix} \frac{2}{3} & \frac{3}{4} \\ \frac{3}{4} & \frac{17}{15} \end{bmatrix} \begin{Bmatrix} \alpha^{(2)} \\ \beta^{(2)} \end{Bmatrix} = \begin{Bmatrix} \frac{1}{3} \\ \frac{1}{4} \end{Bmatrix} \tag{l}$$

from where we can obtain

$$\alpha^{(2)} = +\tfrac{137}{139}, \qquad \beta^{(2)} = -\tfrac{120}{278} \tag{m}$$

We can now compare the results for the first and second approximations:

$x =$	$u^{(1)}$	$\dfrac{du^{(1)}}{dx}$	$u^{(2)}$	$\dfrac{du^{(2)}}{dx}$
0·50	25/64		107/278	
1·00	25/48	0·0	77/139	17/139

Note that, in the second case,

$$\left.\frac{du}{dx}\right|_{x=1} \neq 0$$

but if we increased the number of trial functions the solution would tend to satisfy the boundary condition $du/dx = 0$.

Example 1.9

Consider the case of a prismatic beam on an elastic foundation, for which the equilibrium equation (Figure 1.4) is

$$EI\frac{d^4v}{dx^4} + kv = p \tag{a}$$

where E is the modulus of elasticity, I the moment of inertia, k the foundation constant and p the distributed load on the beam.

One can write—equation (1.51)—the following Galerkin type statement:

$$\int_0^l \left\{ EI\frac{d^4v}{dx^4} + kv - p \right\} \delta v \, dx = 0 \tag{b}$$

We can *integrate by parts* twice the first term of this expression and find that

$$\int_0^l EI\left(\frac{d^2v}{dx^2}\right)\left(\frac{d^2\delta v}{dx^2}\right) dx + \int_0^l kv\,\delta v\,dx - \int_0^l p\,\delta v\,dx$$
$$- \left[EI\frac{d^2v}{dx^2}\frac{d\delta v}{dx} - EI\frac{d^3v}{dx^3}\delta v\right]_{x=0}^{x=l} = 0 \qquad \text{(c)}$$

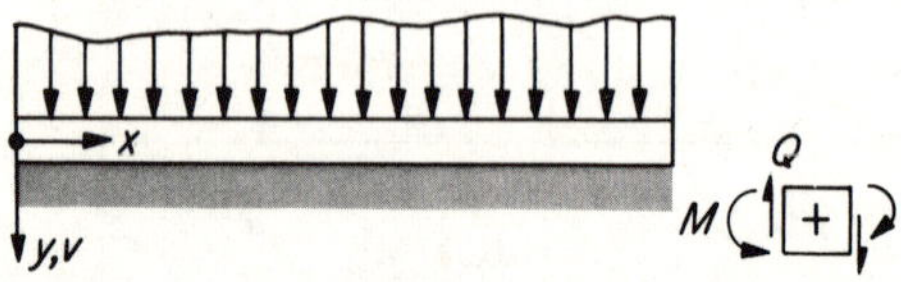

Figure 1.4 Prismatic beam on elastic foundation

The boundary conditions are assumed such that at

$$x = 0, l \qquad M = EI\frac{d^2v}{dx^2} = 0 \quad \text{or} \quad \frac{d\delta v}{dx} = 0$$

and (d)

$$Q = -EI\frac{d^3v}{dx^3} = 0 \quad \text{or} \quad \delta v = 0$$

Hence if the v function satisfies them we have from (c)

$$\int_0^l EI\left(\frac{d^2v}{dx^2}\right)\left(\frac{d^2\delta v}{dx^2}\right) dx + \int_0^l kv\,\delta v\,dx = \int_0^l p\,\delta v\,dx \qquad \text{(e)}$$

Equation (e) is the Principle of Virtual Displacements.

Let us now assume that we have started with a v expression which satisfies the boundary conditions for v and dv/dx (*essential* boundary conditions). Then at a free end (which is the more general case) we need to impose

$$EI\frac{d^2v}{dx^2} = \overline{M} \quad \text{and} \quad -EI\frac{d^3v}{dx^3} = \overline{Q} \qquad \text{(f)}$$

where $\overline{M}$ and $\overline{Q}$ are known applied moment and shear.

At a simply supported end for instance we assume to have imposed $\delta v \equiv 0$, and need to satisfy

$$EI\frac{d^2v}{dx^2} = \overline{M} \qquad \text{(g)}$$

One can now find boundary condition residuals which for the more general free end conditions are

$$\varepsilon_B' = \left\{ EI\frac{d^2v}{dx^2} - \bar{M} \right\}$$

and (h)

$$\varepsilon_B'' = \left\{ -EI\frac{d^3v}{dx^3} - \bar{Q} \right\}$$

The first boundary error has to be weighted by the rotations and the second by the displacements. Thus instead of (b) we now have an *extended* Galerkin statement:

$$\int_0^l \left\{ EI\frac{d^4v}{dx^4} + kv - p \right\} \delta v \, dx = \left[\left(EI\frac{d^3v}{dx^3} + \bar{Q} \right) \delta v - \left(EI\frac{d^2v}{dx^2} - \bar{M} \right) \frac{d\delta v}{dx} \right]_{x=0}^{x=l} \tag{i}$$

Integrating the first term by parts twice we obtain

$$\int_0^l EI\left(\frac{d^2v}{dx^2}\right)\left(\frac{d^2\delta v}{dx^2}\right) dx + \int_0^l kv\,\delta v\, dx = \int_0^l p\,\delta v\, dx + \left[\bar{Q}\,\delta v + \bar{M}\frac{d\delta v}{dx} \right]_{x=0}^{x=l} \tag{j}$$

This new expression allows us to approximate the *natural* boundary conditions (h). Boundary conditions in v and dv/dx are called *essential* and are satisfied exactly (hence δv and $d\delta v/dx$ are identically zero). Note that in our new statement (j), the trial functions can be of an order lower than previously, as we are now working with second-order derivatives instead of fourth-order ones, as in formula (b).

Example 1.10

Next we study the case of seepage in granular soil, which gives a potential problem. Darcy's law for two-dimensional, isotropic soil gives

$$v_x = K\frac{\partial u}{\partial x}, \qquad v_y = K\frac{\partial u}{\partial y} \tag{a}$$

where K is a constant characteristic of the soil and is called the permeability coefficient. u is the head, and v_x and v_y are the velocities in the x and y directions.

If the volume of water in the pores is constant, we have the condition that the 'flow in' must balance the 'flow out'; i.e.

$$\frac{\partial v_x}{\partial x} + \frac{\partial v_y}{\partial y} = 0$$

or

$$\mathscr{L}(u) = K\left(\frac{\partial^2 u}{\partial x^2} + \frac{\partial^2 u}{\partial y^2}\right) = 0 \qquad \text{(b)}$$

The boundary conditions for this problem are basically of two types: (a) *essential* boundary conditions on S_1 (Figure 1.5) such as $u = \bar{u}$, where $\bar{u}$ is a specified value of the potential head; (b) *natural* boundary conditions on S_2, of the type normal velocity equal to a given value, i.e. $\bar{v}_n = K\,\partial u/\partial n$, where n is the normal to the boundary.

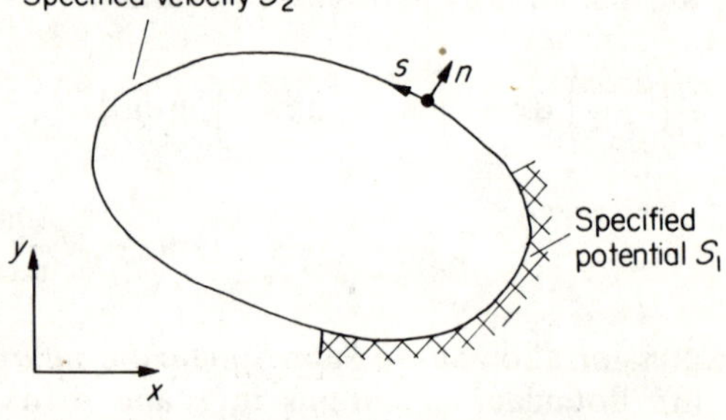

Figure 1.5 Two-dimensional permeable body

If the trial functions satisfy both types of boundary conditions, we only need work with equation (b), which for Galerkin's method can be written

$$\iint K\left\{\frac{\partial^2 u}{\partial x^2} + \frac{\partial^2 u}{\partial y^2}\right\}\delta u \, \mathrm{d}x \, \mathrm{d}y = 0 \qquad \text{(c)}$$

By integrating (c) by parts using Gauss's theorem, one can find the *natural* boundary conditions of the problem. Integrating, we obtain

$$\iint K\left\{\frac{\partial u}{\partial x}\frac{\partial \delta u}{\partial x} + \frac{\partial u}{\partial y}\frac{\partial \delta u}{\partial y}\right\}\mathrm{d}x \, \mathrm{d}y = K\int \frac{\partial u}{\partial x}\delta u \, \mathrm{d}y + K\int \frac{\partial u}{\partial y}\delta u \, \mathrm{d}x \qquad \text{(d)}$$

The right-hand side of (d) is equal to

$$\oint_S K\left\{\frac{\partial u}{\partial x}\frac{\mathrm{d}y}{\mathrm{d}S} - \frac{\partial u}{\partial y}\frac{\mathrm{d}x}{\mathrm{d}S}\right\}\delta u\,\mathrm{d}S = \oint_S K\frac{\partial u}{\partial n}\delta u\,\mathrm{d}S \tag{e}$$

where $S = S_1 + S_2$. Note that on S_1 we have $\delta u \equiv 0$, i.e. the *essential* boundary conditions are assumed to be identically satisfied. This gives (e) equal to

$$K\int_{S_2}\frac{\partial u}{\partial n}\delta u\,\mathrm{d}S \tag{f}$$

Finally (d) can be written as

$$\iint K\left\{\frac{\partial u}{\partial x}\frac{\partial \delta u}{\partial x} + \frac{\partial u}{\partial y}\frac{\partial \delta u}{\partial y}\right\}\mathrm{d}x\,\mathrm{d}y = \int_{S_2} K\frac{\partial u}{\partial n}\delta u\,\mathrm{d}S \tag{g}$$

This means that if one knows certain boundary conditions, such as

$$K\frac{\partial u}{\partial n} = \bar{v}_n \tag{h}$$

which has to be satisfied on the boundary S_2, an expression involving them plus the governing equations inside the body can be written; i.e.

$$\iint K\left(\frac{\partial^2 u}{\partial x^2} + \frac{\partial^2 u}{\partial y^2}\right)\delta u\,\mathrm{d}x\,\mathrm{d}y = \int_{S_2}\left(K\frac{\partial u}{\partial n} - \bar{v}_n\right)\delta u\,\mathrm{d}S \tag{i}$$

Equation (i) is the starting expression for Galerkin's method for the case in which one approximates the equilibrium equation and natural boundary conditions.

One can then write (i) in a different way by integrating by parts as shown before. Hence

$$\iint K\left(\frac{\partial u}{\partial x}\frac{\partial \delta u}{\partial x} + \frac{\partial u}{\partial y}\frac{\partial \delta u}{\partial y}\right)\mathrm{d}x\,\mathrm{d}y = \int_{S_2}\bar{v}_n\,\delta u\,\mathrm{d}S \tag{j}$$

Example 1.11

Another equation which frequently appears in engineering problems is the so called bi-harmonic equation; i.e.

$$\mathscr{L}(u) = \nabla^4 u = p$$

or

$$\nabla^2(\nabla^2 u) - p = \frac{\partial^4 u}{\partial x^4} + 2\frac{\partial^4 u}{\partial x^2\,\partial y^2} + \frac{\partial^4 u}{\partial y^4} - p = 0 \tag{a}$$

In order to define the natural boundary conditions corresponding to the equation, we can weight it by δu and integrate by parts applying Gauss's theorem. This has already been done for Poisson's (harmonic type) equation (Example 1.9), for which case we integrated once to obtain a symmetric operator. For equation (a), the operator becomes symmetric only after two integrations. That is, given

$$\iint (\nabla^4 u - p)\,\delta u\,\mathrm{d}x\,\mathrm{d}y = 0 \tag{b}$$

integrating once one obtains

$$-\iint\left\{\frac{\partial^3 u}{\partial x^3}\frac{\partial \delta u}{\partial x} + \frac{\partial^3 u}{\partial x^2\,\partial y}\frac{\partial \delta u}{\partial y} + \frac{\partial^3 u}{\partial y^2\,\partial x}\frac{\partial \delta u}{\partial x} + \frac{\partial^3 u}{\partial y^3}\frac{\partial \delta u}{\partial y} + p\,\delta u\right\}\mathrm{d}x\,\mathrm{d}y + \oint_S \frac{\partial}{\partial n}\left\{\frac{\partial^2 u}{\partial x^2} + \frac{\partial^2 u}{\partial y^2}\right\}\delta u\,\mathrm{d}S = 0 \tag{c}$$

Integrating again, (c) becomes

$$\iint\left\{\frac{\partial^2 u}{\partial x^2}\frac{\partial^2 \delta u}{\partial x^2} + 2\frac{\partial^2 u}{\partial x\,\partial y}\frac{\partial^2 \delta u}{\partial x\,\partial y} + \frac{\partial^2 u}{\partial y^2}\frac{\partial^2 \delta u}{\partial y^2} - p\,\delta u\right\}\mathrm{d}x\,\mathrm{d}y - \oint_S\left[\frac{\partial}{\partial n}\left(\frac{\partial u}{\partial x}\right)\frac{\partial \delta u}{\partial x} + \frac{\partial}{\partial n}\left(\frac{\partial u}{\partial y}\right)\frac{\partial \delta u}{\partial y}\right]\mathrm{d}S + \oint_S \frac{\partial}{\partial n}\left\{\frac{\partial^2 u}{\partial x^2} + \frac{\partial^2 u}{\partial y^2}\right\}\delta u\,\mathrm{d}S = 0 \tag{d}$$

The last two terms in (d) can be written

$$-\oint_S \frac{\partial}{\partial n}\left(\frac{\partial u}{\partial n}\right)\frac{\partial \delta u}{\partial n}\,\mathrm{d}S + \oint_S \frac{\partial}{\partial n}\left\{\frac{\partial^2 u}{\partial n^2} + \frac{\partial^2 u}{\partial s^2}\right\}\delta u\,\mathrm{d}S$$

The last two integrals correspond to the natural boundary conditions. If their values are known functions, i.e. A and B, we can write

$$\left.\begin{aligned}\mathscr{S}_1(u) &= \frac{\partial}{\partial n}\left(\frac{\partial^2 u}{\partial s^2} + \frac{\partial^2 u}{\partial n^2}\right) = A\\ \mathscr{S}_2(u) &= \frac{\partial}{\partial n}\left(\frac{\partial u}{\partial n}\right) = B\end{aligned}\right\} \tag{e}$$

The final Galerkin expression can be written,

$$\iint \{\mathscr{L}(u) - p\}\,\delta u\,\mathrm{d}x\,\mathrm{d}y = \int \left\{ \{\mathscr{S}_1(u) - A\}\,\delta u - \{\mathscr{S}_2(u) - B\}\frac{\partial \delta u}{\partial n} \right\} \mathrm{d}S \qquad \text{(f)}$$

After integrating (f) twice by parts, we have

$$\iint \{\mathscr{D}_1(u)\mathscr{D}_1(\delta u) + 2\mathscr{D}_2(u)\mathscr{D}_2(\delta u) + \mathscr{D}_3(u)\mathscr{D}_3(\delta u)\}\,\mathrm{d}x\,\mathrm{d}y = \int \left\{ B\frac{\partial \delta u}{\partial n} - A\,\delta u \right\} \mathrm{d}S \qquad \text{(g)}$$

where

$$\mathscr{D}_1(\ ) = \frac{\partial^2(\ )}{\partial x^2}, \qquad \mathscr{D}_2(\ ) = \frac{\partial^2(\ )}{\partial x\,\partial y}, \qquad \mathscr{D}_3(\ ) = \frac{\partial^2(\ )}{\partial y^2}$$

Note the symmetry of the term under the area integral in equation (g), which can now, more conveniently than equation (f), be used in our approximate analysis. The derivatives in equation (g) are of lower order than those in (f), and we now need to use only a second-derivative square integrable function.

Example 1.12

Consider the Laplace equation in cylindrical coordinates:

$$\frac{\mathrm{d}^2 u}{\mathrm{d}r^2} + \frac{1}{r}\frac{\mathrm{d}u}{\mathrm{d}r} = 0 \qquad \text{(a)}$$

which governs the heat flow in a pipe, as shown in Figure 1.6. u is the temperature function at any point. The internal temperature is u_1 and the external temperature is u_2. The external radius is R_2 and the internal one is R_1. The radius r is variable.

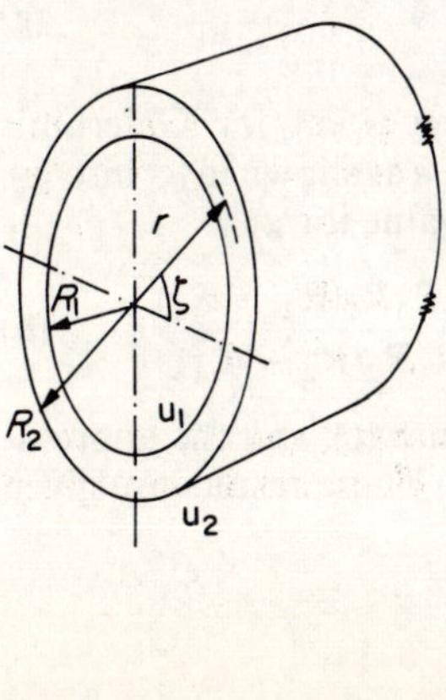

Figure 1.6 Heat flow in a circular pipe

In general we can have temperature or flux boundary conditions but in this case we will only consider the former. Under these conditions the exact solution of equation (a) is

$$u = u_1 + \ln\left(\frac{r}{R_1}\right)\frac{u_2 - u_1}{\ln(R_2/R_1)} \tag{b}$$

Let us now try to approximate the solution of equation (a) by Galerkin's method. One can propose a second-order function as the approximate solution, such as

$$u_{\text{app}} = \left\{\left[\frac{u_1R_2 - u_2R_1}{R_2 - R_1}\right] + \left[\frac{u_2 - u_1}{R_2 - R_1}\right]r\right\} + \alpha\{R_1R_2 - (R_2 + R_1)r + r^2\} \tag{c}$$

which satisfies the temperature boundary conditions, described on the S_1 part of the boundary. The integral for Galerkin's case can in general be written as

$$\iint\left\{\frac{\mathrm{d}^2u}{\mathrm{d}r^2} + \frac{1}{r}\frac{\mathrm{d}u}{\mathrm{d}r}\right\}\delta u\, r\,\mathrm{d}\zeta\,\mathrm{d}r = \int_{S_2}\left(\frac{\mathrm{d}u}{\mathrm{d}r} - g\right)\delta u\,\mathrm{d}S \tag{d}$$

where S_2 is the part of the boundary where flux is prescribed ($\mathrm{d}u/\mathrm{d}r = g$). We can also write (d) as

$$\iint\frac{\mathrm{d}}{\mathrm{d}r}\left(r\frac{\mathrm{d}u}{\mathrm{d}r}\right)\delta u\,\mathrm{d}\zeta\,\mathrm{d}r = \int_{S_2}\left(\frac{\mathrm{d}u}{\mathrm{d}r} - g\right)\delta u\,\mathrm{d}S \tag{e}$$

Integrating with respect to ζ, we obtain

$$2\pi\int\frac{\mathrm{d}}{\mathrm{d}r}\left(r\frac{\mathrm{d}u}{\mathrm{d}r}\right)\delta u\,\mathrm{d}r = \int_{S_2}\left(\frac{\mathrm{d}u}{\mathrm{d}r} - g\right)\delta u\,\mathrm{d}S \tag{f}$$

Integrating now by parts and noting that $\delta u \equiv 0$ on S_1 (note also that S_2 does not exist for this example), we obtain

$$\int_{R_1}^{R_2} r\left(\frac{\mathrm{d}u}{\mathrm{d}r}\right)\left(\frac{\mathrm{d}\delta u}{\mathrm{d}r}\right)\mathrm{d}r = 0 \tag{g}$$

This integral can be taken as the starting point for Galerkin's approximation. If we substitute the function u as shown in (c) into (g), we obtain, after integration, the following value for α:

$$\alpha^{(1)} = \left(\frac{u_2 - u_1}{R_2 - R_1}\right)\left[\frac{-(R_2^3 - R_1^3) + 3R_1R_2(R_2 - R_1)}{(R_2^4 - R_1^4) - 2R_1R_2(R_2^2 - R_1^2)}\right] \tag{h}$$

This value of α can then be substituted into (c), and the approximate values for the temperature are found. These results are given in the table below under the heading *Approximate solution*, where they are compared to the exact solution:

Radius	*Exact solution*	*Approximate solution*
1 (= R_1)	100	100
1·2	73·696	74·666
1·4	51·457	52
1·6	32·192	32
1·8	15·200	14·666
2 (= R_2)	0	0

Example 1.13

Galerkin's method is useful for studying nonlinear problems, most of which do not have a functional in which a Rayleigh–Ritz technique (see Section 1.6) can be used. In general we can propose approximating functions which contain terms associated with linear solutions. Afterwards the function can be improved and the solution iterated until the error in the approximation becomes small.

In order to illustrate the approach, let us study the case of a decay problem governed by the equation

$$v\frac{du}{dx} = D\frac{d^2u}{dx^2} - ku^s \tag{a}$$

where v is the velocity of the fluid, u is the concentration, D the diffusivity and k and s are decay parameters. We will take $s = 2$ for our example.

The boundary conditions for this case are $u = \bar{u}$ at $x = 0$ and $du/dx = 0$ at $x = l$.

In order to apply Galerkin's method we can assume the function

$$u = \bar{u}(1 + \alpha_1 x + \alpha_2 x^2) \tag{b}$$

which satisfies the boundary condition $u = \bar{u}$ at $x = 0$. At $x = l$ we have

$$\left.\frac{du}{dx}\right|_{x=l} = \bar{u}(\alpha_1 + 2\alpha_2 l) = 0 \qquad \therefore \quad \alpha_2 = -\frac{\alpha_1}{2l} \tag{c}$$

Expressing u in terms of $\alpha = \alpha_1$, we can write

$$u = \bar{u}\left(1 + \alpha x - \frac{\alpha}{2l}x^2\right) \tag{d}$$

Linear case ($s = 1$): Let us first solve the problem for which $s = 1$, that is the linear case. The Galerkin expression for this case is

$$\int_0^l \left\{ v\frac{du}{dx} - D\frac{d^2u}{dx^2} + ku \right\} \delta u \, dx = 0 \tag{e}$$

Without bothering about integrating by parts, we can substitute (d) and obtain

$$\int_0^l \left\{ v\alpha\left(1 - \frac{x}{l}\right) + \alpha D\left(\frac{1}{l}\right) + k\left[1 + \alpha\left(x - \frac{x^2}{2l}\right)\right]\right\}\left(x - \frac{x^2}{2l}\right) dx = 0 \tag{f}$$

After integration we have

$$\alpha(\tfrac{3}{24}vl^2 + \tfrac{2}{15}kl^3 + \tfrac{1}{3}Dl) = -\frac{kl^2}{3} \tag{g}$$

$$\therefore \qquad \alpha = -\frac{40kl}{(15vl + 16kl^2 + 40D)} \tag{h}$$

Nonlinear case ($s = 2$): We can now attack the nonlinear problem, starting with the linear solution as the first approximation. For simplicity's sake, we consider $k = v = D = l = \bar{u} = 1$. Hence from (h)

$$\alpha^{(1)} = -0{\cdot}564 \tag{i}$$

Our equilibrium equation becomes

$$\frac{du}{dx} - \frac{d^2u}{dx^2} + u^2 = 0 \tag{j}$$

and we assume that the boundary conditions are the same as before.

Galerkin's expression is now

$$\int_0^1 \left\{ \frac{du}{dx} - \frac{d^2u}{dx^2} + u^2 \right\} \delta u \, dx = 0 \tag{k}$$

Let us now define a ψ function obtained by substituting the $\alpha^{(1)}$ (linear) value of α into (k), with u approximated by equation (d). This function will be different from zero; i.e.

$$\psi^{(1)} = \int_0^1 \left[\alpha^{(1)}(1 - x) + \alpha^{(1)} + \left\{1 + \alpha^{(1)}\left(x - \frac{x^2}{2}\right)\right\}^2 \right]\left(x - \frac{x^2}{2}\right) dx$$

$$= 0{\cdot}333 + 0{\cdot}725\alpha^{(1)} + 0{\cdot}057(\alpha^{(1)})^2 = -0{\cdot}059 \tag{l}$$

Note that the ψ function should be zero for the equilibrium state.

We can now apply the Newton–Raphson method, the recurrence relation for which is

$$d\psi = -\psi$$

or

$$\left(\frac{d\psi}{d\alpha}\right) \Delta\alpha = -\psi \tag{m}$$

where

$$\frac{d\psi}{d\alpha} = \int_0^1 \left[(1 - x) + 1 + 2\left\{1 + \alpha\left(x - \frac{x^2}{2}\right)\right\}\left(x - \frac{x^2}{2}\right)\right]\left(x - \frac{x^2}{2}\right) dx \tag{n}$$

This gives for $\alpha^{(1)}$

$$\left(\frac{d\psi}{d\alpha}\right) = 0{\cdot}725 + 0{\cdot}114\alpha^{(1)} = 0{\cdot}661 \tag{o}$$

We can compute $\Delta\alpha$ as

$$\Delta\alpha = -\frac{\psi^{(1)}}{(d\psi/d\alpha)} = \frac{0{\cdot}059}{0{\cdot}661} = 0{\cdot}088 \tag{p}$$

The new value of α is

$$\alpha^{(2)} = \alpha^{(1)} + \Delta\alpha = -0{\cdot}564 + 0{\cdot}088 = -0{\cdot}476 \tag{q}$$

By computing the $\psi^{(2)}$ function {equation (l)} it is found that

$$\psi^{(2)} = 0{\cdot}333 + 0{\cdot}725\alpha^{(2)} + 0{\cdot}057(\alpha^{(2)})^2 = 0{\cdot}001 \tag{r}$$

which is very nearly equal to zero. If $\psi^{(2)}$ were still significant the above procedure could be repeated.

1.4 INITIAL VALUE PROBLEMS

Galerkin's method is also of interest for the solution of initial value problems. Two different types of initial value equations are important in engineering problems: i.e., the hyperbolic and parabolic equations. First let us discuss the hyperbolic type, usually called the *wave* equation, which for three dimensions can be written

$$\lambda\nabla^2 u = \frac{\partial^2 u}{\partial t^2} \qquad \text{in } V \tag{1.58}$$

where

$$\nabla^2 = \frac{\partial^2}{\partial x^2} + \frac{\partial^2}{\partial y^2} + \frac{\partial^2}{\partial z^2}$$

is a general Laplacian operator and

$$\nabla = \frac{\partial}{\partial x} + \frac{\partial}{\partial y} + \frac{\partial}{\partial z}$$

represents the gradient operator. λ is a positive parameter. The *initial conditions* necessary to solve (1.58) are the values of u and $\partial u/\partial t$ at $t = 0$

$$u(x, 0) = u_0(x), \qquad \frac{\partial u(x, 0)}{\partial t} = \dot{u}_0(x) \tag{1.59}$$

plus the spatial *boundary conditions*

$$u(x, t) = \bar{u}(x, t) \quad \text{on the } S_1 \text{ part of external surface}$$

or

$$\lambda \frac{\partial u}{\partial n}(x, t) = g \qquad \text{on the } S_2 \text{ part of external surface.}$$

The Galerkin statement for this case is

$$\int_0^t \left\{ \iiint_V \left(\lambda \nabla^2 u - \frac{\partial^2 u}{\partial t^2} \right) \delta u \, \mathrm{d}V - \iint_{S_2} \left(\lambda \frac{\partial u}{\partial n} - g \right) \delta u \, \mathrm{d}S \right\} \mathrm{d}t = 0 \tag{1.60}$$

on the understanding that the u functions in (1.60) identically satisfy the initial conditions and the boundary conditions on S_1.

The other type of initial value problem of interest for us is the *diffusion* equation (parabolic type):

$$\nabla^2 u = K \frac{\partial u}{\partial t} \qquad \text{in } V \tag{1.61}$$

which requires only one initial condition; i.e.

$$u(x, 0) = u_0 \tag{1.62}$$

and boundary conditions

$$u(x, t) = \bar{u}(x, t) \quad \text{on } S_1$$

or

$$\frac{\partial u}{\partial n}(x, t) = g \qquad \text{on } S_2 \tag{1.63}$$

Galerkin's statement is now

$$\int_0^t \left\{ \iiint_V \left(\nabla^2 u - K\frac{\partial u}{\partial t} \right) \delta u \, \mathrm{d}V - \int_{S_2} \left(\frac{\partial u}{\partial n} - g \right) \delta u \, \mathrm{d}S \right\} \mathrm{d}t = 0 \tag{1.64}$$

where u satisfies the initial condition (1.62), and $u = \bar{u}$ on S_1. For both cases we will assume that the u function can be approximated by the product of a spatial and a time-dependent function

$$u \simeq \sum_{i=1}^{N} \alpha_i(t)\phi_i(x) \tag{1.65}$$

Note that $\phi_i(x)$ has to satisfy the spatial boundary conditions and $\alpha_i(t)$ the initial conditions.

Example 1.14

Consider the simple equation,

$$\frac{\partial u}{\partial t} + \omega^2 u = 0 \tag{a}$$

with initial conditions $u = u_0$.

We can approximate the solution of (a) using Galerkin's method with

$$u = u_0 + (u_t - u_0)\frac{t}{\Delta t} \tag{b}$$

where Δt is a small time step, u_t the unknown value of the function at the end of Δt, and u_0 the initial value.

Substituting (b) into (a) we obtain

$$\varepsilon = \left\{ \frac{u_t - u_0}{\Delta t} \right\} + \omega^2 \left\{ u_0 + (u_t - u_0)\frac{t}{\Delta t} \right\} \tag{c}$$

This residual has to be orthogonalised only with respect to the function t of equation (b). This is evident as u_0 is fixed and hence the variation δu can be written

$$\delta u = \delta u_t \left(\frac{t}{\Delta t} \right) \tag{d}$$

where $t/\Delta t$ is the only function which orthogonalises (c). Thus

$$\int_0^{\Delta t} \varepsilon t \, \mathrm{d}t = 0 \tag{e}$$

or

$$\int_0^{\Delta t} \left[\left(\frac{u_t - u_0}{\Delta t}\right) + \omega^2\left\{u_0 + (u_t - u_0)\frac{t}{\Delta t}\right\}\right] t\,\mathrm{d}t = 0 \qquad \text{(f)}$$

which after integration becomes

$$\left(\frac{u_t - u_0}{2}\right)\Delta t + \left(\frac{\omega^2 u_0}{2}\right)\Delta t^2 + \omega^2\left(\frac{u_t - u_0}{3}\right)\Delta t^2 = 0 \qquad \text{(g)}$$

$$\therefore \qquad u_t = u_0\left(\frac{3 - \omega^2 \Delta t}{3 + 2\omega^2 \Delta t}\right) \qquad \text{(h)}$$

Equation (h) gives the value of u_t at the end of each small step in function of the initial value (u_0) and the length of the time step. The smaller the time step, the nearer our approximate values will be to the exact solution; i.e.

$$u = u_0\,\mathrm{e}^{-\omega^2 t} \qquad \text{(i)}$$

One can also compare the results obtained using (h) with those corresponding to using the trapezoidal rule in (a), which can be written

$$u = \frac{u_t + u_0}{2}, \qquad \frac{\partial u}{\partial t} = \frac{u_t - u_0}{\Delta t}$$

and gives the recurrence relationship

$$u_t = u_0\left(\frac{2 - \omega^2 \Delta t}{2 + \omega^2 \Delta t}\right) \qquad \text{(j)}$$

The two different approximate solutions for a time step $\Delta t = 0{\cdot}1$ and $\omega = u_0 = 1$ are compared to the exact solution in the following table:

Time	*Exact*	*Galerkin*	*Trapezoidal*
0	1	1	1
1	0·3678	0·3736	0·3675
2	0·1353	0·1392	0·1351
3	$4{\cdot}978 \times 10^{-2}$	$5{\cdot}2172 \times 10^{-2}$	$4{\cdot}9662 \times 10^{-2}$
4	$1{\cdot}8315 \times 10^{-2}$	$1{\cdot}9498 \times 10^{-2}$	$1{\cdot}8254 \times 10^{-2}$
5	$6{\cdot}7379 \times 10^{-3}$	$7{\cdot}2845 \times 10^{-3}$	$6{\cdot}7098 \times 10^{-3}$
6	$2{\cdot}4787 \times 10^{-3}$	$2{\cdot}7219 \times 10^{-3}$	$2{\cdot}4663 \times 10^{-3}$

Example 1.15

Let us now consider a case of temperature diffusion, such as the cooling of an insulated rod of length 2 units (Figure 1.7). The ends of the rod are kept at zero temperature. The cooling is determined by

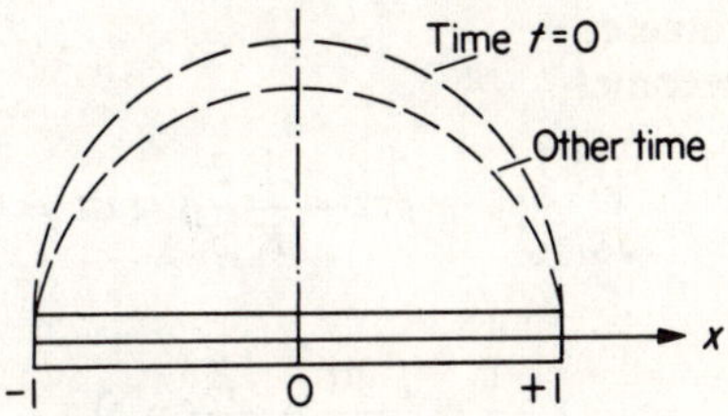

Figure 1.7 Cooling of a rod

the equation

$$\frac{1}{K}\frac{\partial u}{\partial t} = \frac{\partial^2 u}{\partial x^2} \tag{a}$$

where K is the thermal diffusivity. We have the following conditions:

$$\left.\begin{array}{lll} \text{initial condition} & u = u_0(1 - x^2) \quad \text{at } t = 0 & \text{for } -1 \leqslant x \leqslant 1 \\ \text{boundary condition } u = 0 & \text{at } x = \pm 1 & \text{for } t \geqslant 0 \end{array}\right\} \tag{b}$$

As one has taken a quadratic initial variation, the general function for temperature may be written

$$\begin{aligned} u(x, t) &= \phi(x)\alpha(t) \\ &= (1 - x^2)\alpha(t) \end{aligned} \tag{c}$$

where $\alpha(t) = u_0$ when $t = 0$. The Galerkin expression now becomes

$$\int_0^{\Delta t}\left[\int_x \left\{\frac{\partial^2 u}{\partial x^2} - \frac{1}{K}\frac{\partial u}{\partial t}\right\}\phi(x)\,\mathrm{d}x\right]\delta\alpha(t)\,\mathrm{d}t = 0 \tag{d}$$

where Δt is a small time step.

After substituting (c) into (d), one can write

$$\int_0^{\Delta t}\left[\int_{-1}^{+1}\left\{-2\alpha(t) - \frac{1}{K}(1 - x^2)\frac{\partial\alpha(t)}{\partial t}\right\}(1 - x^2)\,\mathrm{d}x\right]\delta\alpha(t)\,\mathrm{d}t = 0$$

which gives

$$\int_0^{\Delta t}\left\{\alpha(t) + \frac{1}{K}\frac{2}{5}\frac{\partial\alpha}{\partial t}\right\}\delta\alpha(t)\,\mathrm{d}t = 0 \tag{e}$$

Let us now assume that the time variation of α within the Δt time step is given by

$$\alpha(t) = u_0(1 + \beta t) \tag{f}$$

$$\therefore \qquad \delta\alpha(t) = u_0\, \delta\beta t$$

where u_0 is the initial value at the centre of the rod and β is an unknown parameter.

Hence (e) becomes

$$\int_0^{\Delta t} \left\{(1 + \beta t) + \frac{1}{K}\frac{2}{5}\beta\right\} t\, \mathrm{d}t = 0 \tag{g}$$

which gives

$$\frac{1}{2} + \frac{\beta\, \Delta t}{3} + \frac{\beta}{5K} = 0 \tag{h}$$

$$\therefore \qquad \beta = -\frac{1}{2}\left(\frac{15K}{5K\, \Delta t + 3}\right) \tag{i}$$

If we have for the sake of illustration, $K = 1$ and $\Delta t = 0{\cdot}01$, then $\beta = -2{\cdot}46$. u_0 will be taken as 100.

Hence for $t = 0$, we have u_c (at the centre) equals 100; for $t = 0{\cdot}01$, $u_c = 100 \times (1 + \beta\Delta t) = 100 \times (1 - 0{\cdot}0246) = 97{\cdot}54$; for $t = 0{\cdot}02$, $u_c = 97{\cdot}54 \times 0{\cdot}9754 = 95{\cdot}14$; for $t = 0{\cdot}03$, $u_c = 95{\cdot}14 \times 0{\cdot}9754 = 92{\cdot}80$; etc. In this way we can integrate on time step by step.

Example 1.16

Let us now investigate the one degree of freedom system defined by the following equation of motion:

$$m\frac{\mathrm{d}^2u}{\mathrm{d}t^2} + c\frac{\mathrm{d}u}{\mathrm{d}t} + ku = p(t) \tag{a}$$

under the two initial conditions

$$u(0) = 0$$

$$\frac{\mathrm{d}u(0)}{\mathrm{d}t} = \omega_d \tag{b}$$

Considering that $m = 1$, $c = 2$, $k = \omega^2$, $\omega_d = \sqrt{(\omega^2 - \eta^2)}$, $p(t) = 0$ and $\eta = c/2\sqrt{(mk)}$. The exact solution for this case is

$$u(t) = \mathrm{e}^{-t} \sin \omega_d t$$

$$\frac{\mathrm{d}u(t)}{\mathrm{d}t} = -\mathrm{e}^{-t} \sin \omega_d t + \omega_d\, \mathrm{e}^{-t} \cos \omega_d t \tag{c}$$

This example will be taken to test the accuracy of Galerkin's method as the hereditary error in the method will tend to become large as time increases.

The simplest function which represents displacements and velocities at the beginning and the end of a time interval Δt is

$$u = \alpha_1 + \alpha_2 t + \alpha_3 t^2 + \alpha_4 t^3 \tag{d}$$

which can be more conveniently rewritten as a function of the values of displacements and velocities at the beginning $(u_0, \dot{u}_0)$ and the end $(u_t, \dot{u}_t)$ of the interval:

$$u = \left(1 - 3\frac{t^2}{\Delta t^2} + 2\frac{t^3}{\Delta t^3}\right)u_0 + \left(t - 2\frac{t^2}{\Delta t} + \frac{t^3}{\Delta t^2}\right)\dot{u}_0 + \left(3\frac{t^2}{\Delta t^2} - 2\frac{t^3}{\Delta t^3}\right)u_t + \left(-\frac{t^2}{\Delta t} + \frac{t^3}{\Delta t^2}\right)\dot{u}_t \tag{e}$$

Since the initial velocity and displacement are known, it is only necessary to orthogonalise the residual with respect to the trial functions corresponding to the final values of displacement and velocity.

One can write the two following Galerkin's expressions:

$$\left.\begin{aligned} \int_0^{\Delta t} \{m\ddot{u} + c\dot{u} + ku - p(t)\}\left\{3\frac{t^2}{\Delta t^2} - 2\frac{t^3}{\Delta t^3}\right\} \mathrm{d}t = 0 \\ \int_0^{\Delta t} \{m\ddot{u} + c\dot{u} + ku - p(t)\}\left\{-\frac{t^2}{\Delta t} + 2\frac{t^3}{\Delta t^2}\right\} \mathrm{d}t = 0 \end{aligned}\right\} \tag{f}$$

Assuming that the forcing function varies linearly throughout the length of the step, we have, after integration,

$$\left(-504\frac{m}{\Delta t} + 210c + 156k\,\Delta t\right)u_t + (462m + 42c\,\Delta t - 22k\,\Delta t^2)\dot{u}_t$$
$$= \left(-504\frac{m}{\Delta t} + 210c - 54k\,\Delta t\right)u_0 + (-42m + 42c\,\Delta t - 13k\,\Delta t^2)\dot{u}_0 + 21\,\Delta t(7p_t + 3p_0)$$

and

$$\left(-42\frac{m}{\Delta t} + 42c + 22k\,\Delta t\right)u_t + (56m + 4\,\Delta t^2 k)\dot{u}_t$$
$$= \left(-42\frac{m}{\Delta t} + 42c - 13k\,\Delta t\right)u_0 + (14m + 7c\,\Delta t - 3\,\Delta t^2 k)\dot{u}_0 + 7\,\Delta t(3p_t + 2p_0)$$

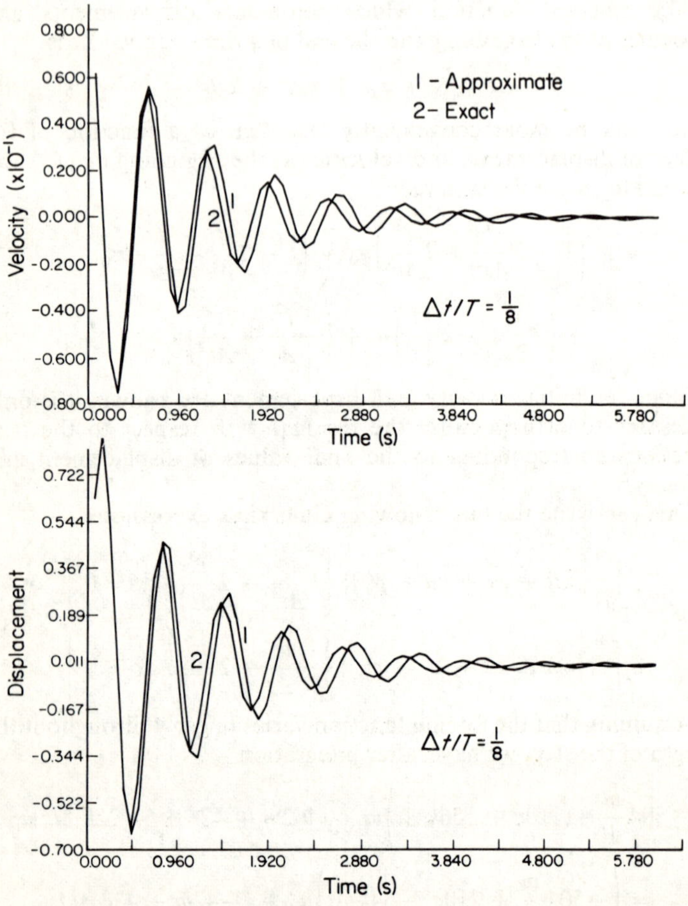

Figure 1.8(a) *Trapezoidal rule results for displacement and velocity*

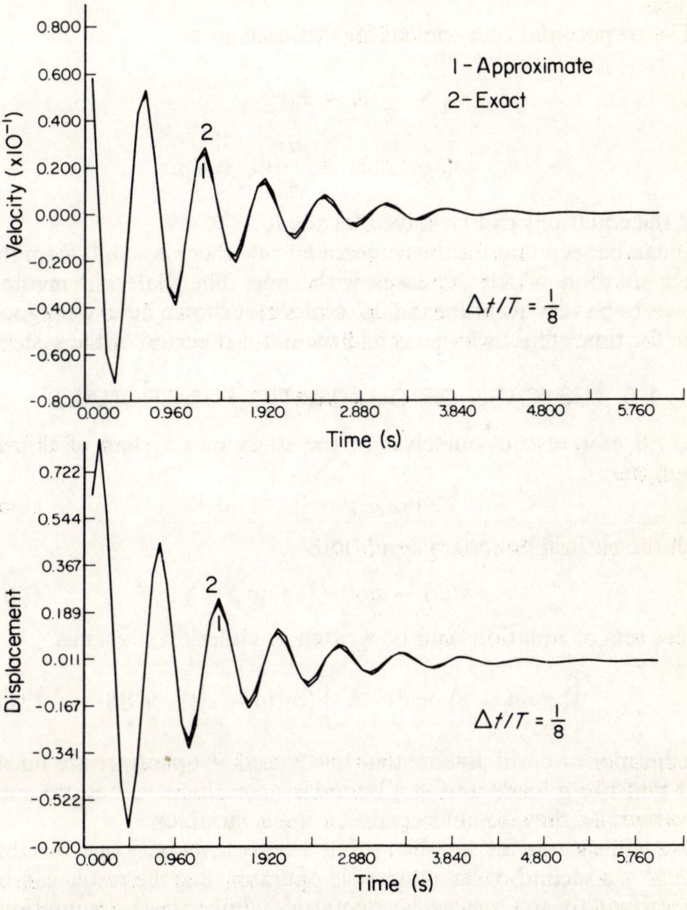

Figure 1.8(b) Galerkin method results for displacement and velocity

In Figure 1.8(a) and (b) the results obtained using the trapezoidal rule and the above Galerkin approach are shown for our case. The time step Δt is, for both cases, $\frac{1}{8}$ of the natural period of the system.

The trapezoidal rule applied for this case gives

$$\begin{aligned} \dot{u}_t &= \dot{u}_0 + \frac{\Delta t}{2}(\ddot{u}_t + \ddot{u}_0) \\ u_t &= u_0 + \Delta t\, \dot{u}_0 + \frac{\Delta t^2}{4}(\ddot{u}_t + \ddot{u}_0) \end{aligned} \qquad \text{(i)}$$

and the equations can be solved for the $\ddot{u}_t$.

It can be seen that for the trapezoidal rule there is a shift from the exact solution which increases with time. The Galerkin method proves to be very accurate and its results (not shown here) were good even for time steps as large as half the natural period of the system.

1.5 THE CASE OF QUADRATIC FUNCTIONALS

We will now restrict ourselves to the study of a system of elliptic equations:

$$\mathscr{L}(u) - p = 0 \qquad \text{in } V \qquad (1.66)$$

with the natural boundary conditions

$$\mathscr{S}(u) - q(s) = 0 \quad \text{on } S \qquad (1.67)$$

These sets of equations can be written in Galerkin's form as

$$\int \{\mathscr{L}(u) - p\}\, \delta u \, \mathrm{d}V = \int \{\mathscr{S}(u) - q(s)\}\, \delta u \, \mathrm{d}S \qquad (1.68)$$

Furthermore we will assume that the $\mathscr{L}$ and $\mathscr{S}$ operators are linear and that the p loads and $q(s)$ boundary condition values are conservative; i.e. they do not depend on the u variables.

We will assume for simplicity that u represents only one variable and $\mathscr{L}$ is a second-order symmetric operator, but the result can be generalised to any *symmetric* operator. Under these assumptions the integration by parts of the first term in equation (1.68) gives

$$\int \mathscr{L}(u)\, \delta u \, \mathrm{d}V = \int \mathscr{S}(u)\, \delta u \, \mathrm{d}S - \int C\mathscr{D}(u)\mathscr{D}(\delta u)\, \mathrm{d}V \qquad (1.69)$$

where C is a positive function which does not depend on u.

Hence equation (1.68) can be written

$$\int C\mathscr{D}(u)\mathscr{D}(\delta u)\, \mathrm{d}V + \int p\, \delta u \, \mathrm{d}V = \int q(s)\, \delta u \, \mathrm{d}S \qquad (1.70)$$

We now write $\delta u = \lambda\eta$, where η is an arbitrary function of x (see Figure 1.9) which satisfies the essential boundary conditions, and λ

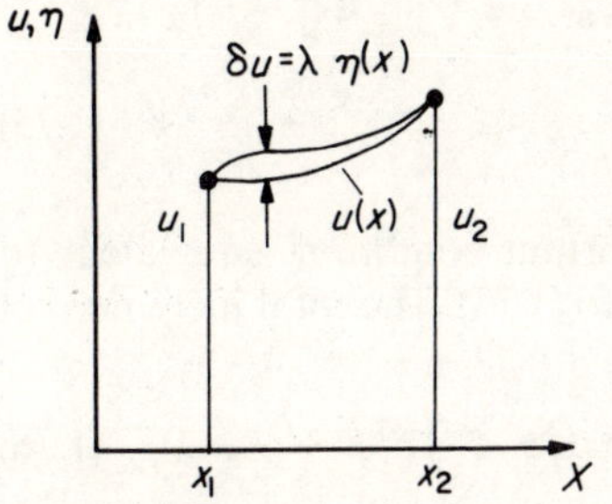

Figure 1.9 Definition of a variation of the u(x) function

is a parameter which does not depend on x. Equation (1.70) can be written

$$\int C\mathscr{D}(u)\mathscr{D}(\lambda\eta)\,\mathrm{d}V + \int p\lambda\eta\,\mathrm{d}V - \int q(s)\lambda\eta\,\mathrm{d}S = 0 \qquad (1.71)$$

As $\mathscr{D}$ is a linear operator and λ is not a function of the spatial co-ordinates, we write

$$\lambda\left\{\int C\mathscr{D}(u)\mathscr{D}(\eta)\,\mathrm{d}V + \int p\eta\,\mathrm{d}V - \int q(s)\eta\,\mathrm{d}S\right\}$$
$$= \lambda\left[\frac{\mathrm{d}}{\mathrm{d}\lambda}\left\{\frac{1}{2}\int C\mathscr{D}^2(u+\lambda\eta)\,\mathrm{d}V + \int p(u+\lambda\eta)\,\mathrm{d}V - \int q(s)(u+\lambda\eta)\,\mathrm{d}S\right\}\right]_{\lambda=0} \equiv 0 \qquad (1.72)$$

Or more briefly

$$\lambda\left\{\frac{\mathrm{d}}{\mathrm{d}\lambda}F(u+\lambda\eta)\right\}_{\lambda=0} = \lambda\left\{\frac{\mathrm{d}}{\mathrm{d}\lambda}F(u+\delta u)\right\}_{\lambda=0} \equiv 0 \qquad (1.73)$$

where $F(u + \lambda\eta)$ is called a quadratic functional (or function of functions) and is equal to

$$F = \frac{1}{2}\int C\mathscr{D}^2(u+\lambda\eta)\,\mathrm{d}V + \int p(u+\lambda\eta)\,\mathrm{d}V - \int q(s)(u+\lambda\eta)\,\mathrm{d}S \qquad (1.74)$$

If the functional can be expanded about the u position, we write

$$F(u + \delta u) = F(u) + \lambda\left(\frac{\mathrm{d}F(u + \delta u)}{\mathrm{d}\lambda}\right)_{\lambda=0} + \frac{\lambda^2}{2!}\left(\frac{\mathrm{d}^2F(u + \delta u)}{\mathrm{d}\lambda^2}\right)_{\lambda=0} + \ldots$$

$$= F(u) + \delta F + \frac{1}{2!}\delta^2 F + \ldots \tag{1.75}$$

The condition $\delta F = 0$ is an equilibrium condition, equivalent to any one of equations (1.68) or (1.70) to (1.73). The total increment of the function F can be written

$$\Delta F = F(u + \delta u) - F(u) = \delta F + \tfrac{1}{2}\delta^2 F + \ldots \tag{1.76}$$

As $\delta F = 0$ we have to investigate the second (or higher) variations of the functional in order to find what type of stationary (or equilibrium) point we have:

$$\Delta F = \tfrac{1}{2}\delta^2 F + \ldots \tag{1.77}$$

The second variation of F gives

$$\delta^2 F = \lambda^2 \int C\mathscr{D}(\eta)\mathscr{D}(\eta)\,\mathrm{d}V = \int C\mathscr{D}^2(\delta u)\,\mathrm{d}V \tag{1.78}$$

If $C > 0$, equation (1.78) is positive definite, which implies that the F functional is a *minimum*. This property can be used to establish bounds in the solution.

Equation (1.66) is usually called the Euler–Lagrange equation corresponding to the F functional. By integrating by parts we have reduced the order of differentiation as F has lower-order derivatives than equation (1.66).

Note that the variational notation used throughout allows us to work with the δs as if they were differentials with u equivalent to independent variables and F equivalent to a function.

Example 1.17

Let us deduce the functional corresponding to the governing equation

$$\mathscr{L}(u) = \frac{\mathrm{d}^2 u}{\mathrm{d}x^2} + u + x = 0 \tag{a}$$

with boundary conditions $u(0) = u(1) = 0$.

One can write the following weighted expression:

$$\delta F = \int_0^1 \left(\frac{d^2u}{dx^2} + u + x \right) \delta u \, dx = 0 \tag{b}$$

Which gives, after integrating by parts,

$$\delta F = \int_0^1 \left(-\frac{du}{dx}\frac{d\delta u}{dx} + u\,\delta u + x\,\delta u \right) dx = 0 \tag{c}$$

from which the F functional can be deduced:

$$F = \tfrac{1}{2} \int_0^1 \left\{ -\left(\frac{du}{dx}\right)^2 + u^2 + 2xu \right\} dx \tag{d}$$

This functional can be easily written because $\mathscr{L}(u)$ is a linear and symmetric operator. If we have at $x = 0$ a condition of the type

$$\left(\frac{du}{dx}\right)_0 = g \tag{e}$$

where g is a given quantity, this condition can be included in (b) in order to obtain the correct functional. That is, we write

$$\delta F = \int_0^1 \left\{ \frac{d^2u}{dx^2} + u + x \right\} \delta u \, dx - \left[\left(\frac{du}{dx} - g \right) \delta u \right]_0 = 0 \tag{f}$$

which gives the following functional:

$$F = \tfrac{1}{2} \int_0^1 \left\{ -\left(\frac{du}{dx}\right)^2 + u^2 + 2ux \right\} dx + [gu]_0 \tag{g}$$

Example 1.18

Consider the case of flow in a channel of unit width when the vertical velocity component is equal to zero ($v = 0$); see Figure 1.10.

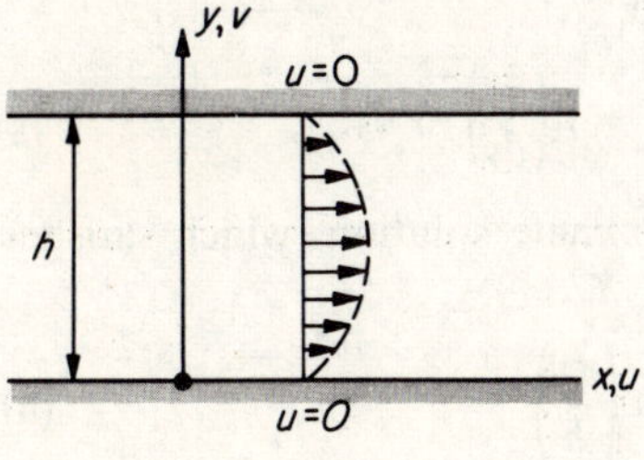

Figure 1.10 Flow in a channel

The continuity equation is

$$\frac{\partial u}{\partial x} + \frac{\partial v}{\partial y} = 0 \tag{a}$$

where u is the velocity component in the x direction, and for $v = 0$ we have $u = u(y)$ only.

For confined time independent laminar flow and forced convection, the momentum equation in the x direction can be written as

$$\rho\left[u\frac{\partial u}{\partial x} + v\frac{\partial u}{\partial y}\right] = -\frac{\partial p}{\partial x} + \mu\frac{\partial^2 u}{\partial y^2} \tag{b}$$

where ρ is the mass density, μ the viscosity and p the pressure. As $u = u(y)$ and $v = 0$, we have

$$0 = -\frac{\partial p}{\partial x} + \mu\frac{\partial^2 u}{\partial y^2} \tag{c}$$

Integrating (c) twice,

$$u = \frac{1}{2\mu}\frac{\partial p}{\partial x}\{y^2\} + C_1 y + C_2 \tag{d}$$

Taking into account the boundary conditions $u = 0$ at $y = 0, h$ we have

$$u = \frac{1}{2}\frac{h^2}{\mu}\left(\frac{\partial p}{\partial x}\right)\left\{\frac{y^2}{h^2} - \frac{y}{h}\right\} \tag{e}$$

Equation (e) gives the Poiseuille flow solution between parallel plates. The variational functional for this problem can be obtained by weighting equation (a) and integrating by parts,

$$\delta F = \int_0^h \left\{-\frac{\partial p}{\partial x} + \mu\frac{\partial^2 u}{\partial y^2}\right\}\delta u \, dy$$

$$= \int_0^h \left\{-\frac{\partial p}{\partial x}\delta u - \mu\frac{\partial u}{\partial y}\frac{\partial \delta u}{\partial y}\right\} dy = 0 \tag{f}$$

The functional can now be deduced,

$$F = -\tfrac{1}{2}\int_0^h \left\{2\frac{\partial p}{\partial x}u + \mu\left(\frac{\partial u}{\partial y}\right)^2\right\} dy \tag{g}$$

One can take the following approximate solution (which satisfies the boundary conditions):

$$u \simeq u_c \sin\left(\frac{\pi y}{h}\right) \tag{h}$$

Thus

$$-F = \frac{1}{2}\int_0^h \left\{2\frac{\partial p}{\partial x}u_c \sin\left(\frac{\pi y}{x}\right) + \mu\left(\frac{\pi}{h}\right)^2 \cos^2\frac{\pi y}{h}u_c^2\right\} dy$$

$$= \frac{\partial p}{\partial x}u_c\left(\frac{2h}{\pi}\right) + \frac{\mu}{2}\left(\frac{\pi}{h}\right)^2\frac{h}{2}u_c^2 \qquad \text{(i)}$$

The variation of F gives

$$\delta F = \left\{\frac{\partial p}{\partial x}\left(\frac{2h}{\pi}\right) + \frac{\mu}{h}\left(\frac{\pi}{h}\right)^2 hu_c\right\}\delta u_c = 0$$

$$\therefore \qquad u_c = -\frac{4h^2}{\pi^3\mu}\frac{\partial p}{\partial x} \qquad \text{(j)}$$

Velocity distributions (h) and (j) are compared in Table 1.1 for the case

$$\frac{h^2}{\mu}\left(\frac{\partial p}{\partial x}\right) = 1$$

Table 1.1 POISEUILLE FLOW RESULTS

y	*Exact*	*Approximate*
0	0	0
0·1	$-4{\cdot}5 \times 10^{-2}$	$-3{\cdot}98648 \times 10^{-2}$
0·2	$-8{\cdot}0 \times 10^{-2}$	$-7{\cdot}58275 \times 10^{-2}$
0·3	−0·105	−0·10436
0·4	−0·12	−0·12269
0·5	−0·125	−0·12900

1.6 RAYLEIGH–RITZ METHOD

The Rayleigh–Ritz method consists in replacing the u variable (or variables) *in an F functional* by an approximate solution of the type

$$u = \sum_{k=1}^{N} \alpha_k\phi_k \qquad (1.79)$$

and afterwards minimising the functional with respect to the α_k variables. Increasing the number of terms in the approximate solution will generally improve the results.

The ϕ_i functions have to satisfy the essential boundary conditions of the problem and to be elements of a given sequence of linearly independent functions, as was the case in Galerkin's method.

Assume a functional of the type

$$F = \int_{x_1}^{x_2} I(u, u_x, x)\,dx \qquad (1.80)$$

where $u_x = \partial u/\partial x$ with boundary conditions $u(x_1) = u(x_2) = 0$. The approximating function for u has to be continuous up to degree $N - 1$ (note N, the highest degree of differentiation in the functional, is here 1) and has to satisfy the corresponding boundary conditions. This implies that each ϕ_k function satisfies the essential boundary conditions exactly; i.e.

$$\phi_k(x_1) = \phi_k(x_2) = 0 \qquad (1.81)$$

Substituting for the trial function and requiring F to be stationary, with respect to $\alpha_1\alpha_2 \ldots \alpha_n$, leads to n equations relating the α_i coefficients

$$\frac{\partial F}{\partial \alpha_i} = 0, \quad i = 1, 2, \ldots n \qquad (1.82)$$

These equations are linear in α_i when F is a quadratic function of u and u_x.

In addition to the above admissibility conditions, the functions ϕ_k must be part of a *complete* set for the solution to converge to the exact solution.

In order to assess the convergence of the method, one has to take two or more trial functions. When the method is applied to a given functional F with a minimum, we can measure convergence by comparing successive values of the functional obtained with the following sequence:

$$\begin{aligned} u^{(1)} &= \alpha_1^{(1)}\phi_1 \\ u^{(2)} &= \alpha_1^{(2)}\phi_1 + \alpha_2^{(2)}\phi_2 \\ \ldots & \\ u^{(i)} &= \alpha_1^{(i)}\phi_1 + \alpha_2^{(i)}\phi_2 + \ldots + \alpha_i^{(i)}\phi_i \end{aligned} \qquad (1.83)$$

where the ith expansion includes all the functions contained in the previous expansions. The functions ϕ_i are normally polynomials or trigonometric functions.

For quadratic functionals, we have, applying the above sequences,

$$F^{(1)} \geqslant F^{(2)} \geqslant \ldots \geqslant F^{(i)} \qquad (1.84)$$

This behaviour is called monotonic, and (1.83) is called a minimising sequence.

Example 1.19

Let us find the approximate solution for the case of a simply supported beam with a concentrated load at $x = l/2$ (Figure 1.11).

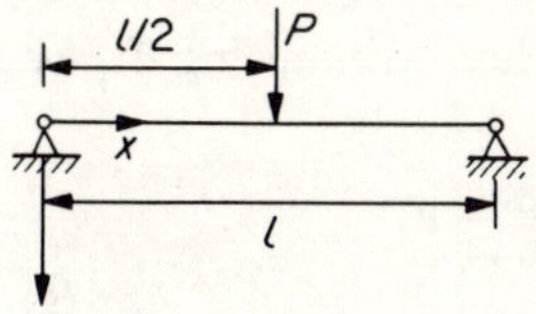

Figure 1.11 Beam under concentrated load

We will take as approximate solutions for the v displacements orthogonal functions of the sinusoidal type. For instance,

$$v = \alpha_1 \sin\left(\frac{x\pi}{l}\right) + \alpha_2 \sin\left(\frac{x}{l}3\pi\right) \qquad \text{(a)}$$
$$= \alpha_1\phi_1 + \alpha_2\phi_2$$

Note that these functions satisfy the boundary conditions ($v = 0$ and moment $M = 0$ at $x = 0, x = l$) and are symmetric. The energy functional is

$$F = \frac{EI}{2}\int_0^l \left(\frac{d^2v}{dx^2}\right)^2 dx - \bar{P}v_c \qquad \text{(b)}$$

where the first term corresponds to the internal energy and the second to the potential of the loads. EI is the flexural rigidity of the beam (E = modulus of elasticity, I = moment of inertia) and v_c is the vertical deflection at the centre.

Now (b) gives, after substituting (a),

$$F = \frac{EI}{2}\int_0^l \left\{\alpha_1\left(\frac{\pi}{l}\right)^2 \sin\frac{x\pi}{l} + \alpha_2\left(\frac{3\pi}{l}\right)^2 \sin\left(\frac{3x\pi}{l}\right)\right\}^2 dx$$
$$- \bar{P}\left\{\alpha_1 \sin\left(\frac{\pi}{2}\right) + \alpha_2 \sin\left(\frac{3\pi}{2}\right)\right\} \qquad \text{(c)}$$
$$= \frac{EI}{2}\int_0^l \left\{\alpha_1^2\left(\frac{\pi}{l}\right)^4 \sin^2\left(\frac{x\pi}{l}\right) + \alpha_2^2\left(\frac{3\pi}{l}\right)^4 \sin^2\left(\frac{3\pi x}{l}\right)\right.$$
$$\left. + 2\alpha_1\alpha_2\left(\frac{\pi}{l}\right)^4 (3)^2 \sin\left(\frac{x\pi}{l}\right) \sin\left(\frac{3\pi x}{l}\right)\right\} dx - \bar{P}\{\alpha_1 - \alpha_2\} \qquad \text{(d)}$$

One can now integrate this equation, taking into account the orthogonality condition

$$\int_0^l \sin\left(\frac{m\pi x}{l}\right) \sin\left(\frac{n\pi x}{l}\right) dx = 0 \quad \text{if } m \neq n$$
$$= l/2 \quad \text{if } m = n \qquad \text{(e)}$$

Thus we can write (d) as

$$F = \frac{EI}{2}\left\{\alpha_1^2\left(\frac{\pi}{l}\right)^4\frac{l}{2} + \alpha_2^2\left(\frac{3\pi}{2}\right)^4\frac{l}{2}\right\} - \bar{P}(\alpha_1 - \alpha_2) \tag{f}$$

where α_1 and α_2 are *independent* of each other. Minimising (f), we have

$$\frac{\partial F}{\partial \alpha_1} = \frac{EI}{2}\left\{2\alpha_1\left(\frac{\pi}{l}\right)^4\frac{l}{2}\right\} - \bar{P} = 0$$

$$\frac{\partial F}{\partial \alpha_2} = \frac{EI}{2}\left\{2\alpha_2\left(\frac{3\pi}{l}\right)^4\frac{l}{2}\right\} + \bar{P} = 0 \tag{g}$$

Solving these equations, we obtain

$$\alpha_1 = \frac{2\bar{P}}{EI}\frac{l^3}{\pi^4}, \qquad \alpha_2 = -\frac{2\bar{P}}{EI}\frac{l^3}{\pi^4}\left(\frac{1}{3}\right)^4 \tag{h}$$

We can generalise this result and obtain α_3, α_4, etc. The n terms solution for the centre deflection is

$$v_c = \frac{2\bar{P}}{EI}\frac{l^3}{\pi^4}\left(1 + \frac{1}{3^4} + \frac{1}{5^4} + \ldots\right) \tag{i}$$

This is a convergent series, thus when $n \to \infty$, we obtain the exact result

$$v_c|_{\text{exact}} = \frac{Pl^3}{EI}\frac{1}{48}$$

1.7 SUBSIDIARY CONDITIONS

In certain cases we want a function to satisfy, in addition to the boundary conditions, certain other relations, called *subsidiary* conditions, which can be introduced using Lagrange multipliers.

Let us review briefly what the Lagrange multipliers are before using them with functionals. Consider a function $f(x, y, z)$ from which we want to obtain the stationary value

$$df = \frac{\partial f}{\partial x}dx + \frac{\partial f}{\partial y}dy + \frac{\partial f}{\partial z}dz = 0 \tag{1.85}$$

subjected to the two subsidiary conditions

$$g_1(x, y, z) = 0, \qquad g_2(x, y, z) = 0 \tag{1.86}$$

We can differentiate (1.86) and obtain

$$dg_1 = \frac{\partial g_1}{\partial x}\,dx + \frac{\partial g_1}{\partial y}\,dy + \frac{\partial g_1}{\partial z}\,dz$$
$$dg_2 = \frac{\partial g_2}{\partial x}\,dx + \frac{\partial g_2}{\partial y}\,dy + \frac{\partial g_2}{\partial z}\,dz \tag{1.87}$$

Let us multiply (1.87) by the parameters λ_1, λ_2 and then add them to (1.85). Thus we have

$$\left(\frac{\partial f}{\partial x} + \lambda_1 \frac{\partial g_1}{\partial x} + \lambda_2 \frac{\partial g_2}{\partial x}\right) dx + \left(\frac{\partial f}{\partial y} + \lambda_1 \frac{\partial g_1}{\partial y} + \lambda_2 \frac{\partial g_2}{\partial y}\right) dy + \left(\frac{\partial f}{\partial z} + \lambda_1 \frac{\partial g_1}{\partial z} + \lambda_2 \frac{\partial g_2}{\partial z}\right) dz = 0 \tag{1.88}$$

We have obtained three equations, which together with (1.86) permit us to determine the five unknowns $x, y, z, \lambda_1, \lambda_2$. The parameters λ_1, λ_2 are the Lagrangian multipliers and they usually have some physical meaning.

One can now write the extremisation problem in another way. Let us define a new function

$$f + \lambda_1 g_1 + \lambda_2 g_2 \tag{1.89}$$

Minimising this with respect to $x, y, z, \lambda_1, \lambda_2$ we obtain

$$\frac{\partial f}{\partial x} + \lambda_1 \frac{\partial g_1}{\partial x} + \lambda_2 \frac{\partial g_2}{\partial x} = 0$$
$$\frac{\partial f}{\partial y} + \lambda_1 \frac{\partial g_1}{\partial y} + \lambda_2 \frac{\partial g_2}{\partial y} = 0 \tag{1.90}$$
$$\frac{\partial f}{\partial z} + \lambda_1 \frac{\partial g_1}{\partial z} + \lambda_2 \frac{\partial g_2}{\partial z} = 0$$

plus $g_1 = 0$ and $g_2 = 0$ conditions. This gives the complete set of equations to be solved.

We can now extend the use of Lagrangian multipliers to the case of functionals. Assume for instance a functional F, a function of u and some of its derivatives under the subsidiary condition

$$G(u) = 0 \tag{1.91}$$

By using the Lagrangian multipliers a new functional can be written

$$F + \lambda G \tag{1.92}$$

Minimising this new functional with respect to the u variables and its derivatives, we obtain

$$\delta(F + \lambda G) = 0 \tag{1.93}$$

or

$$\delta F + \lambda \delta G = 0$$

plus the condition $G = 0$.

Example 1.20

Find the extreme value of the function

$$f(x, y) = x^2 - 2y^2 + xy - 4x + 4 \tag{a}$$

with the subsidiary condition

$$g(x, y) = x + y = 0 \tag{b}$$

One can now write a new function

$$f + \lambda g = (x^2 - 2y^2 + xy - 4x + 4) + \lambda(x + y) \tag{c}$$

and make it stationary with respect to x, y and λ, which gives

$$\begin{aligned} \frac{\partial(f + \lambda g)}{\partial x} &= 2x + y - 4 + \lambda = 0 \\ \frac{\partial(f + \lambda g)}{\partial y} &= -4y + x + \lambda = 0 \\ \frac{\partial(f + \lambda g)}{\partial \lambda} &= x + y = 0 \end{aligned} \tag{d}$$

Solving this system we obtain

$$x = -1, \qquad y = 1, \qquad \lambda = 5 \tag{e}$$

This gives for the stationary value of f

$$f = 6 \tag{f}$$

Note that without the subsidiary conditions we would have

$$x = \tfrac{16}{9}, \qquad y = \tfrac{4}{9} \tag{g}$$

which gives

$$f = \tfrac{36}{81} \tag{h}$$

BIBLIOGRAPHY

AYALA, G. and BREBBIA, C. A., in *Variational Methods in Engineering* (ed. C. A. Brebbia and H. Tottenham), Southampton University Press (1973)

COLLATZ, L., *The Numerical Treatment of Differential Equations*, Springer-Verlag, Berlin (1966)

COURANT, R. and HILBERT, D., *Methods of Mathematical Physics*, Vol. I, Wiley, New York (1953)

DUNCAN, W. J., *Variational Principles in Mechanics and Differential Equations*, Gt. Brit. Aero. Res. Council Dept. R. & M. 1798 (1937)

FINLAYSON, B. A., *The Method of Weighted Residuals and Variational Principles*, Academic Press (1972)

HILDEBRAND, F. B., *Methods of Applied Mathematics*, Prentice-Hall (1952)

KANTOROVICH, L. V. and KRYLOV, V. I., *Approximate Methods of Higher Analysis*, Noordhoff Ltd. (1958)

LANCZOS, C., *The Variational Principles of Mechanics*, University of Toronto Press (1964)

MIKHLIN, S. G., *Variational Methods in Mathematical Physics*, Macmillan, New York (1964)

RAYLEIGH, J. W., *The Theory of Sound*, Dover (1945)

SMIRNOW, V. I., *A Course in Higher Mathematics*, Moscow, Fitzmatgiz (1958)

WEINSTOCK, R., *Calculus of Variations with Applications to Physics and Engineering*, McGraw-Hill, New York (1952)

EXERCISES

1-1 Find the solution for a two-dimensional Poisson's equation applied on a $2a \times 2b$ rectangular domain

$$\frac{\partial^2 u}{\partial x^2} + \frac{\partial^2 u}{\partial y^2} = c \tag{a}$$

with boundary conditions $u = 0$ at $x = \pm a$ and $y = \pm b$. Using Galerkin's method, take as a first approximation

$$u^{(1)} = \alpha_1(a^2 - x^2)(b^2 - y^2) \tag{b}$$

and as a second approximation

$$u^{(2)} = \alpha_1(a^2 - x^2)(b^2 - y^2) + \alpha_2 x^2(a^2 - x^2)(b^2 - y^2) \tag{c}$$

Comment on the choice of functions (b) and (c) and the convergence of the results.

1-2 Find the approximate values of function $u(x)$ which satisfy the following differential equation:

$$\frac{d^2u}{dx^2} + \frac{du}{dx} + x = 0 \tag{a}$$

with $u(0) = u(1) = 0$ and x varying between 0 and 1.

Use the methods of moments, least square error, collocation and Galerkin. Compare the first and second approximations for each of them, using

$$u \simeq x(1 - x)(\alpha_1 + \alpha_2 x) \tag{b}$$

1-3 Assume that a curve is defined by a cubic polynomial $u = \alpha_1 + \alpha_2 x + \alpha_3 x^2 + \alpha_4 x^3$ and find the α_i coefficients by the least squares collocation method for the following data; $u_1 = 3{\cdot}9$ for $x_1 = 0$; $u_2 = 8{\cdot}5$ for $x_2 = 1$; $u_3 = 22$ for $x_3 = 2$; $u_4 = 35$ for $x_4 = 2{\cdot}5$; $u_5 = 52$ for $x_5 = 3$.

1-4 Consider a rectangular domain of dimensions a by b, in which the following governing equation applies:

$$\nabla^4 u = p \tag{a}$$

or

$$\frac{\partial^4 u}{\partial x^4} + 2\frac{\partial^4 u}{\partial x^2\,\partial y^2} + \frac{\partial^4 u}{\partial y^4} = p$$

where p is a constant. The boundary conditions are
Essential conditions:

$$u = 0 \quad \text{on } x = 0, a \text{ and } y = 0, b \text{ boundaries} \tag{b}$$

Natural conditions:

$$\frac{\partial^2 u}{\partial x^2} = 0 \quad \text{on } x = 0, a \text{ boundaries}$$

and

$$\frac{\partial^2 u}{\partial y^2} = 0 \quad \text{on } y = 0, b \text{ boundaries}$$

Solve equation (a) using Galerkin's method and assuming

$$u = \sum_{k=1} \sum_{l=1} \alpha_{kl} \sin\frac{k\pi x}{a} \sin\frac{l\pi y}{b} \tag{c}$$

Compare results with 1, 3 . . . etc., successive terms.

1-5 Given the equation

$$(1 + x)\frac{\mathrm{d}^2 u}{\mathrm{d}x^2} + \frac{\mathrm{d}u}{\mathrm{d}x} + \lambda u = 0 \quad \text{with } u(0) = u(1) = 0 \tag{a}$$

where λ is an eigenvalue, find the two successive approximations to λ using the functions

$$x(1 - x), \qquad x^2(1 - x) \tag{b}$$

1-6 Consider the set of algebraic equations

$$[A]\{X\} = \{B\} \tag{a}$$

where $[A]$ is a symmetrical matrix. Show that equation (a) can be interpreted as the stationary requirement for the functional

$$F = \tfrac{1}{2}\{X\}^T[A]\{X\} - \{X\}^T\{B\} \tag{b}$$

1-7 Deduce the equilibrium equations corresponding to the functional

$$F = \int_{x_1}^{x_2} I(x, u, v, u_x, v_x)\,\mathrm{d}x \tag{a}$$

under the subsidiary constraint

$$J = \int_{x_1}^{x_2} G(x, u, v)\,\mathrm{d}x = 0 \tag{b}$$

1-8 Obtain the equilibrium equation and natural boundary conditions for the functional

$$F = \int_{x_1}^{x_2} I(x, u, u_x)\,\mathrm{d}x - H_2(u) \tag{a}$$

1-9 The beam shown in Figure 1.12 is simply supported at A on a rigid foundation, while the B support is a spring of stiffness k. Assume the v deflection is a polynomial and find the stiffness for which the end C goes neither up nor down when the load $\bar{P}$ is applied. Apply the Rayleigh–Ritz method.

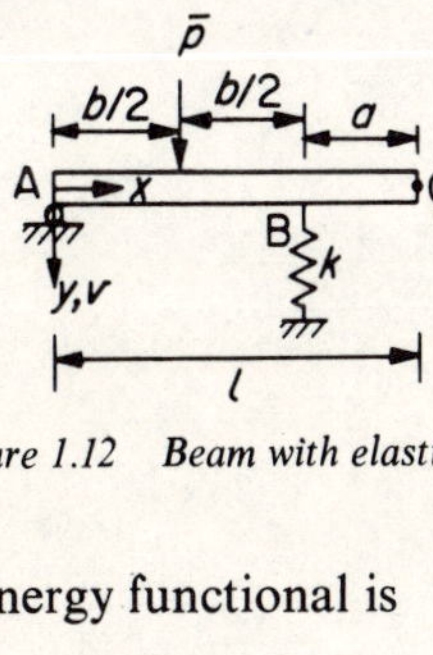

Figure 1.12 Beam with elastic support

The corresponding energy functional is

$$F = \frac{EI}{2}\int_0^l \left(\frac{\mathrm{d}^2 v}{\mathrm{d}x^2}\right)^2 \mathrm{d}x + \frac{k v_b^2}{2} - \bar{P} v_{b/2}$$

1-10 Describe the methods of (a) Galerkin, (b) least squares, (c) collocation and (d) Rayleigh–Ritz, using as an example the equation

$$\nabla^4 u = \frac{\partial^4 u}{\partial x^4} + 2\frac{\partial^4 u}{\partial x^2\,\partial y^2} + \frac{\partial^4 u}{\partial y^4} = p \tag{a}$$

with boundary conditions $u = \partial u/\partial n = 0$ on all the boundaries $x = \pm a$ and $y = \pm b$.

1-11 Minimise the following function:

$$F = f_1^2 + f_2^2 + \ldots + f_n^2 \tag{a}$$

subject to the subsidiary condition

$$G = C_1 f_1 + C_2 f_2 + \ldots C_n f_n = 1 \tag{b}$$

Use Lagrangian multipliers.

2 The Finite Element Technique

2.1 LOCALISED FUNCTIONS

The main difficulty in the direct application of the Galerkin and Rayleigh–Ritz methods is the choice of the global functions. These functions have not only to satisfy the essential boundary conditions of the problem but also to be adequate to describe the geometry, material and other characteristics of the problem. All these conditions are generally very difficult to fulfil and the methods in their 'classical' sense are of limited use. With the advent of high-speed digital computers, the idea of the approximating functions being localised in a small region was developed; in this way simpler functions can be used.

In what follows we discuss the idea of localised Galerkin functions and assume that the same applies for the Rayleigh–Ritz method, an assumption which is valid as it is possible to show that there is always a Galerkin formulation which corresponds to the minimisation of a Rayleigh–Ritz type functional.

The simplest localised function is the one-dimensional one shown in Figure 2.1(a). It has unit value at a particular point i, called a node, and is zero at all other nodes. For elements with external boundaries the function has to satisfy the external essential boundary conditions. In some other cases we may force derivatives of the function to be continuous [Figure 2.1(b)], such as in the case of beams in bending, for which the localised functions must at least be cubic.

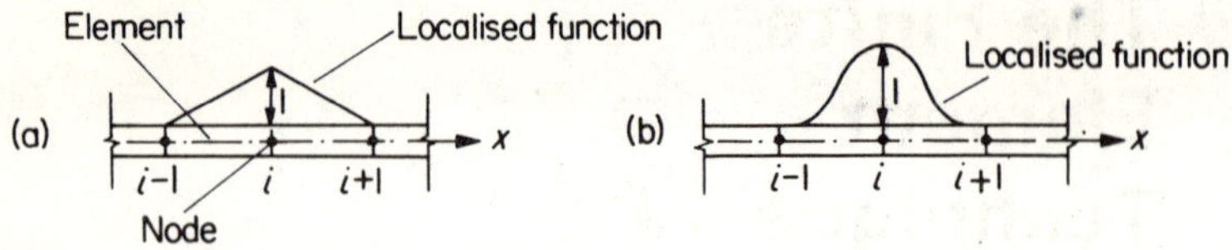

Figure 2.1 Localised one-dimensional functions. (a) Linear case: continuity of the function only. (b) Cubic case: continuity of function and its derivative.

Let us consider as an example the case of a Galerkin type expression which may be written as a function of a symmetric operator $\mathscr{D}$ such that

$$\mu\langle\mathscr{D}(u), \mathscr{D}(\delta u)\rangle = \langle c, \delta u\rangle \tag{2.1}$$

This is the type of operator we obtain for the case of Poiseuille's flow between parallel plates (see Example 1.18)

$$\int_0^h \mu\left(\frac{\mathrm{d}u}{\mathrm{d}x}\right)\left(\frac{\mathrm{d}\delta u}{\mathrm{d}x}\right)\mathrm{d}x = \int_0^h c\,\delta u\,\mathrm{d}x \tag{2.2}$$

One can now divide the total thickness h of the channel into four elements of length l (Figure 2.2(a)) and assume three localised functions at only the interior nodes, as the value of u (velocity) is taken to be zero at nodes 1 and 5. These are the essential boundary conditions for the problem.

The localised functions can now be superimposed to obtain the total distribution of velocities shown in Figure 2.2(b). Hence we can write equation (2.2) as a function of a local coordinate $\bar{x}$ which starts at the lowest point of each element and changes from 0 to 1; equation (2.2) becomes

$$\frac{1}{l}\sum_e \int_0^1 \mu\left(\frac{\mathrm{d}u}{\mathrm{d}\bar{x}}\right)\left(\frac{\mathrm{d}\delta u}{\mathrm{d}\bar{x}}\right)\mathrm{d}\bar{x} = l\sum_e \int_0^1 c\,\delta u\,\mathrm{d}\bar{x}$$

or

$$\begin{aligned}\frac{\mu}{l^2}\Bigg[&\int_0^1 \frac{\mathrm{d}}{\mathrm{d}\bar{x}}(u_2\bar{x})\frac{\mathrm{d}}{\mathrm{d}\bar{x}}(\delta u_2\bar{x})\,\mathrm{d}\bar{x}\\ &+\int_0^1 \frac{\mathrm{d}}{\mathrm{d}\bar{x}}\{u_3\bar{x} + u_2(1-\bar{x})\}\frac{\mathrm{d}}{\mathrm{d}\bar{x}}\{\delta u_3\bar{x} + \delta u_2(1-\bar{x})\}\,\mathrm{d}\bar{x}\\ &+\int_0^1 \frac{\mathrm{d}}{\mathrm{d}\bar{x}}\{u_4\bar{x} + u_3(1-\bar{x})\}\frac{\mathrm{d}}{\mathrm{d}\bar{x}}\{\delta u_4\bar{x} + \delta u_3(1-\bar{x})\}\,\mathrm{d}\bar{x}\\ &+\int_0^1 \frac{\mathrm{d}}{\mathrm{d}\bar{x}}\{u_4(1-\bar{x})\}\frac{\mathrm{d}}{\mathrm{d}\bar{x}}\{\delta u_4(1-\bar{x})\}\,\mathrm{d}\bar{x}\Bigg]\end{aligned}$$

$$= c\left\{\int_0^1 \delta(u_2\bar{x})\,\mathrm{d}\bar{x} + \int_0^1 \delta\{u_3\bar{x} + u_2(1-\bar{x})\}\,\mathrm{d}\bar{x} + \int_0^1 \delta\{u_4\bar{x} + u_3(1-\bar{x})\}\,\mathrm{d}\bar{x}\right\} \tag{2.3}$$

(e is the number of elements).

The left-hand side of equation (2.3) can be written

$$\frac{\mu}{l^2}\left\{\int_0^1 u_2\,\delta u_2\,\mathrm{d}\bar{x} + \int_0^1 (u_3 - u_2)(\delta u_3 - \delta u_2)\,\mathrm{d}\bar{x} + \int_0^1 (u_4 - u_3)(\delta u_4 - \delta u_3)\,\mathrm{d}\bar{x} + \int_0^1 u_4\,\delta u_4\,\mathrm{d}\bar{x}\right\} \tag{2.4}$$

After integration equation (2.4) becomes

$$\frac{\mu}{l^2}\{u_2\,\delta u_2 + (u_3 - u_2)(\delta u_3 - \delta u_2) + (u_4 - u_3)(\delta u_4 - \delta u_3) + u_4\,\delta u_4\} = c(\delta u_2 + \delta u_3 + \delta u_4) \tag{2.5}$$

or

$$\frac{\mu}{cl^2}\{(2u_2 - u_3)\,\delta u_2 + (-u_2 + 2u_3 - u_4)\,\delta u_3 + (-u_3 + 2u_4)\,\delta u_4\} = \delta u_2 + \delta u_3 + \delta u_4 \tag{2.6}$$

As the δu_i are arbitrary variations, we have the following system of equations:

$$\frac{\mu}{cl^2}\begin{bmatrix} 2 & -1 & 0 \\ -1 & 2 & -1 \\ 0 & -1 & 2 \end{bmatrix}\begin{Bmatrix} u_2 \\ u_3 \\ u_4 \end{Bmatrix} = \begin{Bmatrix} 1 \\ 1 \\ 1 \end{Bmatrix} \tag{2.7}$$

which gives (note that $u_2 = u_4$ for symmetry)

$$u_2 = \frac{3}{32}\frac{ch}{\mu^2}, \qquad u_3 = \frac{1}{8}\frac{ch}{\mu} \tag{2.8}$$

These values *coincide* with the exact solution at nodal points, but this does not mean that both solutions are identical, as the approximate one varies linearly from node to node and the exact one quadratically. Furthermore we could compute the exact and approximate values of the quadratic functional associated with (2.1), i.e.

$$F = \tfrac{1}{2}\int\left(\frac{\mathrm{d}u}{\mathrm{d}x}\right)^2\mathrm{d}x - \int cu\,\mathrm{d}x \tag{2.9}$$

and see that they are different, the exact value being the smaller.

The idea of having localised functions can also be applied to two-dimensional problems. In the simplest cases they give for these problems 'pyramidal' (for triangular elements) or 'parabolic' (for rectangular elements) functions; these are shown in Figure 2.3.

The above way of applying the localised functions can be better implemented if each element is considered separately, thus originating the *finite element technique.* In this technique, we consider each individual element and study its properties unrelated to the others. The elements are then joined together, satisfying the necessary continuity and the global boundary conditions.

Let us study separately the element shown in Figure 2.2(c), with a linear variation of velocity between the nodes i and $i + 1$:

$$u = u_{i+1}\bar{x} + (1 - \bar{x})u_i \tag{2.10}$$

We apply, only for this element, the function (2.10) in the variational expression (2.2)

$$\frac{\mu}{l^2}\int_0^1 (u_i - u_{i+1})\,\delta(u_i - u_{i+1})\,\mathrm{d}\bar{x} = c\int_0^1 \{\delta u_{i+1}\bar{x} + \delta u_i(1 - \bar{x})\}\,\mathrm{d}\bar{x} \tag{2.11}$$

We have, after integration,

$$\frac{\mu}{cl^2}\{u_i\,\delta u_i + u_{i+1}\,\delta u_{i+1} - u_i\,\delta u_{i+1} - u_{i+1}\,\delta u_i\} = \delta u_i\tfrac{1}{2} + \delta u_{i+1}\tfrac{1}{2} \tag{2.12}$$

This can be written as

$$\frac{\mu}{cl^2}\begin{bmatrix} 1 & -1 \\ -1 & 1 \end{bmatrix}\begin{Bmatrix} u_i \\ u_{i+1} \end{Bmatrix} = \tfrac{1}{2}\begin{Bmatrix} 1 \\ 1 \end{Bmatrix} \tag{2.13}$$

We can now think of the problem shown in Figure 2.2 as an assemblage of four different elements superimposing their effects at the common nodes. This gives for overall equilibrium

$$\frac{\mu}{cl^2}\begin{bmatrix} 1 & -1 & \cdot & \cdot & \cdot \\ -1 & 2 & -1 & \cdot & \cdot \\ \cdot & -1 & 2 & -1 & \cdot \\ \cdot & \cdot & -1 & 2 & -1 \\ \cdot & \cdot & \cdot & -1 & 1 \end{bmatrix}\begin{Bmatrix} u_1 \\ u_2 \\ u_3 \\ u_4 \\ u_5 \end{Bmatrix} = \begin{Bmatrix} \frac{1}{2} \\ 1 \\ 1 \\ 1 \\ \frac{1}{2} \end{Bmatrix} \tag{2.14}$$

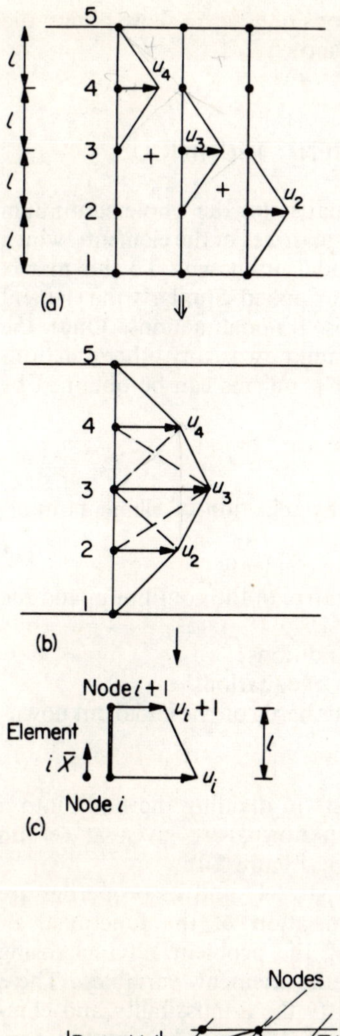

Figure 2.2 Localised functions analysis for Poiseulle's flow: (a) localised functions; (b) distribution of velocity; (c) element

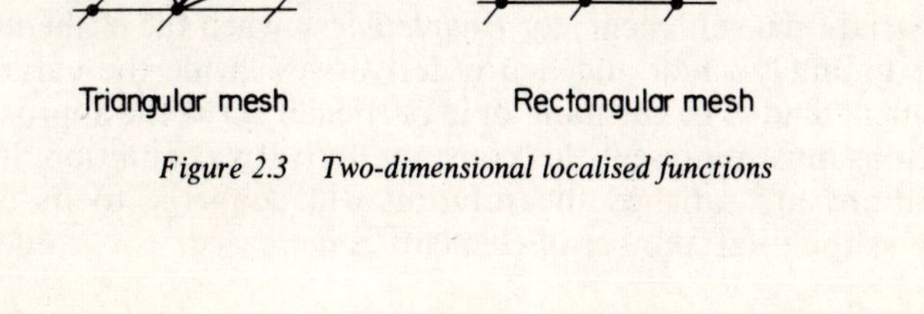

Figure 2.3 Two-dimensional localised functions

After applying the boundary conditions $u_1 = u_5 = 0$, we obtain the system of equations (2.7) previously shown.

2.2 THE FINITE ELEMENT TECHNIQUE

In the finite element technique the matrix for the whole continuum is formed by the contributions of the matrices of the elements, which are expressed as functions of the nodal unknowns. To this matrix the essential boundary conditions are applied. Similarly the element 'inputs' form a vector of generalised nodal actions. Once the system of equations is solved for the unknowns, any other functions of these unknowns necessary in our problems can be obtained by differentiation or integration.

Thus the basic steps of the method are:[1,2,3]

1 Discretisation of the problem by selection of elements interconnected at the nodal points
2 Evaluation of the matrices of the elements
3 Formulation of the complete matrix of the continuum and the input vector
4 Application of the boundary conditions
5 Solution of the resulting system of equations
6 Calculation of any other functions based on the nodal unknowns

The first step of the method consists in dividing the body into a number of elements. The nodal unknowns are given at certain points of these elements, called nodes (Figure 2.4).

Each element is then analysed separately and its properties are generally derived from the minimisation of the functional or Galerkin type expression governing the problem, after choosing some approximate functions for the element variables. These approximate functions have to satisfy the admissibility and completeness conditions for the problem. *Admissibility* implies continuity of the essential variables between elements, and that the order of the expansion is such that the terms are well defined in our variational statement; for *completeness*, when the elements tend to be infinitely small and hence derivatives inside the variational statement tend to be constant (or in particular zero), the approximate functions must represent this constant derivative condition. If these conditions are satisfied the solution will converge to its correct value as the total number of elements is increased.

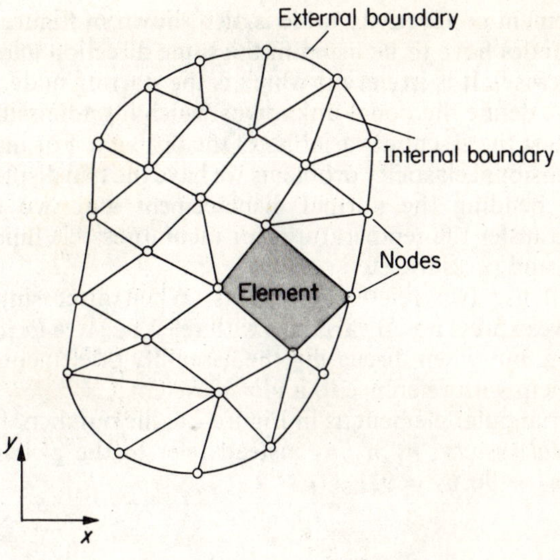

Figure 2.4 Finite element discretisation

The first stage of the method is to divide the body into elements of finite dimensions and to select some points on the boundaries. The elements are referred to as 'finite' elements and the points as nodal points or nodes.

We number the elements and nodes and specify the element node connectivity by listing, for each element, the nodes associated with the element. A typical discretisation of a two-dimensional problem is shown in Figure 2.5. We take the nodes at the corner of the

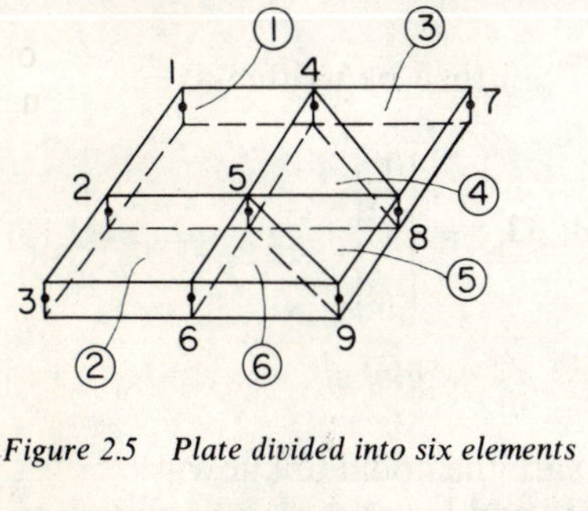

Figure 2.5 Plate divided into six elements

Element	Node n_1	n_2	n_3	n_4	Type
1	1	2	5	4	Rectangular
2	2	3	6	5	
3	4	8	7	–	Triangular
4	4	5	8	–	
5	5	9	8	–	
6	5	6	9	–	

elements on the middle surface of the plate. One could also select additional nodes along the element boundaries.

The element connectivity table is also shown in Figure 2.5. Note that the nodes have to be listed in the same direction (clockwise or anti-clockwise). It is irrelevant which is the starting node.

Next we define the nodal unknowns which for admissibility have to be at least the essential variables of the problem. For instance, for two-dimensional elasticity problems we have the two displacements; for plate bending the normal displacement and two rotations; for heat transfer the temperature; for incompressible fluid flow the velocities and pressure, etc.

We will use two reference systems. When discussing a single element we express nodal variables with reference to a *local* numbering system, but when discussing the assembly of elements we shall express them with reference to a *global* system.

For a triangular element as in Figure 2.6, the numbers 1, 2, 3 refer to the *local* system; n_1, n_2, n_3 instead refer to the *global* one (e.g. $n_1 = 94, n_2 = 96, n_3 = 92$).

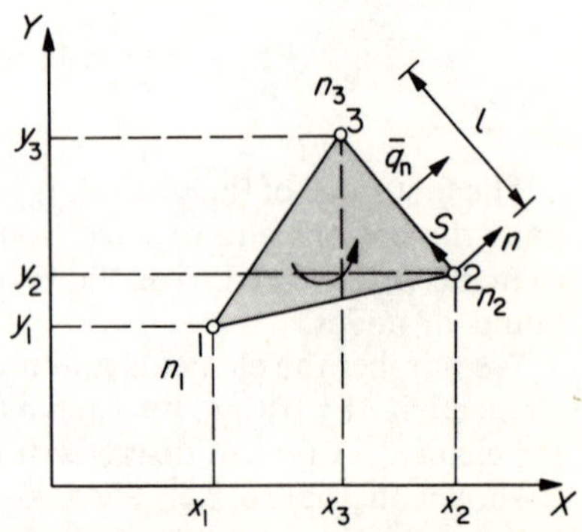

Figure 2.6 Triangular element

The variable vector for a node i can then be written as

$$\mathbf{U}_i = \begin{Bmatrix} u \\ v \\ w \end{Bmatrix}_i \quad \text{or} \quad \mathbf{U}_{n_i} = \begin{Bmatrix} u \\ v \\ w \end{Bmatrix}_{n_i} \tag{2.15}$$

local *global*

The elements of $\mathbf{U}_i$ or $\mathbf{U}_{n_i}$ in (2.15) are the nodal unknowns.

We also introduce upper case ($\mathbf{U}_i$) and lower case ($\mathbf{u}_i$) notation to denote the vector of all nodal unknowns as above or a single nodal unknown vector respectively.

If we assume three nodal unknowns, u, v and w, exist, the vector formed by the vectors of unknowns at the element nodes will be

$$\mathbf{U}^n = \begin{Bmatrix} \mathbf{U}_1 \\ \mathbf{U}_2 \\ \vdots \\ \mathbf{U}_s \end{Bmatrix} \quad \text{or} \quad \begin{Bmatrix} u_1 \\ v_1 \\ w_1 \\ u_2 \\ v_2 \\ w_2 \\ \vdots \\ u_s \\ v_s \\ w_s \end{Bmatrix} \tag{2.16}$$

The superscript n denotes that the vector extends over all nodes of the element, s is the number of nodes in the element, $\mathbf{U}^n$ is called the *element* nodal unknowns vector.

We can now introduce expansions for the $\mathbf{U}$ unknowns over the element domain in terms of the nodal values of the unknowns and interpolation functions (see Section 2.3):

$$\mathbf{U} = \mathbf{\Phi}^T \mathbf{U}^n \tag{2.17}$$

We proceed as for the Rayleigh–Ritz or weighted residual techniques. For simplicity let us assume a quadratic functional F in which the functions $\mathbf{U}$ are substituted:

$$F \simeq F(\mathbf{U}) \tag{2.18}$$

We minimise F with respect to the elements of $\mathbf{U}^n$, which results in a system of linear algebraic equations relating the elements of $\mathbf{U}^n$ with the input vector, which can be written

$$\mathbf{K}\mathbf{U}^n = \mathbf{P} \tag{2.19}$$

Next we have to superimpose all the elements in order to form the original body, an operation which will be explained in Section 2.3.

The critical part of the generation of the $\mathbf{K}$ matrix is in many cases to satisfy inter-element compatibility without resorting to high-order expansions. For functionals with high-order derivatives, formulations which violate inter-element compatibility may be used and they exhibit good convergence (see Chapter 3). It is a necessary condition, for functions of this type to converge, that they should have the terms giving zero derivatives [such as $\alpha_1 + \alpha_2 x$ for a functional with $d^2(\ )/dx^2$] and those that give constant derivatives in the functional (such as $\alpha_3 x^2$ for the above case). In general non-compatible models have to be carefully studied before use, for instance by taking into account the work due to the discontinuities and seeing if it becomes zero when the element size tends to be infinitely small.

2.3 ELEMENT MATRICES

We will now deduce the properties of a single element. We have already mentioned that the expansions for the element unknowns can be written in terms of the nodal values of the unknowns and certain interpolation functions:

$$\mathbf{U} = \boldsymbol{\Phi}^T \mathbf{U}^n \tag{2.20}$$

Hence one can express a u function—where u is one of the unknown variables in the $\mathbf{U}$ vector—as

$$u = \phi_1 u_1 + \phi_2 u_2 + \ldots = \sum_{i=1}^{S} \phi_i u_i = \boldsymbol{\phi}^T \mathbf{u}^n \tag{2.21}$$

where $i = 1, 2, \ldots, S$, etc., are the nodal numbers in the global system, and ϕ_i the corresponding interpolation functions. $\mathbf{u}^n$ is the element vector of nodal unknowns for the u function:

$$\begin{aligned} \mathbf{u}^n &= \{u_1 u_2 \ldots u_S\} \\ \boldsymbol{\phi}^T &= [\phi_1 \phi_2 \ldots \phi_S] \end{aligned} \tag{2.22}$$

If we have more than one variable, let us assume for instance two, u and v, which can be represented by the same interpolation functions, we can write (2.20) as

$$\mathbf{U} = \begin{Bmatrix} u \\ v \end{Bmatrix} = \begin{bmatrix} \boldsymbol{\phi} & \cdot \\ \cdot & \boldsymbol{\phi} \end{bmatrix}^T \begin{Bmatrix} \mathbf{u}^n \\ \mathbf{v}^n \end{Bmatrix} = \boldsymbol{\Phi}^T \mathbf{U}^n \tag{2.23}$$

When there is only one u variable we will write

$$\mathbf{u} = \boldsymbol{\phi}^T \mathbf{u}^n$$

In what follows we will deduce the corresponding finite element matrices for the case of the extended harmonic equation, i.e.

$$\frac{\partial}{\partial x}\left(h\frac{\partial u}{\partial x}\right) + \frac{\partial}{\partial y}\left(h\frac{\partial u}{\partial y}\right) + \lambda u = p \quad \text{in } A \tag{2.24}$$

with boundary conditions

$$u = \bar{u} \quad \text{on } S_1$$

$$h\frac{\partial u}{\partial n} = \bar{q} \quad \text{on } S_2$$

where $u = u(x, y)$; λ is a constant; $h, p, \bar{u}$ and $\bar{q}$ are prescribed functions of x, y; and $S = S_1 + S_2$ is the total boundary.

By requiring $\delta u \equiv 0$ on S_1, where S_1 can be part of the external boundary or the inter-element boundaries shown in Figure 2.4, we can write

$$\delta F = \iint -\left\{\frac{\partial}{\partial x}\left(h\frac{\partial u}{\partial x}\right) + \frac{\partial}{\partial y}\left(h\frac{\partial u}{\partial y}\right) + \lambda u - p\right\} \delta u \, \mathrm{d}A + \int_{S_2} \left(h\frac{\partial u}{\partial n} - \bar{q}\right) \delta u \, \mathrm{d}S = 0 \tag{2.25}$$

One can integrate (2.25) by parts to obtain

$$\delta F = \iint \left\{h\frac{\partial u}{\partial x}\frac{\partial}{\partial x}\delta u + h\frac{\partial u}{\partial y}\frac{\partial}{\partial y}\delta u - \lambda u\,\delta u + p\,\delta u\right\} \mathrm{d}A - \int_{S_2} \bar{q}\,\delta u \, \mathrm{d}S \tag{2.26}$$

Let us now propose an approximation for u over the element as indicated by (2.21)

$$u = \boldsymbol{\phi}^T \mathbf{u}^n \tag{2.27}$$

Note that in order to evaluate (2.26) the derivatives of u must be finite inside the element as well as on the boundaries. Also, to satisfy the condition $\delta u \equiv 0$, u has to be continuous across interior boundaries. This implies that (2.27), when evaluated on a side, must involve the nodal variables for only those nodes which are on the side. We call this requirement 'inter-element' continuity. A linear expansion requires nodes at the vertices; a quadratic expansion requires two end nodes and one interior node, and so on.

We form the spatial derivatives by operating on (2.27)

$$\begin{aligned} \frac{\partial u}{\partial x} &= u_1\frac{\partial \phi_1}{\partial x} + u_2\frac{\partial \phi_2}{\partial x} + \ldots = \boldsymbol{\phi}_{,x}^T \mathbf{u}^n \\ \frac{\partial u}{\partial y} &= u_1\frac{\partial \phi_1}{\partial y} + u_2\frac{\partial \phi_2}{\partial y} + \ldots = \boldsymbol{\phi}_{,y}^T \mathbf{u}^n \end{aligned} \tag{2.28}$$

Then substituting the function u and its derivatives in (2.26) leads to

$$\delta \mathbf{u}^{n,T}\mathbf{K}\mathbf{u}^n - \lambda\, \delta \mathbf{u}^{n,T}\mathbf{M}\mathbf{u}^n - \delta \mathbf{u}^{n,T}\mathbf{P} = 0 \tag{2.29}$$

or

$$(\mathbf{K} - \lambda \mathbf{M})\mathbf{u}^n = \mathbf{P}$$

where **K**, **M**, **P** element matrices are defined by

$$\begin{aligned}
\mathbf{K} &= \iint_{A_e} h(\boldsymbol{\phi}_{,x}\boldsymbol{\phi}_{,x}^T + \boldsymbol{\phi}_{,y}\boldsymbol{\phi}_{,y}^T)\, \mathrm{d}A \\
\mathbf{M} &= \iint_{A_e} \boldsymbol{\phi}\boldsymbol{\phi}^T\, \mathrm{d}A \\
\mathbf{P} &= \iint - p\boldsymbol{\phi}\, \mathrm{d}A + \int_{S_2} \bar{q}\boldsymbol{\phi}\, \mathrm{d}S
\end{aligned} \tag{2.30}$$

Note that **K** and **M** are symmetrical.

Example 2.1

Consider Laplace's equation

$$h_x \frac{\partial^2 u}{\partial x^2} + h_y \frac{\partial^2 u}{\partial y^2} = 0 \tag{a}$$

with boundary conditions

$$u = \bar{u} \quad \text{on } S_1, \qquad q_n = \alpha_{nx} h_x \frac{\partial u}{\partial x} + \alpha_{ny} h_y \frac{\partial u}{\partial y} = \bar{q} \quad \text{on } S_2 \tag{b}$$

α_{nx} and α_{ny} are the direction cosines of the normal n with respect to x and y. One can write (a) and (b) in Galerkin's form as

$$\iint \left(h_x \frac{\partial u}{\partial x}\frac{\partial \delta u}{\partial x} + h_y \frac{\partial u}{\partial y}\frac{\partial \delta u}{\partial y} \right) \mathrm{d}x\, \mathrm{d}y = \int_{S_2} \bar{q}_n\, \delta u\, \mathrm{d}S \tag{c}$$

where the condition $\delta u \equiv 0$ on S_1 is identically satisfied.

Note that (c) is the variation of the functional

$$F = \tfrac{1}{2} \iint \left\{ h_x \left(\frac{\partial u}{\partial x}\right)^2 + h_y \left(\frac{\partial u}{\partial y}\right)^2 \right\} \mathrm{d}x\, \mathrm{d}y - \int_{S_2} \bar{q}_n u\, \mathrm{d}S \tag{d}$$

with essential boundary conditions ($u = \bar{u}$ on S_1) identically satisfied.

Let us now apply (c) on a triangular element, for example the one shown in Figure 2.6. One can propose a function

$$u = \alpha_1 + \alpha_2 x + \alpha_3 y \tag{e}$$

which varies linearly in the triangle and produces three unknowns per element. These unknowns can be related to the nodal values at nodes 1, 2 and 3. Specialising (e) for the corner nodes, we have

$$\begin{Bmatrix} u_1 \\ u_2 \\ u_3 \end{Bmatrix} = \begin{bmatrix} 1 & x_1 & y_1 \\ 1 & x_2 & y_2 \\ 1 & x_3 & y_3 \end{bmatrix} \begin{Bmatrix} \alpha_1 \\ \alpha_2 \\ \alpha_3 \end{Bmatrix} \tag{f}$$

$$\mathbf{u}^n = \mathbf{C}\boldsymbol{\alpha}$$

The inverse of (f) gives the relationship between α_i and u_i

$$\begin{Bmatrix} \alpha_1 \\ \alpha_2 \\ \alpha_3 \end{Bmatrix} = \frac{1}{2A} \begin{bmatrix} 2A_1^0 & 2A_2^0 & 2A_3^0 \\ b_1 & b_2 & b_3 \\ a_1 & a_2 & a_3 \end{bmatrix} \begin{Bmatrix} u_1 \\ u_2 \\ u_3 \end{Bmatrix} \tag{g}$$

$$\boldsymbol{\alpha} = \mathbf{C}^{-1}\mathbf{u}^n$$

where

$$a_i = x_k - x_j \qquad b_i = y_j - y_k \qquad 2A_i^0 = x_j y_k - x_k y_j \tag{h}$$

$$i = 1, 2, 3; \qquad j = 2, 3, 1; \qquad k = 3, 1, 2$$

$$2A = b_1 a_2 - b_2 a_1 \quad \text{(where } A \text{ is the area of the element)}$$

We will arrive at similar expressions in a more direct manner in Chapter 3 by using other types of element functions.

Note that function (e) satisfies compatibility, and has the zero and constant derivative terms which are necessary for completeness when the element size decreases.

From equations (e) and (g) the derivatives of u can be written as

$$\left.\begin{aligned} \frac{\partial u}{\partial x} &= \alpha_2 = \frac{1}{2A}\{b_1 u_1 + b_2 u_2 + b_3 u_3\} = \frac{1}{2A}\mathbf{b}^T\mathbf{u}^n \\ \frac{\partial u}{\partial y} &= \alpha_3 = \frac{1}{2A}\{a_1 u_1 + a_2 u_2 + a_3 u_3\} = \frac{1}{2A}\mathbf{a}^T\mathbf{u}^n \end{aligned}\right\} \tag{i}$$

where

$$\mathbf{a} = \{a_1 a_2 a_3\}, \qquad \mathbf{b} = \{b_1 b_2 b_3\}$$

Similarly the variations are expressed

$$\frac{\partial \delta u}{\partial x} = \frac{1}{2A}\mathbf{b}^T\,\delta\mathbf{u}^n \qquad \frac{\partial \delta u}{\partial y} = \frac{1}{2A}\mathbf{a}^T\,\delta\mathbf{u}^n \tag{j}$$

From (e) and (g), the u function can be written

$$\left.\begin{aligned} u = \phi_1 u_1 + \phi_2 u_2 + \phi_3 u_3 = \boldsymbol{\phi}^T\mathbf{u}^n \\ \delta u = \boldsymbol{\phi}^T\,\delta\mathbf{u}^n \end{aligned}\right\} \tag{k}$$

where

$$\phi_1 = \frac{1}{2A}(2A_1^0 + b_1 x + a_1 y)$$

$$\phi_2 = \frac{1}{2A}(2A_2^0 + b_2 x + a_2 y)$$

$$\phi_3 = \frac{1}{2A}(2A_3^0 + b_3 x + a_3 y)$$

One can now substitute (i), (j) and (k) into the variational statement (c), which becomes

$$\delta\mathbf{u}^{n,T}\frac{1}{4A^2}\iint \{h_x\mathbf{b}\mathbf{b}^T + h_y\mathbf{a}\mathbf{a}^T\}\,\mathrm{d}x\,\mathrm{d}y\,\mathbf{u}^n = \delta\mathbf{u}^{n,T}\int \boldsymbol{\phi}\bar{q}_n\,\mathrm{d}S \tag{l}$$

After integration, the left-hand side of (l) becomes

$$\frac{1}{4A}\{\delta u_1\ \delta u_2\ \delta u_3\}\left\{h_x\begin{bmatrix} b_1^2 & b_1b_2 & b_1b_3 \\ & b_2^2 & b_2b_3 \\ \text{sym.} & & b_3 \end{bmatrix}\right.$$

$$\left. + h_y\begin{bmatrix} a_1^2 & a_1a_2 & a_1a_3 \\ & a_2^2 & a_2a_3 \\ \text{sym.} & & a_3 \end{bmatrix}\right\}\begin{Bmatrix} u_1 \\ u_2 \\ u_3 \end{Bmatrix} \tag{m}$$

Note that integration over x, y only produces a term A (the area of the element) because the integrand is constant. For other interpolation functions the integrand is a function of x and y. The integration is then more complex and it may be necessary to use numerical integration formulae (see Appendix).

Expression (m) can be written

$$\delta\mathbf{u}^{n,T}\mathbf{K}\mathbf{u}^n \tag{n}$$

where $\mathbf{K}$ is the influence coefficient matrix for the element shown in Figure 2.6.

The right-hand side of equation (l) only exists for the boundary S_2 of the body. Let us assume a constant value of $\bar{q}_n$ on side 2–3 for instance (Figure 2.6).

The relationship between the s–n, x–y systems is

$$\begin{Bmatrix} s \\ n \end{Bmatrix} = -\frac{1}{l}\begin{bmatrix} b_1 & -a_1 \\ a_1 & b_1 \end{bmatrix}\left[\begin{Bmatrix} y \\ x \end{Bmatrix} - \begin{Bmatrix} y_2 \\ x_2 \end{Bmatrix}\right]$$
$$\begin{Bmatrix} y \\ x \end{Bmatrix} = -\frac{1}{l}\begin{bmatrix} b_1 & a_1 \\ -a_1 & b_1 \end{bmatrix}\begin{Bmatrix} s \\ n \end{Bmatrix} + \begin{Bmatrix} y_2 \\ x_2 \end{Bmatrix} \tag{o}$$

Hence the integral on S_2, i.e.

$$\int_0^l \boldsymbol{\phi}\bar{q}_n \,\mathrm{d}S = \frac{1}{2A}\int_0^l \begin{Bmatrix} 2A_1^0 + b_1 x + a_1 y \\ 2A_2^0 + b_2 x + a_2 y \\ 2A_3^0 + b_3 x + a_3 y \end{Bmatrix} \bar{q}_n \,\mathrm{d}S \tag{p}$$

after substitution of x and y by s (note that $n = 0$ on side 2–3) and integration, becomes

$$\int_0^l \boldsymbol{\phi}\bar{q}_n \,\mathrm{d}S = \bar{q}_n \begin{Bmatrix} 0 \\ l/2 \\ l/2 \end{Bmatrix} = \bar{q}_n \mathbf{P} \tag{q}$$

From (m) and (q) we have the equilibrium requirement for the element

$$\underset{(3\times 3)}{\mathbf{K}}\;\underset{(3\times 1)}{\mathbf{u}^n} = \underset{(3\times 1)}{\mathbf{P}}$$

2.4 SYSTEM EQUATIONS

In order to obtain the global equilibrium equations for a body formed by different elements, we have to assemble the element matrices and apply the boundary conditions. One first expands the element nodal unknowns vector $\mathbf{U}^n$ in terms of the vector of unknowns at the nodes $\mathbf{U}_{n_i}$, now referred to the *global* numbering system:

$$\mathbf{U}^n = \{\mathbf{U}_{n_i}\} = \begin{Bmatrix} \mathbf{U}_{n_1} \\ \mathbf{U}_{n_2} \\ \vdots \end{Bmatrix} \tag{2.31}$$
$$i = 1, 2, \ldots, s$$

where s is the number of nodes in the element and n_i their number referred to the complete system. We partition $\mathbf{K}$, $\mathbf{M}$ and $\mathbf{P}$ for the element consistently with the partitioning of $\mathbf{U}^n$. They are understood to be resulting from a derivation based on several unknowns for node, i.e. based on $\mathbf{U}^n$.

$$\begin{aligned} \mathbf{K} &= \{\mathbf{k}_{ij}\} \\ \mathbf{M} &= \{\mathbf{m}_{ij}\} \\ \mathbf{P} &= \{\mathbf{p}_i\} \qquad i, j = 1, 2 \ldots x \end{aligned} \tag{2.32}$$

With this notation the terms in (2.29), for instance, take the form

$$\begin{aligned} \delta\mathbf{U}^{n,T}\mathbf{K}\mathbf{U}^n &= \sum_{i=1}^{s} \delta\mathbf{U}_{n_i}^T \left\{ \sum_{j=1}^{s} \mathbf{k}_{ij}\mathbf{U}_{n_j} \right\} \\ \delta\mathbf{U}^{n,T}\mathbf{M}\mathbf{U}^n &= \sum_{i=1}^{s} \delta\mathbf{U}_{n_i}^T \left\{ \sum_{j=1}^{s} \mathbf{m}_{ij}\mathbf{U}_{n_j} \right\} \\ \delta\mathbf{U}^{n,T}\mathbf{P} &= \sum_{i=1}^{s} \delta\mathbf{U}_{n_i}^T \mathbf{p}_i \end{aligned} \tag{2.33}$$

If, in the derivation of the element matrices, the unknowns are referred to the local frame rather than the basic frame, it is necessary to transform the nodal unknowns, $\mathbf{U}_{n_i}, \delta\mathbf{U}_{n_i}$ in (2.33) from the element frame to the basic frame. We use an asterisk to indicate the global frame and obtain

$$\begin{aligned} \mathbf{U}_{n_i} &= \mathbf{R}\mathbf{U}_{n_i}^* \\ \delta\mathbf{U}_{n_i} &= \mathbf{R}\,\delta\mathbf{U}_{n_i}^* \end{aligned} \tag{2.34}$$

where $\mathbf{R}$ contains the direction cosines for the local direction with respect to the global matrices.

Equations (2.33) can now be written

$$\begin{aligned} \delta\mathbf{U}^{n,T}\mathbf{K}\mathbf{U}^n &= \sum_{i=1}^{s} \delta\mathbf{U}_{n_i}^{*,T} \left\{ \sum_{j=1}^{s} \mathbf{k}_{ij}^*\mathbf{U}_{n_j}^* \right\} \\ \delta\mathbf{U}^{n,T}\mathbf{M}\mathbf{U}^n &= \sum_{i=1}^{s} \delta\mathbf{U}_{n_i}^{*,T} \left\{ \sum_{j=1}^{s} \mathbf{m}_{ij}^*\mathbf{U}_{n_j}^* \right\} \\ \delta\mathbf{U}^{n,T}\mathbf{P} &= \sum_{i=1}^{s} \delta\mathbf{U}_{n_i}^{*,T} \mathbf{p}_i^* \end{aligned} \tag{2.35}$$

where

$$\begin{aligned} \mathbf{k}_{ij}^* &= \mathbf{R}^T\mathbf{k}_{ij}\mathbf{R} \\ \mathbf{m}_{ij}^* &= \mathbf{R}^T\mathbf{m}_{ij}\mathbf{R} \\ \mathbf{p}_i^* &= \mathbf{R}^T\mathbf{p}_i \end{aligned} \tag{2.36}$$

The governing equations for the *whole* body can be written as

$$\sum_e \delta \mathbf{U}^{n,T}(\mathbf{K}\mathbf{U}^n - \lambda \mathbf{M}\mathbf{U}^n) = \sum_e \delta \mathbf{U}^{n,T}\,\mathbf{P} \tag{2.37}$$

where $\sum_e$ indicates summation over all the elements. If N denotes the total number of nodes, we can define a system nodal unknown vector (in what follows we assume that $\mathbf{U}_{n_i}$ terms are referred to the global frame, that is we omit the asterisk for simplicity).

$$\mathscr{U} = \{\mathbf{U}_1, \mathbf{U}_2 \ldots \mathbf{U}_N\} \tag{2.38}$$

Expanding (2.37) by summing the contribution of the elements incident on each node, we have

$$\delta\mathscr{U}^T\{\mathscr{K}\mathscr{U} - \lambda\mathscr{M}\mathscr{U}\} = \delta\mathscr{U}^T\mathscr{P} \tag{2.39}$$

or for arbitrary $\delta\mathscr{U}$,

$$\{\mathscr{K} - \lambda\mathscr{M}\}\mathscr{U} = \mathscr{P} \tag{2.40}$$

The partitioned form of (2.40) is

$$\left\{\begin{bmatrix} \mathscr{K}_{11} & \mathscr{K}_{12} & \ldots & \mathscr{K}_{1N} \\ \mathscr{K}_{21} & \mathscr{K}_{22} & \ldots & \mathscr{K}_{2N} \\ \cdots & \cdots & \cdots & \cdots \\ \mathscr{K}_{N1} & \mathscr{K}_{N2} & \ldots & \mathscr{K}_{NN} \end{bmatrix} - \lambda \begin{bmatrix} \mathscr{M}_{11} & \mathscr{M}_{12} & \ldots & \mathscr{M}_{1N} \\ \mathscr{M}_{21} & \mathscr{M}_{22} & \ldots & \mathscr{M}_{2N} \\ \cdots & \cdots & \cdots & \cdots \\ \mathscr{M}_{N1} & \mathscr{M}_{N2} & \ldots & \mathscr{M}_{NN} \end{bmatrix}\right\} \begin{Bmatrix} \mathbf{U}_1 \\ \mathbf{U}_2 \\ \vdots \\ \mathbf{U}_N \end{Bmatrix} = \begin{Bmatrix} \mathscr{P}_1 \\ \mathscr{P}_2 \\ \vdots \\ \mathscr{P}_N \end{Bmatrix} \tag{2.41}$$

We assemble $\mathscr{K}$, $\mathscr{M}$ and $\mathscr{P}$ in *partitioned* form working with successive elements. The contribution for an element is listed below.

In $\mathscr{P}$

$$\mathbf{p}_i \quad \text{in row } i \qquad i = 1, 2 \ldots s$$

In $\mathscr{K}$

$$\mathbf{k}_{ij} \quad \text{in row } i\text{, column } j \qquad i, j = 1, 2 \ldots s \tag{2.42}$$

In $\mathscr{M}$

$$\mathbf{m}_{ij} \quad \text{in row } i\text{, column } j \qquad i, j = 1, 2 \ldots s$$

These operations are carried out for all the elements. Since $\mathbf{M}$ and $\mathbf{K}$ are symmetrical for our example, the $\mathscr{M}$ and $\mathscr{K}$ matrices of (2.40) are going to be symmetrical and only the coefficients on and above the diagonal need to be stored.

Example 2.2

Let us consider a body composed of only four triangular elements (Figure 2.7). Assume that we know the element matrices $\mathbf{K}\mathbf{U}'' = \mathbf{P}$

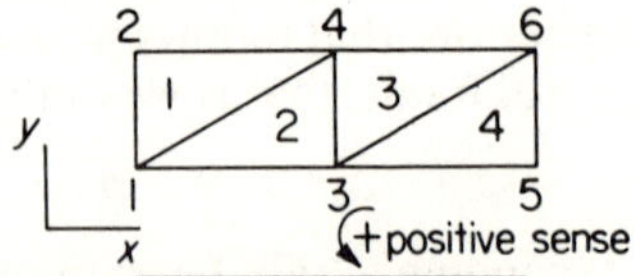

Figure 2.7 Four element body

Element	Nodes n_1	n_2	n_3
1	1	4	2
2	3	4	1
3	3	6	4
4	5	6	3

(see for instance Example 2.1). For instance, for the element 2 the matrices can be written

$$\begin{bmatrix} k_{11}^2 & k_{12}^2 & k_{13}^2 \\ k_{21}^2 & k_{22}^2 & k_{23}^2 \\ k_{31}^2 & k_{32}^2 & k_{33}^2 \end{bmatrix} \begin{Bmatrix} u_1^2 \\ u_2^2 \\ u_3^2 \end{Bmatrix} = \begin{Bmatrix} p_1^2 \\ p_2^2 \\ p_3^2 \end{Bmatrix}$$

or

$$\mathbf{K}\mathbf{U}'' = \mathbf{P} \text{ for element 2}$$

where $\mathbf{K}$ is symmetric.

The unknowns in the above equation are referred to the local numbering system. For the *global* nodal unknowns numbering system, we can write

$$\begin{bmatrix} k_{11}^2 & k_{12}^2 & k_{13}^2 \\ k_{21}^2 & k_{22}^2 & k_{23}^2 \\ k_{31}^2 & k_{32}^2 & k_{33}^2 \end{bmatrix} \begin{Bmatrix} u_3 \\ u_4 \\ u_1 \end{Bmatrix} = \begin{Bmatrix} p_1^2 \\ p_2^2 \\ p_3^2 \end{Bmatrix} \qquad \text{(a)}$$

We are interested in superimposing the effects of all the elements in order to form the system matrices for the whole body. If the structure has six nodes we will finally obtain a 6×6 matrix. A typical element like 2 will have the nine coefficients plus the three right-hand side terms distributed in the global matrix and right-hand side vector, as follows:

$$\mathscr{K} = \begin{bmatrix} k_{33}^{2} & \cdot & k_{31}^{2} & k_{32}^{2} & \cdot & \cdot \\ \cdot & \cdot & \cdot & \cdot & \cdot & \cdot \\ k_{13}^{2} & \cdot & k_{11}^{2} & k_{12}^{2} & \cdot & \cdot \\ k_{23}^{2} & \cdot & k_{21}^{2} & k_{22}^{2} & \cdot & \cdot \\ \cdot & \cdot & \cdot & \cdot & \cdot & \cdot \\ \cdot & \cdot & \cdot & \cdot & \cdot & \cdot \end{bmatrix} \quad \text{and } \mathscr{P} = \begin{bmatrix} p_{3}^{2} \\ \cdot \\ p_{1}^{2} \\ p_{2}^{2} \\ \cdot \\ \cdot \end{bmatrix} \qquad \text{(b)}$$

Once all the elements have been superimposed we will find that the global matrix looks like

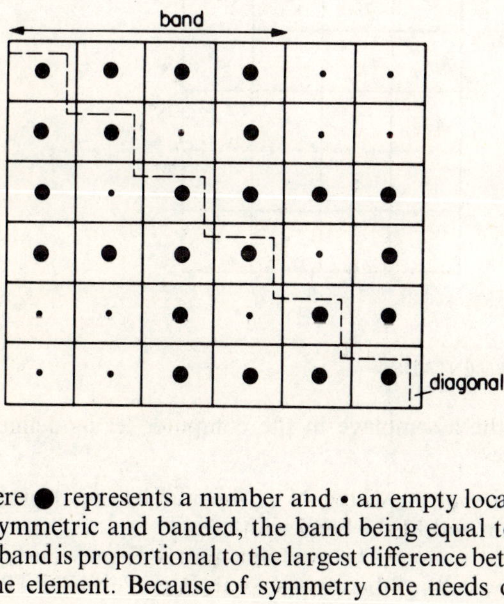

(c)

where ● represents a number and • an empty location. This matrix is symmetric and banded, the band being equal to four. Note that the band is proportional to the largest difference between nodes in the same element. Because of symmetry one needs only to store the diagonal and upper diagonal elements in the form

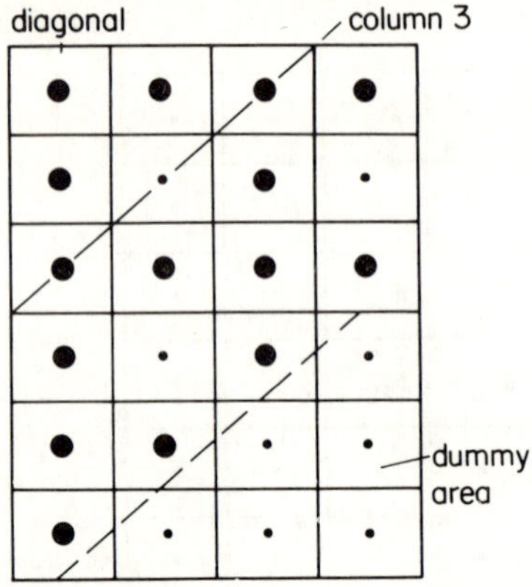

(d)

For instance the coefficients of element 2—equation (b)—would now be stored in the form

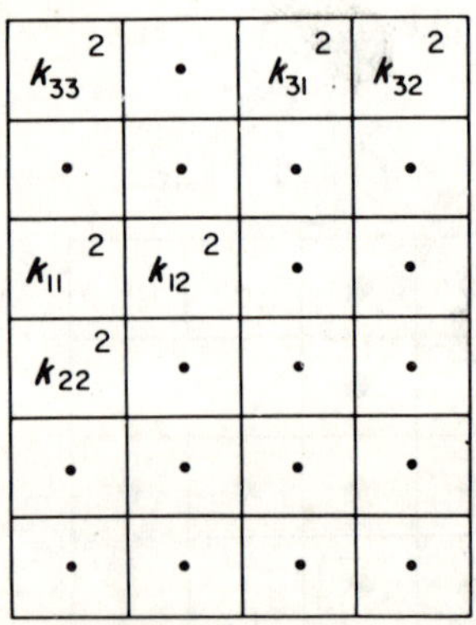

(e)

COMPUTER IMPLEMENTATION

In order to do the assemblage in the computer let us define the following matrices:

A(N,M) Global matrix, where N is the number of degrees of freedom and M the band width.

S(N1,N1) Element matrix. This was our matrix K previously, but we now call it S because K is an integer in FORTRAN.

I(N2,N3) Matrix, where the number of the nodes corresponding to the N2 element are stored. This matrix defines the connectivity table of Figure 2.7. N3 is the number of nodes in each element, which for the above example is equal to N1. In general, however, the number of nodes is different from the number of element unknowns as there could be more than one unknown per node.

We first write the assembler without taking advantage of the symmetric and banded character of the global matrix. We have for each element, NUMEL, the following loops,

```
      DØ 10  IM=1,N3
            IR= I(NUMEL,IM)
            DØ 20    IN=1,N3
                  IC=      I(NUMEL,IN)
                  A(IR,IC)=A(IR,IC)+S(IM,IN)
20          CØNTINUE
10    CØNTINUE
```

In order to reduce the column to the position required in the banded matrix [equations (d) or (c)] we have to modify IC (which defines the columns) as follows:

```
      DØ 10  IM=1,N3
            IR=I(NUMEL,IM)
            DØ20     IN=1,N3
                  IC=I(NUMEL,IN)
                  ID=IC-(IR-1)
                  IF(ID.LT.1) GØTØ 20
                  A(IR,ID)=A(IR,ID)+S(IM,IN)
20                      CØNTINUE
10    CØNTINUE
```

In a similar way the contribution of the nodal actions can be superimposed into a global action vector. Note that if there are no external actions on a node the sum of all the internal actions should be zero.

INTRODUCTION OF BOUNDARY CONDITIONS

The system of equations (2.40) has a solution for the two following cases:

$$\text{(a) } \mathscr{K} - \lambda\mathscr{M} \neq \mathbf{0} \quad \text{and} \quad \mathscr{P} \neq \mathbf{0} \tag{2.43}$$

$$\text{(b) } \mathscr{K} - \lambda\mathscr{M} = \mathbf{0} \quad \text{and} \quad \mathscr{P} = \mathbf{0} \tag{2.44}$$

In the first case we know the value of λ and are interested in finding the $\mathscr{U}$ vector of unknowns, by solving a system of equations,

$$(\mathscr{K} - \lambda\mathscr{M})\mathscr{U} = \mathscr{P}$$

For the second case the unknowns of our problems are a series of λ_i values, called eigenvalues, which satisfy (2.44).

(a) FIRST CASE ($\mathscr{K} - \lambda\mathscr{M} \neq \mathbf{0}$)

Let us assume that we know the values of some of the unknowns $\mathscr{U}$. We could now delete the rows, i.e. governing equations corresponding to these values, and the column elements multiplied by the imposed values can be passed to the right-hand side. This necessitates renumbering rows and columns and results in submatrices of different order. In what follows, we will describe an alternative procedure for introducing displacement boundary conditions which does not require any renumbering. We write the sum $\mathscr{K} - \lambda\mathscr{M}$ as $\mathscr{K}$ for simplicity. We assume that the unknowns at a node r are prescribed,

$$\mathbf{U}_r = \bar{\mathbf{U}}_r \tag{2.45}$$

One can replace the r (matrix) equation by (2.45). If $\bar{\mathbf{U}}_r \neq 0$ we have to operate on the upper diagonal part $\mathscr{K}$ and $\mathscr{P}$ as follows:

(1) $s = 1, 2 \ldots r - 1$

$$\mathscr{P}_s \Rightarrow \mathscr{P}_s - \mathscr{K}_{sr}\bar{\mathbf{U}}_r$$

$$\mathscr{K}_{sr} = \mathbf{0}$$

(2)

$$\begin{aligned} \mathscr{P}_r &= \bar{\mathbf{U}}_r \\ \mathscr{P}_{rr} &= \mathbf{I} \end{aligned} \tag{2.46}$$

(3) $s = r + 1, r + 2 \ldots N$

$$\mathscr{P}_s \Rightarrow \mathscr{P}_s - \mathscr{K}_{rs}^T\bar{\mathbf{U}}_r$$

$$\mathscr{K}_{rs} = \mathbf{0}$$

If the restraint directions of node r do not coincide with the direction of the basic frame, we have first to rotate $\mathscr{K}$ and $\mathscr{P}$. Let $\mathbf{U}_r^*$ denote the displacement matrix at node r referred to the restraint frame, hence

$$\mathbf{U}_r = \mathbf{R}\mathbf{U}_r^* \tag{2.47}$$

We can now write the system of equations as follows:

$$\begin{bmatrix} \mathscr{K}_{11} & \mathscr{K}_{12} & \cdots & \mathscr{K}_{1r}\mathbf{R} & \cdots & \mathscr{K}_{1N} \\ & \mathscr{K}_{22} & \cdots & \mathscr{K}_{2r}\mathbf{R} & \cdots & \mathscr{K}_{2N} \\ & & & \vdots & & \vdots \\ & \text{sym.} & & \mathbf{R}^T\mathscr{K}_{rr}\mathbf{R} & \cdots & \mathbf{R}^T\mathscr{K}_{rN} \\ & & & & & \\ & & & & \cdots & \mathscr{K}_{NN} \end{bmatrix} \begin{Bmatrix} \mathbf{U}_1 \\ \mathbf{U}_2 \\ \vdots \\ \mathbf{U}_r^* \\ \vdots \\ \mathbf{U}_N \end{Bmatrix} = \begin{Bmatrix} \mathscr{P}_1 \\ \mathscr{P}_2 \\ \vdots \\ \mathscr{P}_r^* \\ \vdots \\ \mathscr{P}_N \end{Bmatrix} \quad (2.48)$$

where $\mathscr{P}_r^* = \mathbf{R}^T\mathscr{P}_r$ are the applied actions in the local system.

Once (2.48) has been obtained we can apply the boundary conditions $\mathbf{U}_n^* = \bar{\mathbf{U}}_n$ in the same form as shown before [equations (2.46)].

The above operations are carried out for each restrained node and we obtain a modified system of equations:

$$\mathscr{K}'\mathscr{U}' = \mathscr{P}' \quad (2.49)$$

The size of $\mathscr{U}'$ is the same as $\mathscr{U}$ since we have included the dummy equations.

Example 2.3

In what follows we will assume that we have found the system $\mathscr{K}$ matrix and $\mathscr{P}$ vectors of Example 2.2. We now wish to apply the essential boundary conditions. If we assume that $u_3 = C$ where C is a known value, the third equation can be substituted by this condition as follows:

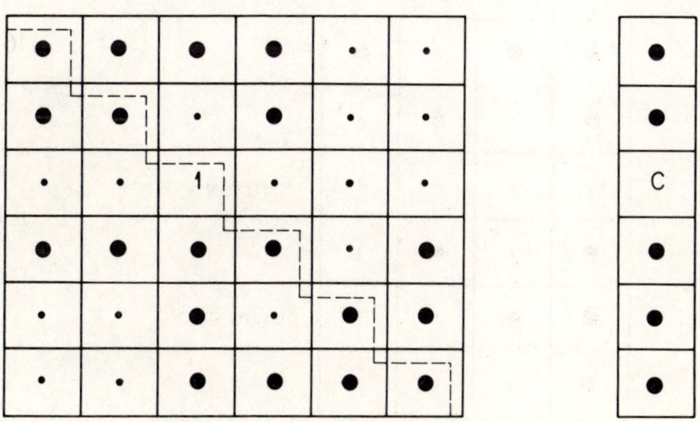

(a)

We operate on the full matrix for simplicity. In practice we will only work with the upper diagonal parts of the matrix.

In order to make this matrix symmetric, we multiply the coefficients of the third column by C and pass them to the right-hand side (with the exception of the coefficient on the diagonal). Thus we have

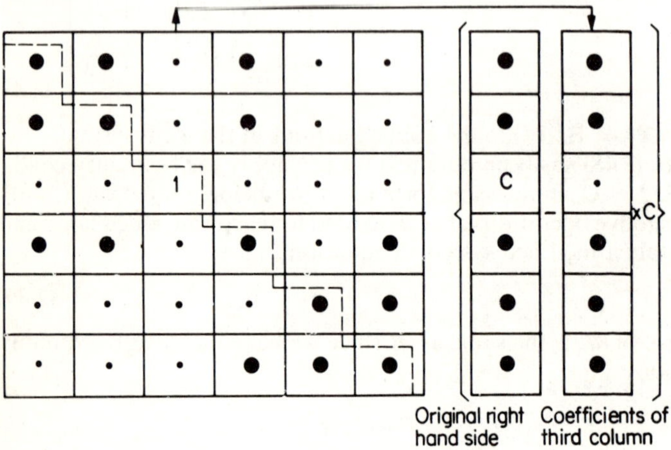

(b)

The new global matrix is now stored as

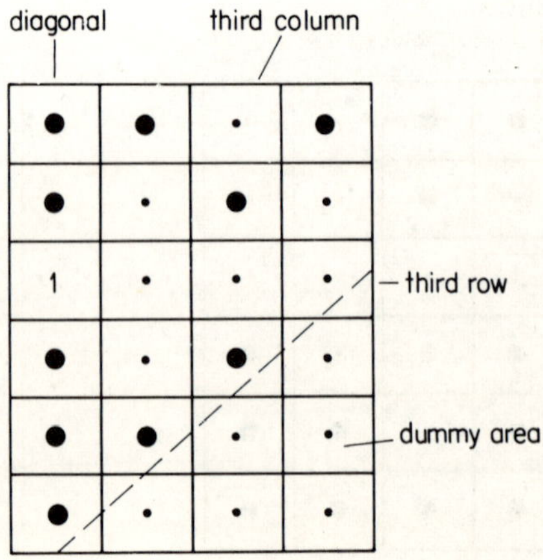

(c)

(b) SECOND CASE ($\mathcal{K} - \lambda\mathcal{M} = \mathbf{0}$)

Let us now study the case for which $\mathcal{P} = 0$. Thus

$$(\mathcal{K} - \lambda\mathcal{M})\mathcal{U} = \mathbf{0} \tag{2.50}$$

The problem reduces to determining the values of λ for which (2.50) has non-trivial solutions. These values are obtained from

$$\mathcal{K} - \lambda\mathcal{M} = \mathbf{0} \tag{2.51}$$

which gives us a series of λ_i values or eigenvalues. Each of these values is associated with a $\mathcal{U}_i$ eigenvector.

The matrix $\mathcal{M}$ in (2.51) is positive definite, but $\mathcal{K}$ is singular due to the zero-derivative type terms. We may impose some full restraint conditions in order to make $\mathcal{K}$ non-singular. For this case we will eliminate the rows and columns corresponding to these unknowns, which implies rearranging the elements of the matrix. Note that, if we did not rearrange the matrices but introduced the boundary conditions as in the first case, each unit value on the diagonal would give us a spurious λ_i eigenvalue.

The final equations can be written

$$(\mathcal{K}' - \lambda\mathcal{M}')\mathcal{U}' = \mathbf{0} \tag{2.52}$$

where the order of the matrices has now been reduced.

2.5 SOLUTION OF THE SYSTEM

(a) FIRST CASE

If $\mathcal{K}'$ is positive definite in equation (2.49), we can use a variety of solution techniques. Gauss elimination which appears to be more efficient in comparison with iterative methods is the only one we will describe here. Solution time in this case is extremely dependent on band width which, in turn, depends on the node numbering scheme. Therefore, one should always try to number the nodes so as to minimise the band width of the system global matrix.

After the $\mathcal{U}'$ vector has been obtained we may be interested in calculating the values of other variables, e.g. the velocities in terms of u, such as

$$v_x = h_x\frac{\partial u}{\partial x}, \qquad v_y = h_y\frac{\partial u}{\partial y} \tag{2.53}$$

For instance, for each of the elements studied in Example 2.1 [see equation (i)], we can write

$$\begin{Bmatrix} v_x \\ v_y \end{Bmatrix} = \begin{bmatrix} h_x b_1 & h_x b_2 & h_x b_3 \\ h_y a_1 & h_y a_2 & h_y a_3 \end{bmatrix} \begin{Bmatrix} u_{n_1} \\ u_{n_2} \\ u_{n_3} \end{Bmatrix} \tag{2.54}$$

One can in this way obtain the $v_x v_y$ values for all the elements, which for this particular case are constants within each of them.

Example 2.4

Solution of the linear system of equations. Once the boundary conditions have been introduced we obtain a positive definite system of equations of the form

$$\mathbf{AX} = \mathbf{B} \tag{a}$$

where a_{ij} terms are the coefficients of the A array and b_i terms the coefficients of the right-hand side vector. We can also write (a) as

$$a_{ij} x_j = b_i; \qquad i, j = 1, 2 \ldots n \tag{b}$$

in indicial notation.

In what follows we will describe how this system can be solved using the Gauss elimination procedure. If the original system is assigned the supercript 'o', we can write

$$\begin{aligned} a^o_{11} x_1 + a^o_{12} x_2 + \ldots + a^o_{1n} x_n &= b^o_1 \\ a^o_{21} x_1 + a^o_{22} x_2 + \ldots + a^o_{2n} x_n &= b^o_2 \\ &\vdots \\ a^o_{n1} x_1 + a^o_{n2} x_2 + \ldots + a^o_{nn} &= b^o_n \end{aligned} \tag{c}$$

One can solve the first equation for x_1 and obtain

$$x_1 = \frac{b^o_1}{a^o_{11}} - \frac{a^o_{12}}{a^o_{11}} x_2 - \frac{a^o_{13}}{a^o_{11}} x_3 - \ldots \frac{a^o_{1n}}{a^o_{11}} x_n \tag{d}$$

We can substitute (d) into the second, third, etc. equations of (c) and obtain a modified system of $(n - 1)$ equations

$$\begin{aligned} a^1_{22} x_2 + a^1_{23} x_3 + \ldots a^1_{2n} x_n &= b^1_2 \\ a^1_{32} x_2 + a^1_{33} x_3 + \ldots a^1_{3n} x_n &= b^1_3 \\ &\vdots \\ a^1_{n2} x_2 + a^1_{n3} x_3 + \ldots a^1_{nn} x_n &= b^1_n \end{aligned} \tag{e}$$

where

$$a_{ij}^1 = a_{ij}^o - a_{i1}^o \frac{a_{1j}^o}{a_{11}^o} \qquad i, j = 2, \ldots, n \qquad \text{(f)}$$
$$b_i^1 = b_i^o - a_{i1}^o \frac{b_1^o}{a_{11}^o}$$

A similar procedure is then used to eliminate x_2 from equations (e) and so on. A general algorithm for the elimination of x_k may then be written

$$x_k = \frac{b_k^{k-1}}{a_{kk}^{k-1}} - \frac{a_{kj}^{k-1}}{a_{kk}^{k-1}} x_j \qquad j = k + 1, \ldots n \qquad \text{(g)}$$

and

$$a_{ij}^k = a_{ij}^{k-1} - a_i^{k-1} \left(\frac{a_{kj}^{k-1}}{a_{kk}^{k-1}} \right) \qquad i, j = k + 1, \ldots n \qquad \text{(h)}$$
$$b_i^i = b_i^{k-1} - a_{ik}^{k-1} \left(\frac{b_k^{k-1}}{a_{kk}^{k-1}} \right)$$

Note that if we divide each equation by the a_{kk} divisor before the elimination we do not need to divide by a_{kk} in (*h*). This is how it will be done in the program.

After the above procedure has been applied $n - 1$ times, the original system of equations is reduced to the following single equation:

$$a_{nn}^{n-1} x_n = b_n^{n-1} \qquad \text{(i)}$$

which is solved directly for x_n

$$x_n = \frac{b_n^{n-1}}{a_{nn}^{n-1}} \qquad \text{(j)}$$

The remaining unknowns can be determined in reverse order by the application of (g); that is

$$x_m = \frac{b_k^{k-1}}{a_{kk}^{k-1}} - \frac{a_{kj}^{k-1}}{a_{kk}^{k-1}} x_j \qquad j = k - 1, n \qquad \text{(k)}$$

Those operations can be written in FORTRAN as follows:

```
      SUBRØUTINE SØLVER(A,B,N)
      DIMENSIØN A(10,10), B(10)
C
C     FØRWARD REDUCTIØN
```

```
C
C       DIVIDE EACH EQUATIØN BY A(K,K)
C
        DØ 10    K=1,N
              K1=K+1
              B(K)=B(K)/A(K,K)
C
C       CHECK FØR LAST EQUATIØN
C
              IF (K.EQ.N) GØTØ 100
C
C       FØRM A(K,J)/A(K,K)
C
              DØ 20    J=K1,N
                  IF(A(K,J).EQ.0)GØTØ 20
                  A(K,J)=A(K,J)/A(K,K)
                  DØ 30    I=K1,N
                      A(I,J)=A(I,J)-A(I,K)*A(K,J)
   30             CØNTINUE
C
C       MØDIFY B(J)
C
                  B(J)=B(J)-A(J,K)*B(K)
   20         CØNTINUE
   10   CØNTINUE
C
C       BACK SUBSTITUTIØN
C
  100   K1=K
        K=K-1
        IF(K.EQ.0) GØTØ 200
              DØ 40    J=K1,N
                  B(K)=B(K)-A(K,J)*B(J)
   40         CØNTINUE
        GØTØ 100
  200   RETURN
        END
```

Symmetric and banded system of equations. The systems of equations obtained in most practical problems are not only symmetric but also banded. This property means that the matrix **A** has the rectangular form shown in Example 2.3. For this case we can modify the above solver and work only with **A**(N, M) where M is the band width, and a_{ij} terms are the coefficients on and above the diagonal.

```
      SUBRØUTINE BANDSØL (A,B,N,M)
      DIMENSIØN A(20,8),B(20),C(8)
C
C     FØRWARD REDUCTIØN
C
C     DIVIDE EACH EQUATIØN BY A(K,1)
      DØ 10   K=1,N
         B(K)=B(K)/A(K,1)
C
C     CHECK FØR LAST EQUATIØN
C
         IF (K.EQ.N) GØTØ 100
C
C     FØRM A(K,J)/A(K,1) UNTIL M
C
         DØ 20     J=2,M
            C(J)=A(K,J)
            A(K,J)=A(K,J)/A(K,1)
   20    CØNTINUE
         DØ 30     L=2,M
            I=K+L-1
            IF(N.LT.I)GØTØ 30
            J=0
            DØ 40     LL=L,M
               J=J+1
               A(I,J)=A(I,J)-C(L)*A(K,LL)
   40       CØNTINUE
            B(I)=B(I)-C(L)*B(K)
   30    CØNTINUE
   10 CØNTINUE
C
C     BACK SUBSTITUTIØN
C
  100 K=K-1
      IF(K.EQ.0) GØTØ 200
         DØ 50     J=2,M
            L=K+J-1
            IF(N.LT.L) GØTØ 50
            B(K)=B(K)-A(K,J)*B(L)
   50    CØNTINUE
      GØTØ 100
  200 RETURN
      END
```

The reader can verify by comparing this listing with that for the full matrix that we are carrying out the same types of operations but taking into account the symmetry and banded character of the matrix and the different location of the column elements, which are now stored in a 'skew' form. The vector C is simply a working space. A check statement like IF(N.LT.I) GØTØ 30 stops the program from working outside the storage space for the real matrix, i.e. in the 'dummy' area [see equation (c) of Example 2.3].

(b) SECOND CASE

For this case we have the classical eigenvalue problem,

$$\mathcal{K}\mathcal{U} = \lambda\mathcal{M}\mathcal{U} \tag{2.55}$$

where $\mathcal{K}$ and $\mathcal{M}$ are assumed to be symmetrical and positive definite.

For each λ_i value we obtain a corresponding $\mathcal{U}_i$ such that

$$\begin{aligned} \mathcal{U}_i^T\mathcal{M}\mathcal{U}_i &= 1 \\ \mathcal{U}_i^T\mathcal{K}\mathcal{U}_i &= \lambda_i \end{aligned} \tag{2.56}$$

One can also show that the solution vectors $\mathcal{U}_i, \mathcal{U}_j$ are *orthogonal* with respect to $\mathcal{M}$ and $\mathcal{K}$; i.e.

$$\left.\begin{aligned} \mathcal{U}_j^T\mathcal{K}\mathcal{U}_i &= 0 \\ \mathcal{U}_j^T\mathcal{M}\mathcal{U}_i &= 0 \end{aligned}\right\} \quad \text{for } i \neq j \tag{2.57}$$

The final solution is generally a linear combination of the eigenvectors

$$\mathcal{U} = \sum_{i=1}^{s} q_i\mathcal{U}_i \tag{2.58}$$

where s is less than or equal to the number of total degrees of freedom of the system, and the qs are called generalised coordinates.

Let us assume that the original system equations were of the type

$$\mathcal{K}\mathcal{U} + \mathcal{M}\ddot{\mathcal{U}} = \mathcal{P} \tag{2.59}$$

where for harmonic motion $\ddot{\mathcal{U}} = \mathcal{U}\,e^{i\omega t}$, where ω is the rotating frequency, and if $\mathcal{P} = 0$ we obtain equation (2.55); i.e.

$$\begin{aligned} \mathcal{K}\mathcal{U} - \lambda\mathcal{M}\mathcal{U} &= \mathbf{0} \\ \lambda &= \omega^2 \end{aligned} \tag{2.60}$$

Substituting (2.58) into the equilibrium equations (2.59), we obtain

$$\sum_{i=1}^{s} \{q_i \mathscr{K} \mathscr{U}_i + \ddot{q}_i \mathscr{M} \mathscr{U}_i\} = \mathscr{P} \tag{2.61}$$

Premultiplying it by $\mathscr{U}_j^T$ we can reduce (2.61) to s uncoupled differential equations:

$$\ddot{q}_i + \lambda_i q_i = P_i \qquad i = 1, 2 \ldots s \tag{2.62}$$

where $P_i = \mathscr{U}_i^T \mathscr{P}$, $\lambda_i = \omega_i^2$.

The solution of equations (2.62) gives the contribution of each q_i in formula (2.58).

The essential difficulty with modal superposition is the choice of s, that is how many generalised coordinates one should take. In general, to make s the same as the number of degrees of freedom is expensive. Also, interpretation of the relative importance of the higher modes is difficult since the higher frequencies tend to be closely spaced. In selecting which modes to include for a particular case, one should compare the modal participation factors for $\mathscr{P}$ (that is the P_is).

To simplify the presentation, we have not included damping terms, although it is fairly straightforward to incorporate proportional damping in the formulation, but this introduces additional complications as regards finding the solution.

Example 2.5

Let us now study the solution of a linear system of equations of the form

$$\begin{bmatrix} a_{11} & a_{12} & \ldots & a_{1n} \\ a_{21} & a_{22} & \ldots & a_{2n} \\ \vdots & \vdots & & \vdots \\ a_{n1} & a_{n2} & \ldots & a_{nn} \end{bmatrix} \begin{Bmatrix} x_1 \\ x_2 \\ \vdots \\ x_n \end{Bmatrix} = \begin{bmatrix} \lambda & 0 & \ldots & 0 \\ 0 & \lambda & \ldots & 0 \\ \vdots & \vdots & & \vdots \\ 0 & 0 & & \end{bmatrix} \begin{Bmatrix} x_1 \\ x_2 \\ \vdots \\ x_n \end{Bmatrix} \tag{a}$$

or

$$\mathbf{AX} = \lambda \mathbf{IX}$$

where $\mathbf{A}$ is a real symmetric matrix, $\mathbf{X}$ is a vector of variables and λ the scalar parameter known as the eigenvalue. The problem consists in finding the values of λ and $\mathbf{X}$ which satisfy (a). For a matrix of order N there are N eigenvalues and N eigenvectors. We will now propose a transformation $\mathbf{R}$, such as

$$\mathbf{X} = \mathbf{R}\mathbf{X}' \tag{b}$$

Equation (a) becomes

$$\mathbf{A}\mathbf{R}\mathbf{X}' = \lambda\mathbf{R}\mathbf{X} \tag{c}$$

or

$$\mathbf{R}^T\mathbf{A}\mathbf{R}\mathbf{X}' = \lambda\mathbf{I}\mathbf{X}' \tag{d}$$

(where $\mathbf{R}^T\mathbf{R} = 1$). We will now assume that an $\mathbf{R}$ exists such that we can write (d) as

$$\mathbf{A}'\mathbf{X}' = \lambda\mathbf{X}' \tag{e}$$

$$\begin{bmatrix} a'_{11} & 0 & 0 & \dots & 0 \\ 0 & a'_{22} & 0 & \dots & 0 \\ 0 & 0 & a'_{33} & \dots & 0 \\ \vdots & \vdots & \vdots & & \vdots \\ 0 & 0 & 0 & \dots & a'_{nn} \end{bmatrix} \begin{Bmatrix} x'_1 \\ x'_2 \\ \vdots \\ x'_n \end{Bmatrix} = \begin{bmatrix} \lambda_1 & 0 & 0 & \cdots \\ 0 & \lambda_2 & 0 & \cdots \\ 0 \,. & 0 & \lambda_3 & \cdots \\ \vdots & \vdots & \vdots & \ddots \\ & & & \lambda_n \end{bmatrix} \begin{Bmatrix} x'_1 \\ x'_2 \\ \vdots \\ x'_n \end{Bmatrix}$$

where $\mathbf{A}'$ is a diagonal matrix which gives the N values of λ_n.

For each λ_n, we will have a corresponding eigenvector $\mathbf{X}_n$

$$\mathbf{A}'\mathbf{X}'_n = \lambda_n\mathbf{I}\mathbf{X}'_n \tag{f}$$

We can now write all the eigenvectors (each as a column) in a square matrix $\mathbf{V}$, in the same order as the eigenvectors, such that

$$\mathbf{A}\mathbf{V} = \mathbf{V}\lambda \tag{g}$$

If we premultiply by $\mathbf{V}^{-1}$,

$$\mathbf{V}^{-1}\mathbf{A}\mathbf{V} = \mathbf{V}^{-1}\mathbf{V}\lambda = \lambda\mathbf{I} \tag{h}$$

Comparing (h) with (d) we deduce that

$$\mathbf{V} = \mathbf{R} \tag{i}$$

The problem is thus reduced to finding the matrix $\mathbf{R}$ which diagonalises $\mathbf{A}$.

Jacobi's method. We will discuss here only Jacobi's method, which finds the $\mathbf{R}$ matrix by doing a series of simple two-dimensional transformations. These transformations are applied to the off-diagonal terms of $\mathbf{A}$ in order to reduce them to zero.

Let us first consider a two-dimensional system; the matrix **A** is simply

$$\mathbf{A} = \begin{bmatrix} a_{11} & a_{12} \\ a_{21} & a_{22} \end{bmatrix}$$

The transformation for an orthogonal system gives

$$\begin{Bmatrix} X_1 \\ X_2 \end{Bmatrix} = \begin{bmatrix} \cos\theta & -\sin\theta \\ \sin\theta & \cos\theta \end{bmatrix} \begin{Bmatrix} X'_1 \\ X'_2 \end{Bmatrix}$$

Thus,

$$\mathbf{A}' = \mathbf{R}^T\mathbf{AR}$$

$$= \begin{bmatrix} a_{11}\cos^2\theta + 2a_{12}\sin\theta\cos\theta + a_{22}\sin^2\theta & a_{12}(\cos^2\theta - \sin^2\theta) + \sin\theta\cos\theta(a_{22} - a_{11}) \\ \text{sym.} & a_{11}\sin^2\theta - 2a_{11}\sin\theta\cos\theta + a_{22}\cos^2\theta \end{bmatrix}$$

In order to eliminate the off-diagonal term, we require

$$a_{12}(\cos^2\theta - \sin^2\theta) + \sin\theta\cos\theta(a_{22} - a_{11}) = 0$$

or

$$a_{12}\tan^2\theta + (a_{11} - a_{22})\tan\theta - a_{12} = 0$$

Thus,

$$\tan\theta = \frac{-(a_{11} - a_{22}) \pm \sqrt{\{(a_{11} - a_{22})^2 + 4a_{12}^2\}}}{2a_{12}}$$

Let us restrict ourselves to one of the roots, for instance the one with the positive sign in the numerator:

$$\tan\theta = \frac{-(a_{11} - a_{22}) + \sqrt{\{(a_{11} - a_{22})^2 + 4a_{12}^2\}}}{2a_{12}} \tag{j}$$

Note that the other root will be 90° out of phase and that this is equivalent to working only in the $-\pi/2 < \theta < \pi/2$ interval.

We can write

$$\begin{aligned} \cos\theta &= (1 + \tan^2\theta)^{-\frac{1}{2}} \\ \sin\theta &= \cos\theta\tan\theta \end{aligned} \tag{k}$$

Jacobi's method consists in applying the above transformation to all the off-diagonal terms until all of them are, to a small error, equal to zero. We can start with the largest (in absolute value) off-diagonal term and rotate to make this term zero, etc. Note that, if the term is in the location I,J, we have to interchange 1 and 2 by I and J in formula (j).

After the matrix **A** is rotated r times, we have

$$\mathbf{R}_r^T \ldots \mathbf{R}_2^T \mathbf{R}_1 \mathbf{A} \mathbf{R}_1 \mathbf{R}_2 \ldots \mathbf{R}_r = \mathbf{A}'$$

or (l)

$$\mathbf{V}^T \mathbf{A} \mathbf{V} = \mathbf{A}'$$

The r different eigenvalues will be given by

$$\mathbf{V} = \mathbf{R}_1 \mathbf{R}_2 \ldots \mathbf{R}_r$$

Computer subroutine for the Jacobi method.

```
      SUBRØUTINE JACØB(A,V,ERR,N)
C
C     THE ORIGINAL MATRIX IS GIVEN IN A. ERR
C     GIVES THE REQUIRED PRECISION. THE
C     ITERATION STOPS WHEN THE LARGEST OFF
C     DIAGONAL ELEMENT IS LESS THAN ERR*
C     (LARGEST OFF-DIAG. ELEM. AT THE
C     BEGINNING). N, ORDER OF THE MATRIX.
C     EIGENVALUES ARE FINALLY WRITTEN ON
C     DIAGONAL OF A IN DECREASING ORDER,
C     EIGENVECTORS IN V. IF NUMBER OF ITER
C     IS LARGER THAN ITM THE SUBROUTINE
C     ALSO STOPS.
C
      DIMENSIØN   A(10,10), V(10,10)
C
C     INITIALIZATIØN
      ITM=200
      IT=0
      DØ 10   I=1,N
      DØ 10   J=1,N
      V(I,J)=0
   10 IF(I.EQ.J)V(I,J)=1
C
C     SEARCH FØR LARGEST ØFF-DIAG. ELEMENT.
C
   13 T=0
      M=N-1
      DØ 20   I=1,M
      J1=I+1
```

```
      DØ 20     J=J1,N
            IF(ABS(A(I,J)).LE.T) GØTØ20
            T=ABS(A(I,J))
            IR=I
            IC=J
   20 CØNTINUE
C
C     FØRM ACCURACY REQUIRED AND CHECK IF
C     EXCEEDED
C
      IF (IT.EQ.0) T1=T*ERR
      IF (T.LE.T1) GØTØ 999
C
C     FØRM CØS AND SIN
C
      PS=A(IR,IR)-A(IC,IC)
      TA = (-PS + SQRT(PS*PS + 4*T*T))/(2*A(IR,IC))
      C=1./SQRT(1.+TA*TA)
      S=C*TA
C
C     ØBTAIN NEW V MATRIX FØR EINVECTØRS
C
      DØ 50     I = 1,N
      P=V(I,IR)
      V(I,IR)=C*P+S*V(I,IC)
   50 V(I,IC)=C*V(I,IC)-S*P
C
C     RØTATE MATRIX A
C
      I=1
  100 IF(I.EQ.IR) GØTØ 200
      P=A(I,IR)
      A(I,IR)=C*P+S*A(I,IC)
      A(I,IC)=C*A(I,IC)-S*P
      I=I+1
      GØTØ 100
  200 I=IR+1
  300 IF(I.EQ.IC)GØTØ 400
      P=A(IR,I)
      A(IR,I)=C*P+S*A(I,IC)
      A(I,IC)=C*A(I,IC)-S*P
      I=I+1
      GØTØ 300
  400 I=IC+1
```

```
  500 IF(I.GT.N) GØTØ 600
      P=A(IR,I)
      A(IR,I)=C*P+S*A(IC,I)
      A(IC,I)=C*A(IC,I)-S*P
      I=I+1
      GØTØ 500
C
C     DEFINE DIAGØNAL TERMS AND ØFF-DIAG.
C     ELEMENT ØF A
C
  600 P=A(IR,IR)
      A(IR,IR)=C*C*P+2.*C*S*A(IR,IC)+S*S*A(IC,IC)
      A(IC,IC)=C*C*A(IC,IC)+S*S*P-2.*C*S*A(IR,IC)
      A(IR,IC)=0
C
C     CHECK IF MAX NUMBER ØF ITER EXCEEDED
C
      IT=IT+1
      IF(IT.LT.ITM) GØTØ 13
  999 RETURN
      END
```

Note: ERR is normally in the order 10^{-6} to 10^{-8}.

General case. Consider that we have, instead of equation (a), the following system:

$$\mathbf{AX} = \lambda\mathbf{BX} \tag{m}$$

where $\mathbf{B}$ is a symmetric and positive definite matrix. We would like to reduce (m) to the form of equation (a).

One can start with $\mathbf{B}$ and apply the Jacobi method to it. We will then obtain the eigenvectors ($\mathbf{V}$ matrix) and a diagonal ($\mathbf{D}$) matrix with the eigenvalues. We can write

$$\mathbf{V}^T\mathbf{BV} = \mathbf{D}$$

$$\mathbf{B} = \mathbf{VDV}^T \tag{n}$$

Thus equation (n) becomes

$$\mathbf{AX} = \lambda[\mathbf{VDV}^T]\mathbf{X}$$

or

$$\mathbf{V}^T\mathbf{AX} = \lambda\mathbf{DV}^T\mathbf{X}$$

$$\mathbf{V}^T\mathbf{AVV}^T\mathbf{X} = \lambda\mathbf{DV}^T\mathbf{X}$$

Let us now call $\mathbf{H} = \mathbf{V}^T\mathbf{A}\mathbf{V}$ and $\mathbf{X}' = \mathbf{V}^T\mathbf{X}$, we then have

$$\mathbf{H}\mathbf{X}' = \lambda\mathbf{D}\mathbf{X}' \tag{o}$$

note that $\mathbf{D}$ is a diagonal matrix which can be easily written as

$$\mathbf{D} = \mathbf{G}\mathbf{G} \tag{p}$$

where $\mathbf{G}$ is another diagonal matrix with element $g_{ii} = \sqrt{d_{ii}}$. Thus

$$\mathbf{H}\mathbf{X}' = \lambda\mathbf{G}\mathbf{G}\mathbf{X}'$$

$$\mathbf{G}^{-1}\mathbf{H}\mathbf{X}' = \lambda\mathbf{G}\mathbf{X}'$$

or

$$\mathbf{G}^{-1}\mathbf{H}\mathbf{G}^{-1}\mathbf{G}\mathbf{X}' = \lambda\mathbf{G}\mathbf{X}'$$

We can now define

$$\mathbf{X}'' = \mathbf{G}\mathbf{X}'$$

to obtain

$$(\mathbf{G}^{-1}\mathbf{H}\mathbf{G}^{-1})\mathbf{X}'' = \lambda\mathbf{I}\mathbf{X}''$$

or

$$\mathbf{Q}\mathbf{X}'' = \lambda\mathbf{I}\mathbf{X}'' \tag{q}$$

Equation (q) has the same form as (a). We can solve it and afterwards obtain $\mathbf{X}$ from $\mathbf{X}''$, by remembering that

$$\mathbf{X}'' = \mathbf{G}\mathbf{X}', \qquad \mathbf{X}' = \mathbf{V}^T\mathbf{X} \tag{r}$$

The eigenvalues λ are the same for (m) as for (q).

2.6 THE GENERAL PROGRAM

We can now summarise the different stages of the finite element method for the case of solving the Laplace equation

$$h_x\frac{\partial^2 u}{\partial x^2} + h_y\frac{\partial^2 u}{\partial y^2} = 0 \tag{2.63}$$

with boundary conditions

$$u = \bar{u} \text{ on } S_1 \quad \text{and} \quad v_n = h_x\frac{\partial u}{\partial x}\alpha_{nx} + h_y\frac{\partial u}{\partial x}\alpha_{ny}$$

α_{nx}, α_{ny} are the direction cosines of the outward normal to the boundary with respect to x and y axis.

$h_x h_y$ are assumed to be constant over each element for simplicity, and the element matrix of Example 2.1 will be used.

The finite element computer program consists of the following steps:

1 Reading and checking of the data
2 Generation of element matrices
3 Assemblage of global matrix
4 Application of boundary conditions
5 Solution of the system of equations
6 Printing of nodal u values [also the v_x, v_y values may be computed as shown in (2.54) and afterwards listed, though this has not been done in the following program]

Example 2.6

Computer program for Laplace's equations. In order to become independent of FORTRAN formats we will assume the data are read in a subroutine DATR and the results are printed in a subroutine OUTW, which are not included in the following listing.

The subroutine ZER sets all the elements of a given matrix or vector equal to zero, and is assumed to have been written by the user.

```
          DIMENSIØN A(20,8), S(3,3), C(8), B(20), I(12,3),
         1X(20), Y(20), HX(12), HY(12), IBC(13), VIBC(13)

          CALL DATR(N,N2,N4,I,X,Y,HX,HY,IBC,VIBC)
C
C         N, NUMBER ØF NØDES: N2, NUMBER ØF
C         ELEMENTS: N4, NUMBER ØF APPLIED U
C         VALUES: I(12,3) IS THE CØNNECTIVITY
C         MATRIX FØR THE ELEMENTS: X(20), Y(20) ARE
C         THE CØØRDINATES ØF EACH NØDE: HX(12),
C         HY(12) ARE THE 'DIFFUSIVITY' CØEFF. IN X
C         AND Y DIRECTIØNS: IBC(13) GIVES THE
C         NUMBER ØF EQUATIØN FØR WHICH U IS
C         EQUAL TØ AVIBC(13) VALUE.
C
C
C         DETERMINE BAND WIDTH MB
C
          IX=0
          DØ 100    K=1,N2
                 IY1=IABS(I(K,1)-I(K,2))
                 IY2=IABS(I(K,2)-I(K,3))
                 IY3=IABS(I(K,3)-I(K,1))
```

```
          IF (IY1.GE.IY2. AND.IY1.GE.IY3) IY=IY1
          IF (IY2.GE.IY3. AND.IY2.GE.IY1) IY=IY2
          IF (IY3.GE.IY1. AND.IY3.GE.IY2) IY=IY3
  100 IF(IY.GT.IX)IX=IY
      MB=IX+1
      IF(MB.GT.8) GØTØ 999
      CALL ZER(A,N,MB)
C
C     FØRM GLØBAL MATRIX
C
      DØ 101    K=1,N2
      DELT=(X(I(K,2))*Y(I(K,3))+X(I(K,1))*Y(I(K,2))+
     1X(I(K,3))*Y(I(K,1)))-(X(I(K,2))*Y(I(K,1))+
     2X(I(K,3))*Y(I(K,2))+X(I(K,1))*Y(I(K,3)))

      B1=(Y(I(K,2))-Y(I(K,3)))/DELT
      B2=(Y(I(K,3))-Y(I(K,1)))/DELT
      B3=(Y(I(K,1))-Y(I(K,2)))/DELT

      A1=(X(I(K,3))-X(I(K,2)))/DELT
      A2=(X(I(K,1))-X(I(K,3)))/DELT
      A3=(X(I(K,2))-X(I(K,1)))/DELT

      S(1,1)=(HX(K)*B1*B1+HY(K)*A1*A1)*DELT/2
      S(1,2)=(HX(K)*B2*B1+HY(K)*A2*A1)*DELT/2
      S(1,3)=(HX(K)*B3*B1+HY(K)*A3*A1)*DELT/2

      S(2,2)=(HX(K)*B2*B2+HY(K)*A2*A2)*DELT/2
      S(2,3)=(HX(K)*B3*B2+HY(K)*A3*A2)*DELT/2

      S(3,3)=(HX(K)*B3*B3+HY(K)*A3*A3)*DELT/2

      S(2,1)=S(1,2)
      S(3,1)=S(1,3)
      S(3,2)=S(2,3)
C
C     ASSEMBLE ELEMENT MATRIX INTØ A
C
      DØ 200    IM=1,3
         IR=I(K,IM)
         DØ 200    IN=1,3
         IC=I(K,IN)
         ID=IC-(IR-1)
         IF(ID.LT.1)GØTØ 200
```

```
          A(IR,ID)=A(IR,ID)+S(IM,IN)
  200 CØNTINUE
  101 CØNTINUE
C
C     APPLY BØUNDARY CØNDITIØNS
C
      CALL ZER(B,N,1)
          KF=MB-1
      DØ 300    K=1,N4
          N5=IBC(K)
         DØ 400    K1=1,KF
          IR=N5-K1
          IF(IR.LT.1) GØTØ 450
          B(IR)=B(IR)-A(IR,(K1+1))*VIBC(K)
          A(IR,(K1+1))=0
  450     IR=N5+K1
          IF(IR.GT.N)GØTØ 400
          B(IR)=B(IR)-A(N5,K1+1)*VIBC(K)
          A(N5,(K1+1))=0
  400    CØNTINUE
      A(N5,1)=1
      B(N5)=VIBC(K)
  300 CØNTINUE
C
C     CALL SØLVER
C
      CALL BANDSØL(A,B,N,MB).
C
C     CALL SUBRØUTINE TØ PRINT VALUES ØF U
C
      CALL ØUTW(B,X,Y,N)
  999 STØP
      END
```

REFERENCES

1. DESAI, C. S. and ABEL, J. F., *Introduction to the Finite Element Method*, Van Nostrand Reinhold Co (1972)
2. BREBBIA, C. A. and CONNOR, J. J., *Fundamentals of Finite Element Techniques for Structural Engineers*, Butterworths (1973)
3. MARTIN, H. C. and CAREY, G. F., *Introduction to Finite Element Analysis*, McGraw-Hill (1973)
4. NORRIE, D. H. and DE VRIES, G., *The Finite Element Method, Fundamentals and Applications*, Academic Press (1973)

EXERCISES

2-1 Compare the conventional Galerkin method with the finite element technique. Point out the differences.

2-2 Write a macro flow chart for a computer program to find the eigenvalues and eigenvectors corresponding to the equation

$$h\frac{\partial^2 u}{\partial x^2} + h\frac{\partial^2 u}{\partial y^2} + \lambda u = 0$$

with boundary conditions $u = \bar{u}$ on S_1 and $\partial u/\partial n = 0$ on S_2. Assume the Jacobi subroutine of Example 2.6 is used.

2-3 Check if the function

$$u = \alpha_1 + \alpha_2\xi + \alpha_3\eta + \alpha_4\xi\eta \tag{a}$$

satisfies admissibility and completeness for the rectangular element shown in Figure 2.8, when solving the Laplace equation.

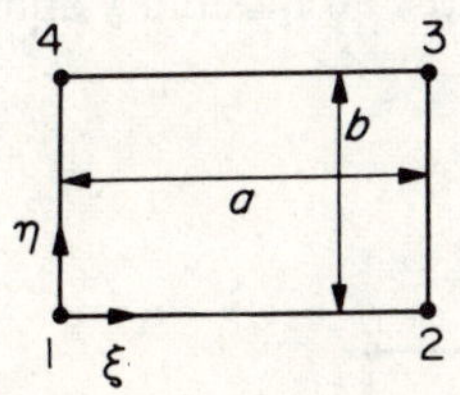

Figure 2.8 Quadrilateral element

2-4 Take the approximate u function on the rectangular element shown in Figure 2.8 as

$$u = \alpha_1 + \alpha_2\xi + \alpha_3\eta + \alpha_4\xi\eta \tag{a}$$

Deduce the corresponding **K** and **P** matrices for the Laplace equation shown in Example 2.1.

2-5 Discuss the difference and relative merits of the triangular element of Example 2.1 and the rectangular one deduced in Exercise 2.4.

2-6 The Laplace equation

$$\frac{\partial^2 u}{\partial x^2} + \frac{\partial^2 u}{\partial y^2} = 0 \tag{a}$$

can be expressed in centred finite difference form for node 5 (Figure 2.9) as

$$\frac{1}{l^2}(u_2 + u_4 + u_6 + u_8 - 4u_5) = 0 \tag{b}$$

Compare it with the equation one can obtain using the finite element matrix of Example 2.1 for the two grids shown in Figure 2.9(b).

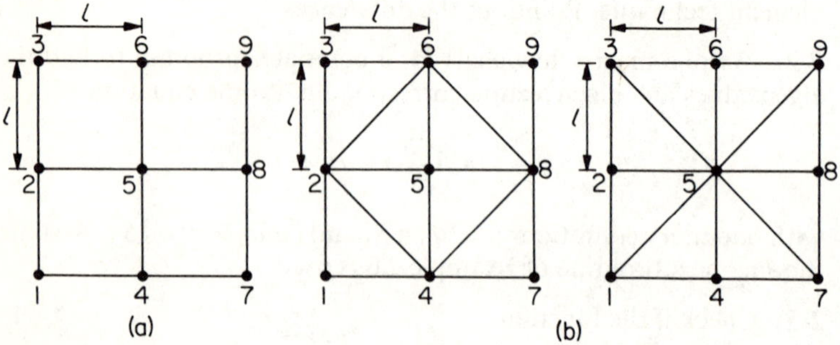

Figure 2.9 Finite difference and finite element grids. (a) *Finite difference grid*; (b) *Finite element grids*

2-7 What type of u function would you propose for a six node model as shown in Figure 2.10? Discuss how the admissibility and completeness conditions are satisfied.

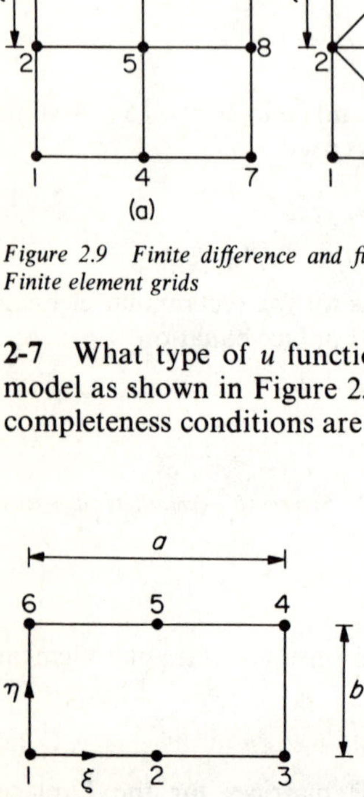

Figure 2.10 Six node model

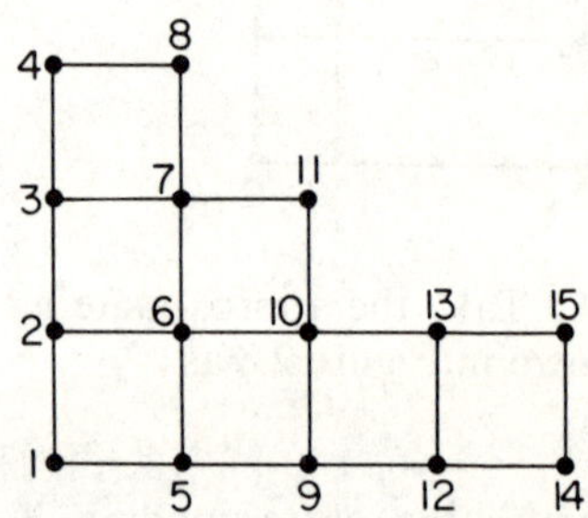

Figure 2.11 Twelve element discretisation

2-8 Assemble the elements shown in Figure 2.11 using an element matrix

$$\mathbf{K}^i = \begin{bmatrix} 4 & 1 & 1 & 1 \\ 1 & 4 & 1 & 1 \\ 1 & 1 & 4 & 1 \\ 1 & 1 & 1 & 4 \end{bmatrix}$$

with one degree of freedom per node. Assume boundary conditions $u_{15} = u_{14} = 0$.

Can the band width be improved by renumbering?

2-9 Assemble four triangular elements as shown in Figure 2.12 using the matrix deduced in Example 2.1. Once the matrix is assembled as

$$\underset{(5\times 5)}{\mathscr{K}}\ \underset{(5\times 1)}{\mathscr{U}} = \underset{(5\times 1)}{\mathscr{P}}$$

where $\mathscr{U}^T = \{u_1 u_2 u_3 u_4 u_5\}$, $\mathscr{P}^T = \{p_1 p_2 p_3 p_4 0\}$ write u_5 as a function of the other four variables and compare the resulting (4×4) matrix with the one developed in Exercise 2.4.

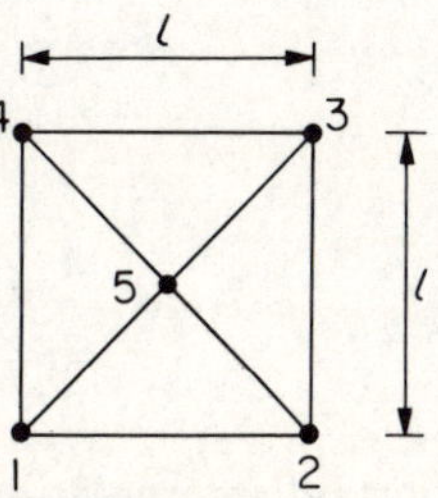

Figure 2.12 Quadrilateral formed by four triangular elements

2-10 Discretise into 10 to 20 triangular elements the two-dimensional shape shown in Figure 2.13 using symmetry considerations. The example is a heat transfer problem and one can assume $\partial u/\partial n = 0$ on certain sections. The thermal diffusivity coefficient is the same for all the section.

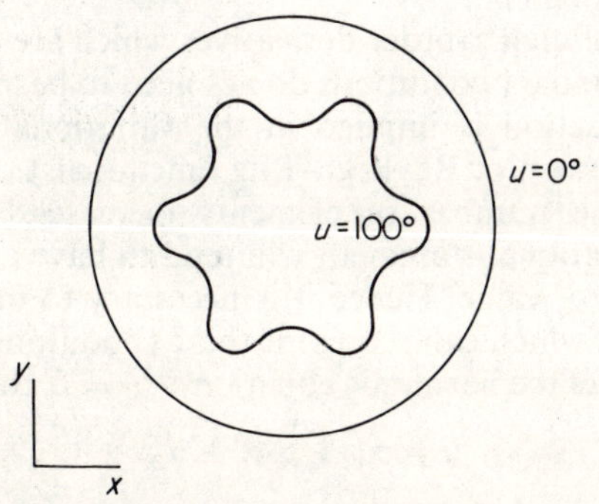

Figure 2.13 Temperature distribution problem

2-11 Assume that the diffusion coefficients of equation (a), Example 2.1 are given as $h_x' h_y'$ in a principal reference system $X'Y'$ inclined at θ degrees with respect to the XY system. Express the diffusion coefficients $(h_x h_y)$ in the XY system as a function of h_x', h_y' and θ.

3 Interpolation Functions

3.1 INTRODUCTION

In this chapter we will deduce the more important interpolation functions for the case of the harmonic and bi-harmonic equations. In general these functions have to satisfy the following conditions:

(1) They have to be continuous inside and between elements up to the order $n - 1$, where n is the order of the highest derivative in the functional. For the harmonic equation (e.g. $\nabla^2 u = 0$) this implies continuity of u, and for the bi-harmonic (e.g. $\nabla^4 u = 0$) it means continuity of u and $\partial u/\partial n$.

Continuity of higher-order derivatives which are associated with the natural boundary conditions do not need to be imposed, as their eventual satisfaction is implied in the variational statement (i.e. Galerkin expression or Rayleigh–Ritz functionals).

(2) When the number of elements increases, the derivatives within the variational statement will tend to have a constant, or in particular a zero, value. Hence it is necessary to include, in our u function, terms which can represent these conditions.

For the case of the harmonic equation $\nabla^2 u = 0$, these terms imply

$$u = \alpha_1 + \alpha_2 x + \alpha_3 y \tag{3.1}$$

the reason being that the highest derivatives in the variational statement are now $\partial u/\partial x$ and $\partial u/\partial y$, the α_1 represents the 'zero derivative' type term and the α_2, α_3 the 'constant derivative' terms.

For the bi-harmonic equation $\nabla^4 u = 0$, the highest-order derivatives in the functional are $\partial^2 u/\partial x^2, \partial^2 u/\partial x\, \partial y, \partial^2 u/\partial y^2$. Hence the u functions must have the following terms:

$$u = \alpha_1 + \alpha_2 x + \alpha_3 y + \alpha_4 x^2 + \alpha_5 xy + \alpha_6 y^2 \tag{3.2}$$

where the α_1 represents a translation-type term, and α_2, α_3 are rotation terms. α_4, α_5 and α_6 satisfy the constant derivatives condition.

As already mentioned an interpolation function can be expressed as

$$u = \mathbf{A}(x, y)\boldsymbol{\alpha} \tag{3.3}$$

where $\boldsymbol{\alpha}$ is a set of parameters. The number of α_i parameters has to equal the number of nodal unknowns. Specialising (3.3) for the element nodes, we have

$$\mathbf{u}^n = \mathbf{C}(x_i y_i)\boldsymbol{\alpha} \tag{3.4}$$

where $\mathbf{u}^n$ is the nodal unknowns vector and $\mathbf{C}$ is a square matrix. Hence

$$\boldsymbol{\alpha} = \mathbf{C}^{-1}\mathbf{u}^n \tag{3.5}$$

and we can express (3.3) as

$$u = (\mathbf{A}\mathbf{C}^{-1})\mathbf{u}^n = \boldsymbol{\phi}^T\mathbf{u}^n \tag{3.6}$$

where the terms in $\boldsymbol{\phi}$ are the interpolation functions ϕ_i, each of them associated with a u_i unknown; i.e.

$$u = \phi_1 u_1 + \phi_2 u_2 + \ldots = \boldsymbol{\phi}^T\mathbf{u}^n \tag{3.7}$$

If we have two variables, for instance u and v, which can be represented by the same interpolation function, we can write

$$\mathbf{U} = \begin{Bmatrix} u \\ v \end{Bmatrix} = \begin{bmatrix} \boldsymbol{\phi} & \cdot \\ \cdot & \boldsymbol{\phi} \end{bmatrix}^T \begin{Bmatrix} \mathbf{u}^n \\ \mathbf{v}^n \end{Bmatrix} = \boldsymbol{\Phi}^T\mathbf{U}^n \tag{3.8}$$

In the following we will deduce the interpolation functions for triangular and rectangular elements for

1 *First-order continuity* or functions for which only inter-element continuity of the function itself is required. Such are those functions used in the formulation of the harmonic equation.
2 *Second-order continuity* or functions with inter-element continuity of the function and its derivatives. These functions are used for the bi-harmonic equation.

3.2 FIRST-ORDER CONTINUITY FUNCTIONS FOR TRIANGULAR ELEMENTS

RELATIONSHIP BETWEEN CARTESIAN AND TRIANGULAR COORDINATES

The generation of interpolation functions for a triangular element is simplified considerably if one works with oblique coordinates.

Figure 3.1 shows the relative node numbering scheme and sense for the oblique directions $\bar{\xi}_1\bar{\xi}_2$. The nodes are numbered in an anti-clockwise direction, and the side opposite to node i is defined as side i. The oblique coordinates can be made dimensionless with respect to side length and vary from 0 to 1.

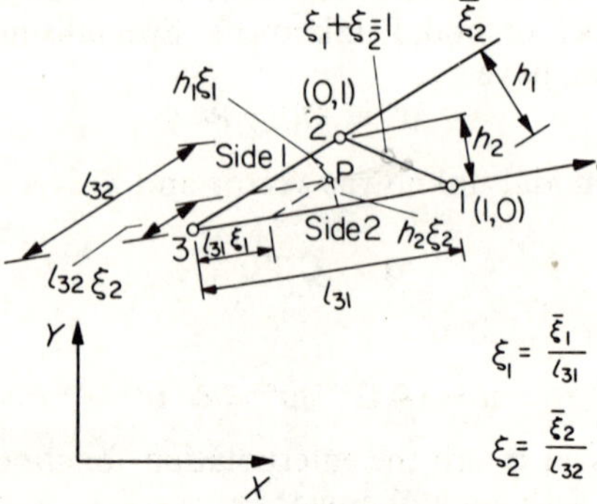

Figure 3.1 Oblique coordinates

In order to establish the relationship between the Cartesian and oblique coordinates, we consider an arbitrary $P(x, y) = P(\xi_1, \xi_2)$ point. Thus,

$$
\begin{aligned}
x &= x_3 + (x_1 - x_3)\xi_1 + (x_2 - x_3)\xi_2 \\
y &= y_3 + (y_1 - y_3)\xi_1 + (y_2 - y_3)\xi_2
\end{aligned}
\tag{3.9}
$$

Also ξ_1 and ξ_2 are constant along side 3;

$$
\xi_1 + \xi_2 = 1 \quad \text{on side 3} \tag{3.10}
$$

We write (3.9) as

$$
\begin{aligned}
x &= \xi_1 x_1 + \xi_2 x_2 + (1 - \xi_1 - \xi_2)x_3 \\
y &= \xi_1 y_1 + \xi_2 y_2 + (1 - \xi_1 - \xi_2)y_3
\end{aligned}
\tag{3.11}
$$

and its inverted form gives

$$
\begin{aligned}
\xi_1 &= \frac{1}{2A}(2A_1^0 + b_1 x + a_1 y) \\
\xi_2 &= \frac{1}{2A}(2A_2^0 + b_2 x + a_2 y)
\end{aligned}
\tag{3.12}
$$

where

$$a_i = x_k - x_j \qquad b_i = y_j - y_k$$

$$2A_i^0 = x_j y_k - x_k y_j \tag{3.13}$$

$$A = \text{area of triangle} = \tfrac{1}{2}(b_1 a_2 - b_2 a_1)$$

where $i = 1, 2$ for $j = 2, 3$ and $k = 3, 1$.

The dimensionless coordinates can also be interpreted as area ratios. We see from Figure 3.1 that the distance from side 1 along the normal towards node 1 is equal to $h_1 \xi_1$. Then

$$\xi_1 = \frac{\text{area (32P)}}{\text{area (321)}} = \frac{A_1}{A} \tag{3.14}$$

where A_1 is defined as the area bound by side 1 and P (Figure 3.2).

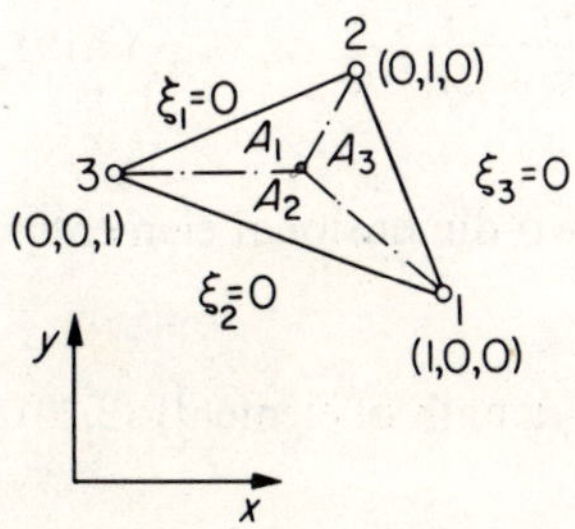

Figure 3.2 Area coordinates

We have then three area coordinates $(\xi_1 \xi_2 \xi_3)$ but only two are independent since

$$\xi_1 + \xi_2 + \xi_3 = 1$$

Although they are not independent, it is convenient to use all these coordinates when generating interpolation functions.

We can find ξ_3 in terms of x and y from (3.12)

$$\xi_3 = 1 - \xi_1 - \xi_2 = \frac{1}{2A}(2A_3^0 + b_3 x + a_3 y)$$

$$A_3^0 = A - A_1^0 - A_2^0 \tag{3.15}$$

$$a_3 = -a_1 - a_2 \qquad b_3 = -b_1 - b_2$$

Hence the relationship between the Cartesian and triangular coordinates ξ_i becomes [see equation (3.11)]

$$x = x_1\xi_1 + x_2\xi_2 + x_3\xi_3$$
$$y = y_1\xi_1 + y_2\xi_2 + y_3\xi_3 \tag{3.16}$$

Since the expressions for the element matrices generally involve Cartesian derivatives, we will need to apply

$$\begin{aligned}\frac{\partial}{\partial x}\{f(\xi_1\xi_2\xi_3)\} &= \sum_{i=1}^{3}\frac{\partial f}{\partial \xi_i}\frac{\partial \xi_i}{\partial x} = \frac{1}{2A}\sum_{i=1}^{3} b_i\frac{\partial f}{\partial \xi_i}\\ \frac{\partial}{\partial y}\{f(\xi_1\xi_2\xi_3)\} &= \sum_{i=1}^{3}\frac{\partial f}{\partial \xi_i}\frac{\partial \xi_i}{\partial y} = \frac{1}{2A}\sum_{i=1}^{3} a_i\frac{\partial f}{\partial \xi_i}\end{aligned} \tag{3.17}$$

Higher derivatives are generated by repeated application of (3.17).

In order to evaluate the integral

$$\iint f(\xi_1\xi_2\xi_3)\,\mathrm{d}A \tag{3.18}$$

we can use the formula

$$\iint \xi_1^i\xi_2^j\xi_3^k\,\mathrm{d}A = \frac{i!j!k!}{(i+j+k+2)!}2A \tag{3.19}$$

where ! indicates factorial.

Integration formula (3.19) is valid for two-dimensional elements. For the one-dimensional case we have

$$\int \xi_1^i\xi_2^j\,\mathrm{d}S = \frac{i!j!}{(i+j+1)!}L \qquad (L = \text{length of element}) \tag{3.20}$$

and, for three dimensions,

$$\int \xi_1^i\xi_2^j\xi_3^k\xi_4^l\,\mathrm{d}S = \frac{i!j!k!l!}{(i+j+k+l+3)!}6V \tag{3.21}$$

(V = volume of the element)

Care should be taken in the transformation of coordinates as the new system is such that the ξ_i terms are not all independent. Thus in two-dimensional transformations for instance it is advisable to work with the oblique system of Figure 3.1, where $\xi_1\xi_2$ are independent. For this system,

$$\iint f(\xi_1\xi_2)\,\mathrm{d}A = 2A\int_0^1\left[\int_0^{1-\xi_2} f(\xi_1\xi_2)\,\mathrm{d}\xi_1\right]\mathrm{d}\xi_2 \tag{3.22}$$

LINEAR FUNCTION

The simplest two-dimensional expansion is obtained for a three node triangular element (Figure 3.2):

$$u = \xi_1 u_1 + \xi_2 u_2 + \xi_3 u_3 \tag{3.23}$$

where u_i terms are the nodal values of the function. We can express u as

$$u = \boldsymbol{\phi}^T \mathbf{u}^n$$
$$u^n = \{u_1 u_2 u_3\}, \qquad \boldsymbol{\phi}^T = [\xi_1 \xi_2 \xi_3] \tag{3.24}$$

The derivatives of u are obtained by operating on (3.23)

$$\frac{\partial u}{\partial x} = \frac{1}{2A} \sum_{i=1}^{3} b_i \frac{\partial u}{\partial \xi_i} = \frac{1}{2A} \sum_{i=1}^{3} b_i u_i = \frac{1}{2A} [b_1 b_2 b_3] \mathbf{u}^n$$
$$\frac{\partial u}{\partial y} = \frac{1}{2A} \sum_{i=1}^{3} a_i u_i = \frac{1}{2A} [a_1 a_2 a_3] \mathbf{u}^n \tag{3.25}$$

Note that after substitution of ξ_i as a function of x, y coordinates, one can write u as

$$u = \alpha_1 + \alpha_2 x + \alpha_3 y \tag{3.26}$$

or in oblique coordinates,

$$u = \alpha'_1 + \alpha'_2 \xi_1 + \alpha'_2 \xi_2$$

The triangular ξ_i coordinates formulation is, however, more convenient as u is directly expressed as a function of the nodal unknowns and not as a function of the α_i terms which are a combination of nodal values. The triangular coordinates formulation permits the use of simple integration rules and is generally suitable for higher-order elements.

Example 3.1

Consider the variational statement corresponding to the Laplace equation seen in Example 2.1; i.e.

$$\iint \left(h_x \frac{\partial u}{\partial x} \frac{\partial \delta u}{\partial x} + h_y \frac{\partial u}{\partial y} \frac{\partial \delta u}{\partial y} \right) \mathrm{d}x\, \mathrm{d}y = \int_{S_2} \bar{q}_n\, \delta u\, \mathrm{d}S \tag{a}$$

Now instead of using

$$u = \alpha_1 + \alpha_2 x + \alpha_3 y$$

as an approximate function we will apply (3.23), i.e.

$$u = \xi_1 u_1 + \xi_2 u_2 + \xi_3 u_3 = \boldsymbol{\phi}^T \mathbf{u}^n \tag{b}$$

over the triangular element shown in Figure 3.3. We will see that the determination of **K** and **P** becomes easier.

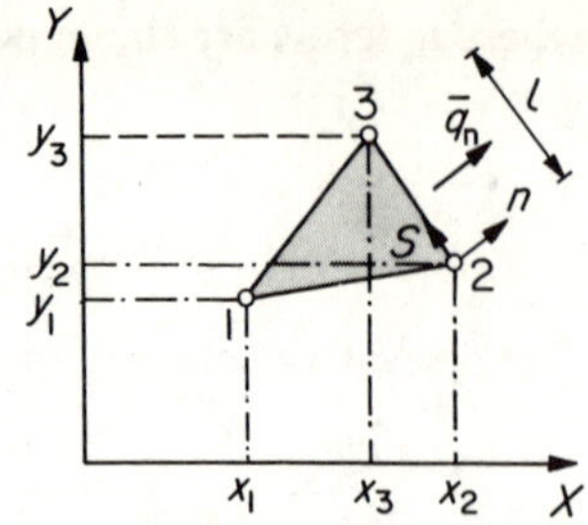

Figure 3.3 Triangular element

We now have [see equation (3.25)]

$$\frac{\partial u}{\partial x} = \frac{\partial u}{\partial \xi_1}\frac{\partial \xi_1}{\partial x} + \frac{\partial u}{\partial \xi_2}\frac{\partial \xi_2}{\partial x} + \frac{\partial u}{\partial \xi_3}\frac{\partial \xi_3}{\partial x} = u_1\frac{\partial \xi_1}{\partial x} + u_2\frac{\partial \xi_2}{\partial x} + u_3\frac{\partial \xi_3}{\partial x}$$
$$= \frac{1}{2A}\{u_1 b_1 + u_2 b_2 + u_3 b_3\} \tag{c}$$

$$\frac{\partial u}{\partial y} = \frac{\partial u}{\partial \xi_1}\frac{\partial \xi_1}{\partial y} + \frac{\partial u}{\partial \xi_2}\frac{\partial \xi_2}{\partial y} + \frac{\partial u}{\partial \xi_3}\frac{\partial \xi_3}{\partial y} = u_1\frac{\partial \xi_1}{\partial y} + u_2\frac{\partial \xi_2}{\partial y} + u_3\frac{\partial \xi_3}{\partial y}$$
$$= \frac{1}{2A}\{u_1 a_1 + u_2 a_2 + u_3 a_3\}$$

These relationships can be written in matrix form as

$$\frac{\partial u}{\partial x} = \frac{1}{2A}\mathbf{b}^T\mathbf{u}^n \quad \text{similarly} \quad \frac{\partial \delta u}{\partial x} = \frac{1}{2A}\mathbf{b}^T\,\delta\mathbf{u}^n$$
$$\frac{\partial u}{\partial y} = \frac{1}{2A}\mathbf{a}^T\mathbf{u}^n \quad \text{similarly} \quad \frac{\partial \delta u}{\delta y} = \frac{1}{2A}\mathbf{a}^T\,\delta\mathbf{u}^n \tag{d}$$

where $\mathbf{b} = \{b_1 b_2 b_3\}$, $\mathbf{a} = \{a_1 a_2 a_3\}$ and $\mathbf{u}^n = \{u_1 u_2 u_3\}$.

One can substitute (b) and (d) into the variational statement (a), which becomes

$$\frac{1}{4A^2}\delta\mathbf{u}^{n,T}\iint\{h_x\mathbf{b}\mathbf{b}^T + h_y\mathbf{a}\mathbf{a}^T\}\,\mathrm{d}x\,\mathrm{d}y\,\mathbf{u}^n = \delta\mathbf{u}^{n,T}\int\boldsymbol{\phi}\bar{q}_n\,\mathrm{d}S \tag{e}$$

After integrating the left-hand side of this equation we obtain the same result as in Example 2.1, that is expression (m). For the right-hand side the integration is now simplified by using triangular coordinates. Assume that $\bar{q}_n$ only applies on side 2–3 (Figure 3.3). Hence

$$\mathbf{P} = \int_{\text{node 2}}^{\text{node 3}}\boldsymbol{\phi}\bar{\mathbf{q}}_n\,\mathrm{d}S = \int_{\text{node 2}}^{\text{node 3}}\begin{Bmatrix}0\\ \xi_2\\ \xi_3\end{Bmatrix}\bar{q}_n\,\mathrm{d}S \tag{f}$$

Next one can apply formula (3.20) which gives, for a constant $\bar{q}_n$,

$$\mathbf{P} = \bar{q}_n \begin{Bmatrix} 0 \\ l/2 \\ l/2 \end{Bmatrix} \qquad \text{(g)}$$

This result is of course the same as in Example 2.1, but its derivation is simpler.

QUADRATIC FUNCTION

In order to obtain accurate results using elements which employ linear interpolation functions, in some cases one has to use a large number of elements. Therefore, considerable effort has been devoted to developing higher-order functions, of which the most used in practical applications is the quadratic.[2]

The quadratic function can be expressed in oblique coordinates as

$$u = \alpha_1 + \alpha_2\xi_1 + \alpha_3\xi_2 + \alpha_4\xi_1^2 + \alpha_5\xi_1\xi_2 + \alpha_6\xi_2^2 \qquad (3.27)$$

and specialising for the nodal values of $\xi_1\xi_2$ we can solve the $\boldsymbol{\alpha}$ as a function of $\mathbf{u}^n$, nodal values. The expression can be written as (see Figure 3.4)

$$u = \boldsymbol{\phi}^T\mathbf{u}^n \qquad (3.28)$$

where

$$\boldsymbol{\phi}^T = \{\xi_1(2\xi_1 - 1), \xi_2(2\xi_2 - 1), \xi_3(2\xi_3 - 1), 4\xi_1\xi_2, 4\xi_2\xi_3, 4\xi_3\xi_1\}$$
$$\mathbf{u}^n = \{u_1u_2u_3u_4u_5u_6\}$$

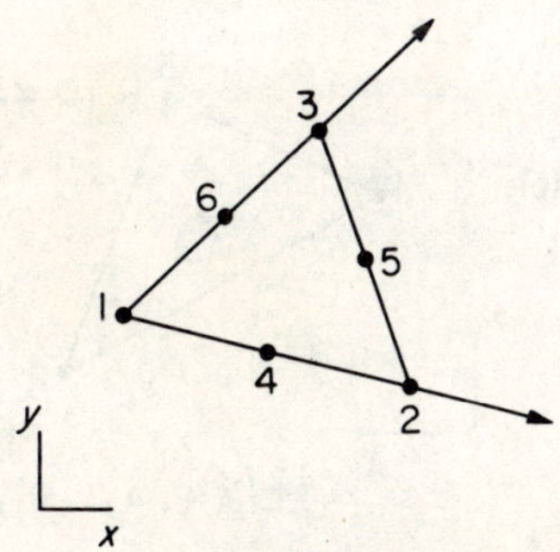

Node	ξ_1	ξ_2	ξ_3
1	1	0	0
2	0	1	0
3	0	0	1
4	$\frac{1}{2}$	$\frac{1}{2}$	0
5	0	$\frac{1}{2}$	$\frac{1}{2}$
6	$\frac{1}{2}$	0	$\frac{1}{2}$

Figure 3.4 Quadratic function triangular element

The derivatives are given by

$$\frac{\partial u}{\partial x} = \frac{1}{2A}\sum b_i \frac{\partial u}{\partial \xi_i} = \frac{1}{2A}\mathbf{b}^T\boldsymbol{\psi}\mathbf{u}^n$$
$$\frac{\partial u}{\partial y} = \frac{1}{2A}\sum a_i \frac{\partial u}{\partial \xi_i} = \frac{1}{2A}\mathbf{a}^T\boldsymbol{\psi}\mathbf{u}^n \tag{3.29}$$

with

$$\boldsymbol{\psi} = \begin{bmatrix} 4\xi_1 - 1 & 0 & 0 & 4\xi_2 & 0 & 4\xi_3 \\ 0 & 4\xi_2 - 1 & 0 & 4\xi_1 & 4\xi_3 & 0 \\ 0 & 0 & 4\xi_3 - 1 & 0 & 4\xi_2 & 4\xi_1 \end{bmatrix} \tag{3.30}$$

Evaluation of integrals is now more complex and it can be more convenient to use some numerical integration formulae (see Appendix) than equations (3.19) and (3.20).

CUBIC FUNCTION

For this case u needs to be expressed in terms of ten parameters. In order to satisfy inter-element compatibility, the function for a side must depend only on the nodal quantities for that side. Since the function is cubic, four nodal quantities are required to define the distribution on a side.

One possibility is to work with corner nodes and two interior nodes per side, taking as nodal quantities the values of the function at the nodes. This model is shown in Figure 3.5(a). The side nodes are located at the third points. An additional interior node is needed to

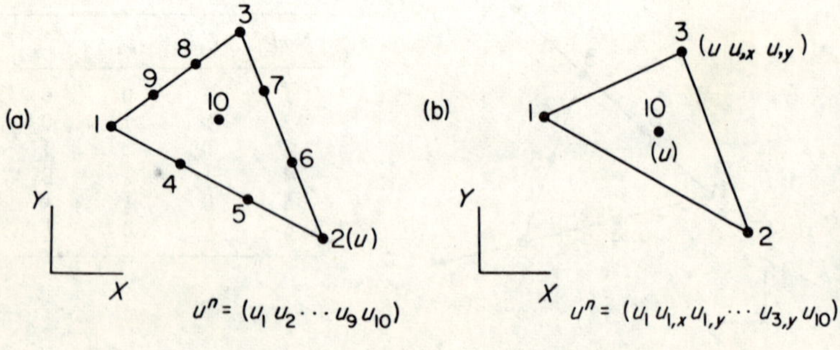

Figure 3.5 Cubic function. (a) *The function as a nodal variable*; (b) *The function and its derivatives as nodal variables*

maintain completeness of the polynomial since, if the polynomial is not complete, the resulting matrices will have a preferred direction, which is not desirable. It is convenient to take the interior node at the centroid.

The function has the form

$$u = \sum_{i=1}^{9} \phi_i u_i + \phi_{10} u_{10} \tag{3.31}$$

where the ϕ terms are given in Table 3.1.

Table 3.1 SOME ELEMENT FUNCTIONS FOR TWO DIMENSIONS

Triangle	*Rectangle*
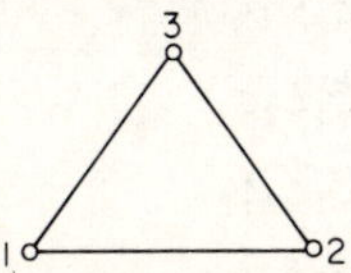 $\phi_i = \xi_i, i = 1, 2, 3$	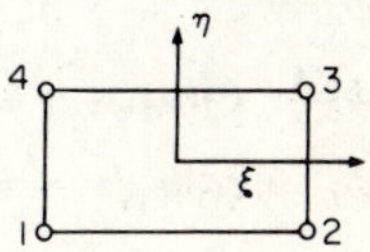 $\phi_i = \frac{1}{4}(1 + \xi\xi_i)(1 + \eta\eta_i), i = 1, \ldots, 4$
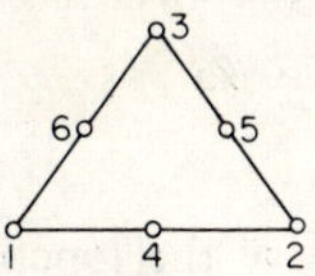 *At corners* $\phi_i = (2\xi_i - 1)\xi_i, i = 1, 2, 3$ *At mid-sides* $\phi_j = 4\xi_k\xi_l$ $j = 4, 5, 6; k = 1, 2, 3; l = 2, 3, 1$	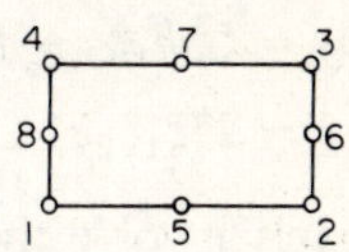 *At corners* $\phi_i = \frac{1}{4}(1 + \xi\xi_i)(1 + \eta\eta_i), i = 1, 2, 3, 4$ *At* $i = 6, 8$ $\phi_i = \frac{1}{2}(1 + \xi\xi_i)(1 - \eta^2)$ *At* $i = 5, 7$ $\phi_i = \frac{1}{2}(1 - \xi^2)(1 + \eta\eta_i)$
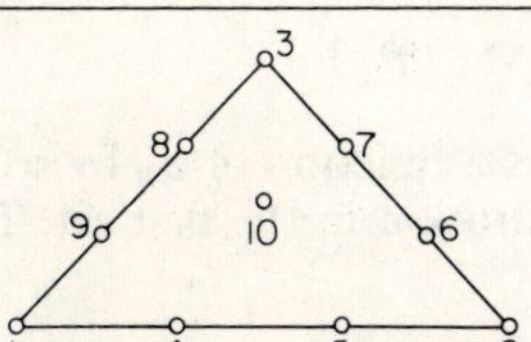 *At corners* $\phi_i = \frac{1}{2}(3\xi_i - 1)(3\xi_i - 2)\xi_i, i = 1, 2, 3$ *On sides,* $i = 4, 5, 6, 7, 8, 9$ $\phi_i = \frac{9}{2}\xi_k\xi_l(3\xi_k - 1), k = 1, 2, 2, 3, 3, 1$ $l = 2, 1, 3, 2, 1, 3$ *On centre,* $i = 10$ $\phi_{10} = 27\xi_1\xi_2\xi_3$	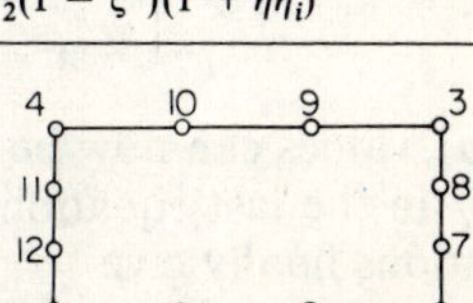 *At corners* $\phi_i = \frac{1}{32}(1 + \xi_i)(1 + \eta\eta_i)$ $\times [-10 + 9(\xi^2 + \eta^2)], i = 1, 2, 3, 4$ *At* $i = 7, 8, 11, 12$ $\phi_i = \frac{9}{32}(1 + \xi\xi_i)(1 - \eta^2)(1 + 9\eta_i\eta)$ *At* $i = 5, 6, 9, 10$ $\phi_i = \frac{9}{32}(1 + \eta\eta_i)(1 - \xi^2)(1 + 9\xi\xi_i)$

In order to reduce the band width, one can attempt to work only with corner nodes. Since four quantities are necessary for inter-element compatibility, we can include the function derivatives as nodal quantities.

The model is shown in Figure 3.5(b).

The appropriate interpolation polynomials in terms of triangular coordinates are,[2]

$$
\begin{aligned}
u &= \phi_1 u_1 + \phi_2 u_{1,x} + \phi_3 u_{1,y} + \ldots + \phi_9 u_{3,y} + \phi_c u_c \\
\phi_1 &= \xi_1^2(\xi_1 + 3\xi_2 + 3\xi_3) - 7\xi_1\xi_2\xi_3 \\
\phi_2 &= \xi_1^2(a_3\xi_2 - a_2\xi_3) + (a_2 - a_3)\xi_1\xi_2\xi_3 \\
\phi_3 &= \xi_1^2(b_2\xi_3 - b_3\xi_2) + (b_3 - b_2)\xi_2\xi_1\xi_3 \\
\phi_4 &= \xi_2^2(\xi_2 + 3\xi_3 + 3\xi_1) - 7\xi_1\xi_2\xi_3 \\
\phi_5 &= \xi_2^2(a_1\xi_3 + a_3\xi_1) + (a_3 - a_1)\xi_1\xi_2\xi_3 \\
\phi_6 &= \xi_2^2(b_3\xi_1 - b_1\xi_3) + (b_1 - b_3)\xi_1\xi_2\xi_3 \\
\phi_7 &= \xi_3^2(\xi_3 + 3\xi_1 + 3\xi_2) - 7\xi_1\xi_2\xi_3 \\
\phi_8 &= \xi_3^2(a_2\xi_1 - a_1\xi_2) + (a_1 - a_2)\xi_1\xi_2\xi_3 \\
\phi_9 &= \xi_3^2(b_1\xi_2 - b_2\xi_1) + (b_2 - b_1)\xi_1\xi_2\xi_3 \\
\phi_{10} &= 27\xi_1\xi_2\xi_3
\end{aligned}
\tag{3.32}
$$

Both models involve the centroidal value of the function as the unknown. In certain cases we can substitute this value as a function of the other unknowns by matrix partition and condensation. Let us assume we have a system of equations such as

$$
\begin{bmatrix} \mathbf{K}_B & \mathbf{K}_{BC} \\ \mathbf{K}_{BC}^T & \mathbf{K}_C \end{bmatrix} \begin{Bmatrix} \mathbf{u}_B \\ \mathbf{u}_C \end{Bmatrix} = \begin{Bmatrix} \mathbf{P}_B \\ \mathbf{P}_C \end{Bmatrix} \tag{3.33}
$$

The $\mathbf{u}_C$ values can now be expressed as a function of $\mathbf{u}_B$ by solving for $\mathbf{u}_C$ in the last equations and substitution in the first set. These operations finally give

$$
\mathbf{K}_B^* \mathbf{u}_B = \mathbf{P}_B^* \tag{3.34}
$$

where

$$
\begin{aligned}
\mathbf{K}_B^* &= \mathbf{K}_B - \mathbf{K}_{BC}\mathbf{K}_C^{-1}\mathbf{K}_{BC}^T \\
\mathbf{P}_B^* &= \mathbf{P}_B - \mathbf{K}_{BC}\mathbf{K}_C^{-1}\mathbf{P}_C
\end{aligned}
$$

After such a process the element matrix is said to be ‘condensed’.

Example 3.2

We will now deduce the element matrices for the variational statement

$$\iint \left(h_x \frac{\partial u}{\partial x}\frac{\partial \delta u}{\partial x} + h_y \frac{\partial u}{\partial y}\frac{\partial \delta u}{\partial y} \right) \mathrm{d}x\,\mathrm{d}y = \int_{S_2} \bar{q}_n\, \delta u\, \mathrm{d}S \qquad \text{(a)}$$

using the quadratic function of equation (3.28).

$$u = \boldsymbol{\phi}^T \mathbf{u}^n \qquad \text{(b)}$$

One has now [see (3.39)]

$$\frac{\partial u}{\partial x} = \frac{1}{2A}\mathbf{b}\boldsymbol{\psi}\mathbf{u}^n$$

$$\frac{\partial u}{\partial y} = \frac{1}{2A}\mathbf{a}\boldsymbol{\psi}\mathbf{u}^n \qquad \text{(c)}$$

Hence (a) becomes

$$\delta\mathbf{u}^{n,T}\frac{1}{4A^2}\iint (h_x\boldsymbol{\psi}^T\mathbf{b}\mathbf{b}^T\boldsymbol{\psi} + h_y\boldsymbol{\psi}^T\mathbf{a}\mathbf{a}^T\boldsymbol{\psi})\,\mathrm{d}x\,\mathrm{d}y\,\mathbf{u}^n = \delta\mathbf{u}^{n,T}\int_{S_2}\boldsymbol{\phi}\bar{q}_n\,\mathrm{d}S \qquad \text{(d)}$$

The matrix multiplication and integration involved in the two terms on the left-hand side gives two 6 × 6 matrices, which can be written

$$\mathbf{I}_b = \iint \boldsymbol{\psi}^T\mathbf{b}\mathbf{b}^T\boldsymbol{\psi}\,\mathrm{d}A, \qquad \mathbf{I}_a = \iint \boldsymbol{\psi}^T\mathbf{a}\mathbf{a}^T\boldsymbol{\psi}\,\mathrm{d}A \qquad \text{(e)}$$

We have assumed that h_x and h_y are constants.

The integration of the first of these integrals using rule (3.19) gives

$$\mathbf{I}_b = A\begin{bmatrix} b_1^2 & -\frac{1}{3}b_1b_2 & -\frac{1}{3}b_1b_3 & \frac{4}{3}b_1b_2 & 0 & \frac{4}{3}b_1b_3 \\ & b_2^2 & -\frac{1}{3}b_2b_3 & \frac{4}{3}b_1b_2 & \frac{4}{3}b_2b_3 & 0 \\ & & b_3^2 & 0 & \frac{4}{3}b_2b_3 & \frac{4}{3}b_1b_3 \\ & & & \frac{8}{3}(b_1^2+b_1b_2+b_2^2) & \frac{4}{3}(b_1b_2+b_2^2+b_2b_3)+\frac{8}{3}b_1b_3 & \frac{4}{3}(b_1b_3+b_1^2+b_1b_2)+\frac{8}{3}b_2b_3 \\ & & & & \frac{8}{3}(b_2^2+b_2b_3+b_3^2) & \frac{4}{3}(b_2b_3+b_3^2+b_1b_3)+\frac{8}{3}b_1b_2 \\ & \text{sym.} & & & & \frac{8}{3}(b_3^2+b_1b_3+b_1^2) \end{bmatrix} \qquad \text{(f)}$$

A similar result is obtained for $\mathbf{I}_a$ integral but with a_i instead of b_i.

The matrix **K** can now be assembled, as

$$\mathbf{K} = \frac{1}{4A^2}[h_x \mathbf{I}_b + h_y \mathbf{I}_a] \tag{g}$$

Next, we form the **P** matrix which can be defined for $\bar{q}_n$ acting only on the 2–3 side ($\xi_1 \equiv 0$), as

$$\mathbf{P} = \int_{\text{node 2}}^{\text{node 3}} \begin{Bmatrix} 0 \\ \xi_2(2\xi_2 - 1) \\ \xi_3(2\xi_3 - 1) \\ 0 \\ 4\xi_2\xi_3 \\ 0 \end{Bmatrix} \bar{q}_n \, \mathrm{d}S \tag{h}$$

Let us assume $\bar{q}_n$ is constant over that side for simplicity, though it would not be difficult to consider the $\bar{q}_n$ varying, for instance, linearly, if necessary; i.e.

$$\bar{q}_n = \bar{q}_{n2}\xi_2 + \bar{q}_{n3}\xi_3 \tag{i}$$

After integrating (h) using formula (3.20), we have

$$\mathbf{P} = \begin{Bmatrix} 0 \\ \frac{1}{6}l \\ \frac{1}{6}l \\ 0 \\ \frac{2}{3}l \\ 0 \end{Bmatrix} \bar{q}_n \tag{j}$$

(l = length of the side.)

3.3 FIRST-ORDER CONTINUITY FUNCTIONS FOR RECTANGULAR ELEMENTS

SYSTEM OF COORDINATES

For rectangular elements (Figure 3.6) we can define a set of dimensionless coordinates ξ, η, such as

$$\xi = \frac{1}{a}(x - x_c), \qquad \eta = \frac{1}{b}(y - y_c) \tag{3.35}$$

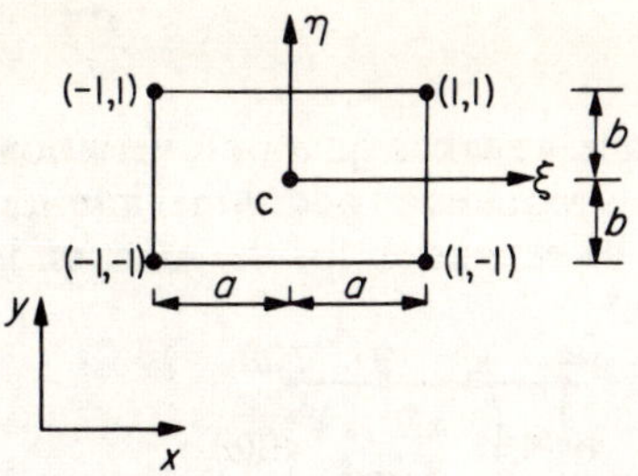

Figure 3.6 Rectangular elements

with limits ± 1. The derivatives and area integral in terms of the dimensionless coordinates are

$$\frac{\partial}{\partial x} = \frac{1}{a}\frac{\partial}{\partial \xi}$$
$$\frac{\partial}{\partial y} = \frac{1}{b}\frac{\partial}{\partial \eta} \tag{3.36}$$

and

$$\iint_A f(x, y)\,\mathrm{d}A = 4ab\int_{-1}^{+1}\int_{-1}^{+1} f(\xi, \eta)\,\mathrm{d}\xi\,\mathrm{d}\eta \tag{3.37}$$

SIMPLEST RECTANGLE

For the simplest rectangle we have only corner nodal unknowns, hence the function must vary linearly on the boundaries. Appropriate two-dimensional interpolation functions for the u function can be generated by evaluating

$$u = \alpha_1 + \alpha_2\xi + \alpha_3\eta + \alpha_4\xi\eta \tag{3.38}$$

at the nodal points and solving for α. The result is

$$\begin{aligned}
u &= \phi_1 u_1 + \phi_2 u_2 + \phi_3 u_3 + \phi_4 u_4\\
\phi_1 &= \tfrac{1}{4}(1 - \xi)(1 - \eta)\\
\phi_2 &= \tfrac{1}{4}(1 + \xi)(1 - \eta)\\
\phi_3 &= \tfrac{1}{4}(1 + \xi)(1 + \eta)\\
\phi_4 &= \tfrac{1}{4}(1 - \xi)(1 + \eta)
\end{aligned} \tag{3.39}$$

REFINED RECTANGULAR ELEMENTS

To improve on the first-order rectangle, we take a quadratic variation of the function on the boundary. This requires an additional interior node per side [see Figure 3.7(a)]. The expansion for the function u

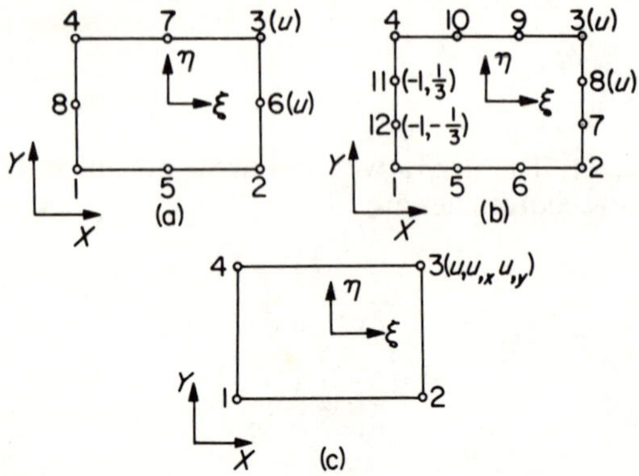

Figure 3.7 Higher-order rectangular elements. (a) *Second order*; (b) *Third order—translations*; (c) *Third order—translations and first derivatives*

involves eight parameters. Specialising

$$u = \alpha_1 + \alpha_2\xi + \alpha_3\eta + \alpha_4\xi^2 + \alpha_5\xi\eta + \alpha_6\eta^2 + \alpha_7\xi^2\eta + \alpha_8\xi\eta^2 \tag{3.40}$$

at the nodes leads to

$$u = \sum_{i=1}^{8} \phi_i u_i \tag{3.41}$$

where

$$4\phi_1 = (1 - \xi)(1 - \eta)(-\xi - \eta - 1)$$
$$4\phi_2 = (1 + \xi)(1 - \eta)(\xi - \eta - 1)$$
$$4\phi_3 = (1 + \xi)(1 + \eta)(\xi + \eta - 1)$$
$$4\phi_4 = (1 - \xi)(1 + \eta)(-\xi + \eta - 1)$$
$$2\phi_5 = (1 - \xi^2)(1 - \eta)$$
$$2\phi_6 = (1 - \eta^2)(1 + \xi)$$
$$2\phi_7 = (1 - \xi^2)(1 + \eta)$$
$$2\phi_8 = (1 - \eta^2)(1 - \xi)$$

A rectangular element corresponding to the cubic triangle is obtained by taking a cubic variation of the function on the boundary. Two possibilities exist: we can include two additional nodes on each side, as shown in Figure 3.7(b), and work only with translations as nodal quantities. An alternative approach is to work only with corner nodes and include derivatives as nodal quantities. For the latter model [Figure 3.7(c)] there are three variables per node

$$u, \frac{\partial u}{\partial x}, \frac{\partial u}{\partial y}$$

In both cases, we obtain the interpolation formulation by evaluating

$$u = \alpha_1 + \alpha_2\xi + \alpha_3\eta + \alpha_4\xi^2 + \alpha_5\xi\eta + \alpha_6\eta^2 + \alpha_7\xi^3 + \alpha_8\xi^2\eta + \alpha_9\xi\eta^2 + \alpha_{10}\eta^3 + \alpha_{11}\xi^3\eta + \alpha_{12}\xi\eta^3 \quad (3.42)$$

and the dimensionless derivatives $(\partial u/\partial\xi, \partial u/\partial\eta)$ at the node points. The resulting expressions are listed below:

Model A: Nodal variables are values of the function

$$u = \sum_{i=1}^{12} \phi_i u_i \quad (3.43)$$

where

$$\phi_1 = \tfrac{1}{32}(1 - \xi)(1 - \eta)\{-10 + 9(\xi^2 + \eta^2)\}$$
$$\phi_2 = \tfrac{1}{32}(1 + \xi)(1 - \eta)\{-10 + 9(\xi^2 + \eta^2)\}$$
$$\phi_3 = \tfrac{1}{32}(1 + \xi)(1 + \eta)\{-10 + 9(\xi^2 + \eta^2)\}$$
$$\phi_4 = \tfrac{1}{32}(1 - \xi)(1 + \eta)\{-10 + 9(\xi^2 + \eta^2)\}$$
$$\phi_5 = \tfrac{9}{32}(1 - \xi^2)(1 - \eta)(1 - 3\xi)$$
$$\phi_6 = \tfrac{9}{32}(1 - \xi^2)(1 - \eta)(1 + 3\xi)$$
$$\phi_7 = \tfrac{9}{32}(1 - \eta^2)(1 + \xi)(1 - 3\eta)$$
$$\phi_8 = \tfrac{9}{32}(1 - \eta^2)(1 + \xi)(1 + 3\eta)$$
$$\phi_9 = \tfrac{9}{32}(1 - \xi^2)(1 + \eta)(1 + 3\xi)$$
$$\phi_{10} = \tfrac{9}{32}(1 - \xi^2)(1 + \eta)(1 - 3\xi)$$
$$\phi_{11} = \tfrac{9}{32}(1 - \eta^2)(1 - \xi)(1 + 3\eta)$$
$$\phi_{12} = \tfrac{9}{32}(1 - \eta^2)(1 - \xi)(1 - 3\eta)$$

Model B: Nodal variables are the function and its first derivatives[3]

$$u = \boldsymbol{\phi}^T \mathbf{u}^n \tag{3.44}$$

$$u^n = \{u_1 u_{1,x} u_{1,y} u_2 u_{2,x} u_{2,y} u_3 u_{3,x} u_{3,y} u_4 u_{4,x} u_{4,y}\}$$

where

$$\phi_1 = \tfrac{1}{8}(\eta - 1)(\xi - 1)\{\tfrac{1}{2}(\eta - 1)(\xi - 1) - \tfrac{1}{2}(\eta + 1)(\xi + 1) - (\eta + 1)(\eta - 1) - (\xi + 1)(\xi - 1)\}$$

$$\phi_2 = \frac{a}{8}(\eta - 1)(\xi - 1)^2(\xi + 1)$$

$$\phi_3 = -\frac{b}{8}(\eta - 1)^2(\xi - 1)(\eta + 1)$$

$$\phi_4 = \tfrac{1}{8}(\eta - 1)(\xi + 1)\{\tfrac{1}{2}(\eta - 1)(\xi + 1) - \tfrac{1}{2}(\eta + 1)(\xi - 1) + (\eta + 1)(\eta - 1) + (\xi + 1)(\xi - 1)\}$$

$$\phi_5 = \frac{a}{8}(\eta - 1)(\xi + 1)^2(\xi - 1)$$

$$\phi_6 = \frac{b}{8}(\eta - 1)^2(\xi + 1)(\eta + 1)$$

$$\phi_7 = \tfrac{1}{8}(\eta + 1)(\xi + 1)\{\tfrac{1}{2}(\eta + 1)(\xi + 1) - \tfrac{1}{2}(\xi - 1)(\eta - 1) - (\eta + 1)(\eta - 1) - (\xi + 1)(\xi - 1)\}$$

$$\phi_8 = \frac{a}{8}(\eta + 1)(\xi + 1)^2(\xi - 1)$$

$$\phi_9 = \frac{b}{8}(\eta + 1)^2(\xi + 1)(\eta - 1)$$

$$\phi_{10} = \tfrac{1}{8}(\eta + 1)(\xi - 1)\{\tfrac{1}{2}(\eta + 1)(\xi - 1) - \tfrac{1}{2}(\eta - 1)(\xi + 1) + (\eta + 1)(\eta - 1) + (\xi + 1)(\xi - 1)\}$$

$$\phi_{11} = \frac{a}{b}(\eta + 1)(\xi - 1)^2(\xi + 1)$$

$$\phi_{12} = -\frac{b}{8}(\eta + 1)^2(\xi - 1)(\eta - 1)$$

The interpolation functions corresponding to some two-dimensional rectangular element functions are shown in Table 3.1.

Table 3.2 SOME ELEMENT FUNCTIONS FOR THREE DIMENSIONS

Tetrahedron	*Cube*
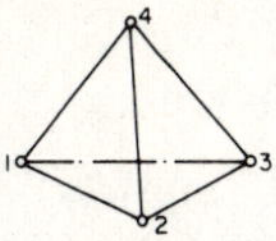 $\phi_i = \xi_i,$ $i = 1, 2, 3, 4$	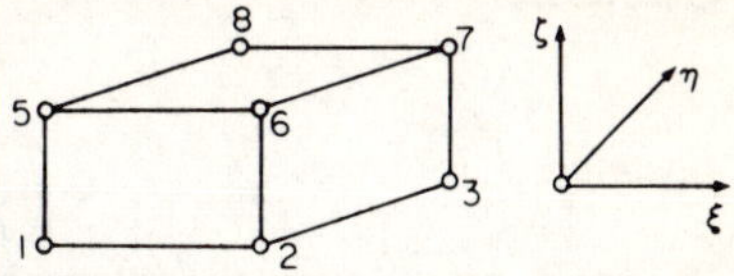 $\phi_i = \frac{1}{8}(1 + \xi\xi_i)(1 + \eta\eta_i)(1 + \zeta\zeta_i)$
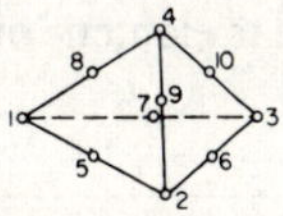 *Corner nodes* $\phi_i = (2\xi_i - 1)\xi_i, i = 1, 2, 3, 4$ *Midside,* $i = 5, 6, 7, 8, 9, 10$ $\phi_i = 2\xi_k\xi_l$ $k = 1, 2, 3, 1, 2, 3; l = 2, 3, 1, 4, 4, 4$	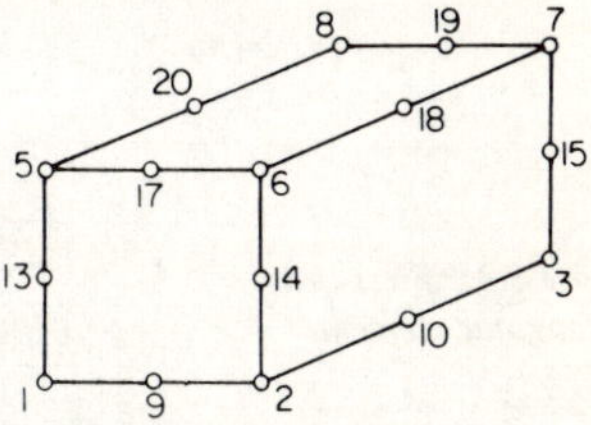 *Corner nodes,* $i = 1, 2, \ldots, 8$ $\phi_i = \frac{1}{8}(1 + \xi\xi_i)(1 + \eta\eta_i)(1 + \zeta\zeta_i)$ $\times (\xi\xi_i + \eta\eta_i + \zeta\zeta_i - 2)$ *On sides,* $i = 9, 12, 17, 19$ $\phi_i = \frac{1}{4}(1 - \xi^2)(1 + \eta\eta_i)(1 + \zeta\zeta_i)$ *On sides,* $i = 10, 11, 18, 20$ $\phi_i = \frac{1}{4}(1 - \eta^2)(1 + \xi\xi_i)(1 + \zeta\zeta_i)$ *On sides,* $i = 13, 14, 15, 16$ $\phi_i = \frac{1}{4}(1 - \zeta^2)(1 + \xi\xi_i)(1 + \eta\eta_i)$
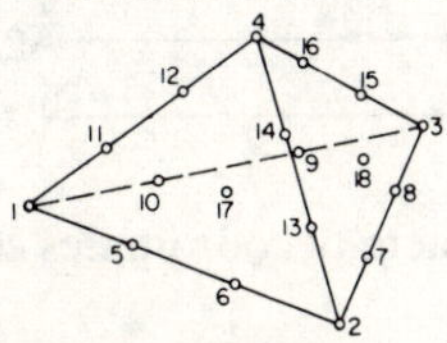 *Corner nodes,* $i = 1, 2, 3, 4$ $\phi_i = \frac{1}{2}(3\xi_i - 1)(2\xi_i - 2)\xi_i$ *On sides* $i = 5, 6, 7, 8, 9, 10, 11, 12, 13, 14, 15, 16$ $\phi_i = \frac{9}{2}(3\xi_i - 1)\xi_k\xi_l$ $k = 1, 2, 2, 3, 3, 1, 1, 4, 2, 4, 3, 4$ $l = 2, 1, 3, 2, 1, 3, 4, 1, 4, 2, 4, 3$ *At mid-faces,* $i = 17, 18, 19$ $\phi_i = 27\xi_j\xi_k\xi_l$ $j = 1, 2, 3; k = 2, 3, 1; l = 4, 4, 4$	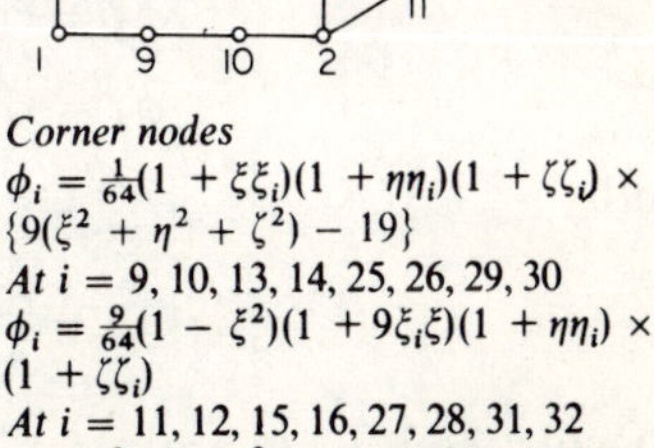 *Corner nodes* $\phi_i = \frac{1}{64}(1 + \xi\xi_i)(1 + \eta\eta_i)(1 + \zeta\zeta_i) \times$ $\{9(\xi^2 + \eta^2 + \zeta^2) - 19\}$ *At* $i = 9, 10, 13, 14, 25, 26, 29, 30$ $\phi_i = \frac{9}{64}(1 - \xi^2)(1 + 9\xi_i\xi)(1 + \eta\eta_i) \times$ $(1 + \zeta\zeta_i)$ *At* $i = 11, 12, 15, 16, 27, 28, 31, 32$ $\phi_i = \frac{9}{64}(1 - \eta^2)(1 + 9\eta_i\eta)(1 + \xi\xi_i) \times$ $(1 + \zeta\zeta_i)$ *At* $i = 17, 18, 19, 20, 21, 22, 23, 24$ $\phi_i = \frac{9}{64}(1 - \zeta^2)(1 + 9\zeta_i\zeta)(1 + \xi\xi_i) \times$ $(1 + \eta\eta_i)$
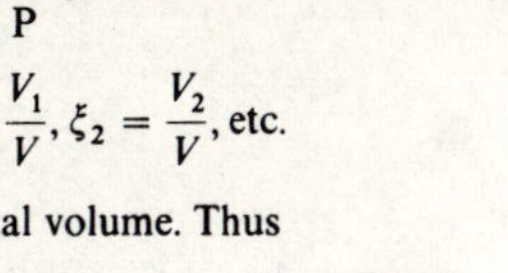 *Coordinate system* The system is defined in terms of the volumes generated by an arbitrary point P $\xi_1 = \frac{V_1}{V}, \xi_2 = \frac{V_2}{V}$, etc. V total volume. Thus $\begin{Bmatrix} 1 \\ x \\ y \\ z \end{Bmatrix} = \begin{bmatrix} 1 & 1 & 1 & 1 \\ x_1 & x_2 & x_3 & x_4 \\ y_1 & y_2 & y_3 & y_4 \\ z_1 & z_2 & z_3 & z_4 \end{bmatrix} \begin{Bmatrix} \xi_1 \\ \xi_2 \\ \xi_3 \\ \xi_4 \end{Bmatrix}$	Coordinates ξ, η, ζ vary between $+1$, -1

These functions can be easily generalised to three dimensions (Table 3.2).

Example 3.3

We will now form the element matrices corresponding to the Laplace equation for the simplest rectangular element of Figure 3.8.

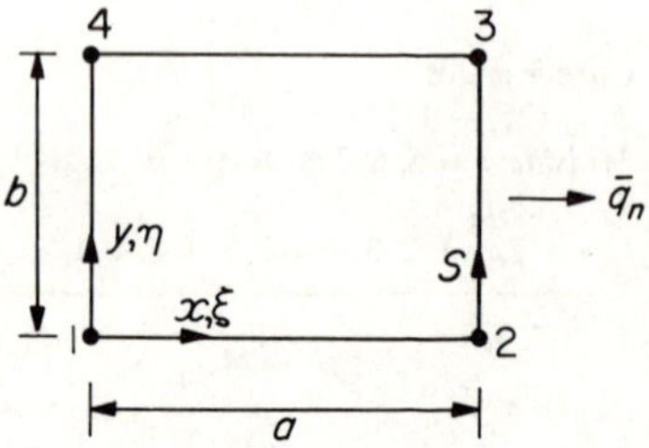

Figure 3.8 First-order rectangular element

The element is referred to a Cartesian system of coordinates centred on node 1 for simplicity.

We have

$$\begin{aligned} u &= \phi_1 u_1 + \phi_2 u_2 + \phi_3 u_3 + \phi_4 u_4 \\ &= \boldsymbol{\phi}^T \mathbf{u}^n \end{aligned} \tag{a}$$

where

$$\begin{aligned} \phi_1 &= (1 - \xi)(1 - \eta) \\ \phi_2 &= \xi(1 - \eta) \\ \phi_3 &= \xi\eta \\ \phi_4 &= (1 - \xi)\eta \end{aligned}$$

with

$$\xi = \frac{x}{a}, \quad \eta = \frac{y}{b}$$

The first derivatives are

$$\begin{Bmatrix} \dfrac{\partial u}{\partial x} \\ \dfrac{\partial u}{\partial y} \end{Bmatrix} = \begin{Bmatrix} \dfrac{1}{a} \dfrac{\partial u}{\partial \xi} \\ \dfrac{1}{b} \dfrac{\partial u}{\partial \eta} \end{Bmatrix} = \begin{Bmatrix} \boldsymbol{\phi}_{,x} \\ \boldsymbol{\phi}_{,y} \end{Bmatrix} \mathbf{u}^n \tag{b}$$

where

$$\begin{Bmatrix} \boldsymbol{\phi}_{,x} \\ \boldsymbol{\phi}_{,y} \end{Bmatrix} = \begin{bmatrix} -\dfrac{(1-\eta)}{a} & \dfrac{1-\eta}{a} & \dfrac{\eta}{a} & -\dfrac{\eta}{a} \\ -\dfrac{1-\xi}{b} & -\dfrac{\xi}{b} & \dfrac{\xi}{b} & \dfrac{1-\xi}{b} \end{bmatrix} \tag{c}$$

Substituting (b) into the variational expression shown in equation (a) of Example 3.2, we obtain

$$\delta \mathbf{u}^{n,T} ab \int_0^1 \int_0^1 (h_x \boldsymbol{\phi}_{,x} \boldsymbol{\phi}_{,x}^T + h_y \boldsymbol{\phi}_{,y} \boldsymbol{\phi}_{,y}^T)\, d\xi\, d\eta\, \mathbf{u}^n = \delta \mathbf{u}^{n,T} \int_{S_2} \boldsymbol{\phi} \bar{q}_n\, dS \tag{d}$$

We can integrate the first term on the left-hand side, which gives, for constant h_x,

$$h_x ab \int_0^1 \int_0^1 \boldsymbol{\phi}_{,x} \boldsymbol{\phi}_{,x}^T\, d\xi\, d\eta$$

$$= h_x \frac{b}{a} \int_0^1 \int_0^1 \begin{bmatrix} (1-\eta)^2 & -(1-\eta)^2 & -\eta(1-\eta) & \eta(1-\eta) \\ & (1-\eta)^2 & \eta(1-\eta) & -\eta(1-\eta) \\ & \text{sym.} & \eta^2 & -\eta^2 \\ & & & \eta^2 \end{bmatrix} d\xi\, d\eta \tag{e}$$

$$= h_x \left(\frac{b}{a}\right) \frac{1}{6} \begin{bmatrix} 2 & -2 & -1 & 1 \\ & 2 & 1 & -1 \\ & & 2 & -2 \\ & \text{sym.} & & 2 \end{bmatrix} \tag{f}$$

Similarly,

$$h_y ab \int_0^1 \int_0^1 \boldsymbol{\phi}_{,y} \boldsymbol{\phi}_{,y}^T\, d\xi\, d\eta = h_y \left(\frac{a}{b}\right) \frac{1}{6} \begin{bmatrix} 2 & 1 & -1 & -2 \\ & 2 & -2 & -1 \\ & \text{sym.} & 2 & 1 \\ & & & 2 \end{bmatrix} \tag{g}$$

To determine the **P** matrix it is necessary to integrate on the boundary on which $\bar{q}_n$ is applied. Let us assume that this needs to be done only on side 2–3, and that $\bar{q}_n$ is constant over it:

$$\mathbf{P} = b \int_2^3 \begin{Bmatrix} 0 \\ 1-\eta \\ \eta \\ 0 \end{Bmatrix} \bar{q}_n \, d\eta = \bar{q}_n b \begin{Bmatrix} 0 \\ \frac{1}{2} \\ \frac{1}{2} \\ 0 \end{Bmatrix} \tag{h}$$

COMPARISON STUDY

Emery and Carson[4] have carried out some comparison studies of different element matrices for the case of steady state and transient heat transfer. The solution of the steady state one-dimensional diffusion equation was taken as a test case, i.e.

$$\frac{d^2u}{dx^2} = p \tag{a}$$

where $p = x^5$, with boundary conditions,

$$\left.\frac{du}{dx}\right|_{x=1} = 0 \qquad u|_{x=0} = 0 \tag{b}$$

They used linear, quadratic and the cubic triangular element corresponding to equations (3.32). The results are presented in Figure 3.9, where they are compared with a finite difference solution. l indicates the distance between nodes, noting that the total length is

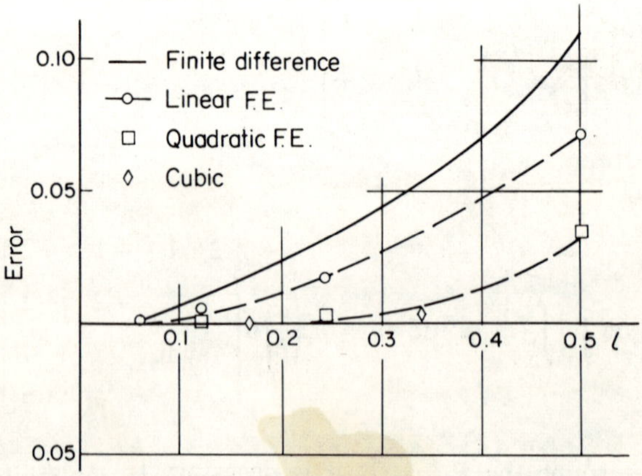

Figure 3.9 Error in u

equal to 1. The results represent the error in the temperature u at the point $x = 1$. The exact value of u at that point is obtained by integrating (a) with boundary conditions (b) and is equal to 0·1429.

The results point out the convenience of using finite elements when trying to represent complex source functions of the type $p = x^5$. In finite elements these functions are distributed in accordance with the interpolation functions; i.e. for equations (a) and (b) we have the following variational statement:

$$\int_0^L \left\{ \frac{d^2u}{dx^2} - p \right\} \delta u \, dx = \left. \frac{\partial u}{\partial x} \delta u \right|_L \tag{c}$$

Integrating by parts, this gives

$$\int_0^L \left\{ \left(\frac{du}{dx} \right) \left(\frac{d\delta u}{dx} \right) + p \, \delta u \right\} dx = 0 \tag{d}$$

where the p function is distributed in accordance with the interpolation function for u. In finite difference instead, the contributions of the p function are 'lumped' at the nodes.

The results for this one-dimensional case seem also to show that it is generally more efficient to use higher-order elements. Among these elements the quadratic gives good results with very little extra computational work and without the complications present in other higher-order elements, such as having large elements for which the properties and geometry may be cumbersome to define. Taking into consideration these and other tests, such as those carried out for stress analysis problems,[1] the quadratic model is the one we generally recommend.

3.4 ISOPARAMETRIC ELEMENTS

It is more accurate to represent bodies with curved boundaries using elements with curved sides (Figure 3.10). The transformation from straight to curved sides is done as follows. Consider that the x, y coordinates can be expressed in terms of curvilinear coordinates

$$x = x(\xi, \eta), \qquad y = y(\xi, \eta) \tag{3.45}$$

The choice of ξ, η depends on the element geometry and will be discussed later. We generate interpolation functions in terms of

ξ, η and evaluate curvilinear derivatives:

$$\boldsymbol{\phi} = \boldsymbol{\phi}(\xi, \eta) = \{\phi_1, \phi_2, \ldots\}$$
$$\boldsymbol{\phi}_{,\xi} = \left\{\frac{\partial \phi_1}{\partial \xi}, \frac{\partial \phi_2}{\partial \xi} \ldots\right\}, \qquad \boldsymbol{\phi}_{,\eta} = \left\{\frac{\partial \phi_1}{\partial \eta}, \frac{\partial \phi_2}{\partial \eta}, \ldots\right\} \tag{3.46}$$

Since the equation for the element matrices involves Cartesian derivatives, we have to transform curvilinear to Cartesian derivatives before finding the element matrices. The general transformation for a function ϕ is

$$\frac{\partial \phi}{\partial x} = \frac{\partial \phi}{\partial \xi}\frac{\partial \xi}{\partial x} + \frac{\partial \phi}{\partial \eta}\frac{\partial \eta}{\partial x}$$
$$\frac{\partial \phi}{\partial y} = \frac{\partial \phi}{\partial \xi}\frac{\partial \xi}{\partial y} + \frac{\partial \phi}{\partial \eta}\frac{\partial \eta}{\partial y} \tag{3.47}$$

Usually explicit expressions for ξ, η in terms of x and y are not easily available. Thus, we consider ϕ to be a function of x and y, form the derivatives with respect to ξ and η and then invert:

$$\begin{Bmatrix} \dfrac{\partial \phi}{\partial \xi} \\ \dfrac{\partial \phi}{\partial \eta} \end{Bmatrix} = \begin{bmatrix} \dfrac{\partial x}{\partial \xi} & \dfrac{\partial y}{\partial \xi} \\ \dfrac{\partial x}{\partial \eta} & \dfrac{\partial y}{\partial \eta} \end{bmatrix} \begin{Bmatrix} \dfrac{\partial \phi}{\partial x} \\ \dfrac{\partial \phi}{\partial y} \end{Bmatrix} = \mathbf{J} \begin{Bmatrix} \dfrac{\partial \phi}{\partial x} \\ \dfrac{\partial \phi}{\partial y} \end{Bmatrix} \tag{3.48}$$

Thus

$$\begin{Bmatrix} \dfrac{\partial \phi}{\partial x} \\ \dfrac{\partial \phi}{\partial y} \end{Bmatrix} = \mathbf{J}^{-1} \begin{Bmatrix} \dfrac{\partial \phi}{\partial \xi} \\ \dfrac{\partial \phi}{\partial \eta} \end{Bmatrix} \tag{3.49}$$

and

$$\frac{\partial \phi}{\partial x} = \frac{1}{|\mathbf{J}|}\left(\frac{\partial y}{\partial \eta}\frac{\partial \phi}{\partial \xi} - \frac{\partial y}{\partial \xi}\frac{\partial \phi}{\partial \eta}\right)$$
$$\frac{\partial \phi}{\partial y} = \frac{1}{|\mathbf{J}|}\left(-\frac{\partial x}{\partial \eta}\frac{\partial \phi}{\partial \xi} + \frac{\partial x}{\partial \xi}\frac{\partial \phi}{\partial \eta}\right) \tag{3.50}$$

where $|\mathbf{J}|$ is the determinant of $\mathbf{J}$ and is called the Jacobian.

$$|\mathbf{J}| = \frac{\partial x}{\partial \xi}\frac{\partial y}{\partial \eta} - \frac{\partial x}{\partial \eta}\frac{\partial y}{\partial \xi} \tag{3.51}$$

The Jacobian must be finite if the transformation is to be unique. One can show that the differential area transforms to

$$dA = (\text{absolute value of } |\mathbf{J}|)\, d\xi\, d\eta \tag{3.52}$$

The presence of variable coefficients in the derivative transformations necessitates the use of numerical integration to evaluate the element matrices. Formulae for triangular and rectangular domains are given in the Appendix.

If the coordinate expansions are identical with the variable expansions the element is called 'isoparametric'.[5] We must ensure that the variable expansions in terms of curvilinear coordinates satisfy the constant function and constant derivatives (completeness) condition. Hence by suitably specialising the function u

$$u = \boldsymbol{\phi}^T \mathbf{u}^n \tag{3.53}$$

the condition

$$u = \alpha_1 + \alpha_2 x + \alpha_3 y \tag{3.54}$$

must be satisfied.

To do this let us consider the nodal variables are prescribed as

$$u_i = \alpha_1 + \alpha_2 x_i + \alpha_3 y_i \qquad (i = \text{node under consideration}) \tag{3.55}$$

Next substitute into (3.53)

$$\begin{aligned} u &= \sum_{i=1}^{n} (\alpha_1 + \alpha_2 x_i + \alpha_3 y_i)\phi_i(\xi, \eta) \\ &= \alpha_1 \sum_{i=1}^{n} \phi_i(\xi, \eta) + \alpha_2 \sum_{i=1}^{n} x_i \phi_i(\xi, \eta) + \alpha_3 \sum_{i=1}^{n} y_i \phi_i(\xi, \eta) \end{aligned} \tag{3.56}$$

Hence the interpolation functions must satisfy

$$\begin{aligned} \sum_{i=1}^{n} \phi_i &= 1 \\ \sum_{i=1}^{n} x_i \phi_i &= x \\ \sum_{i=1}^{n} y_i \phi_i &= y \end{aligned} \tag{3.57}$$

The last two relationships are the definition of an isoparametric element. The first is satisfied by the interpolation functions.

Example 3.4

We will discuss the transformation of coordinates shown above for the simplest rectangular element, which, as an isoparametric element, transforms into a general quadrilateral [Figure 3.10(a)].

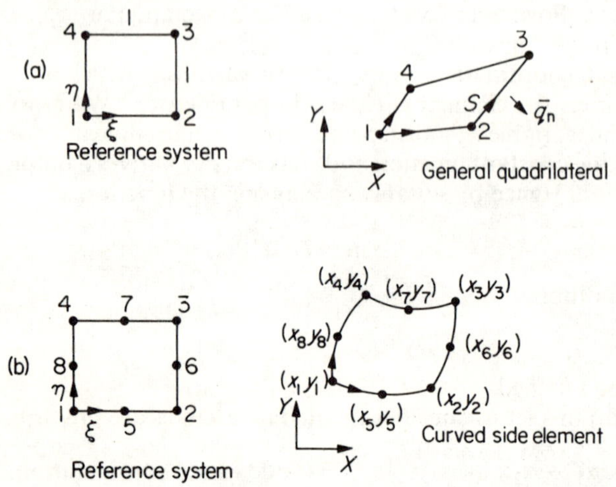

Figure 3.10 Isoparametric elements. (a) *Linear element*; (b) *Quadratic element*

One can express the x, y coordinates in terms of their nodal values as

$$x = \sum_{i=1}^{4} \phi_i x_i = \boldsymbol{\phi}^T \mathbf{x}^n$$
$$y = \sum_{i=1}^{n} \phi_i y_i = \boldsymbol{\phi}^T \mathbf{y}^n \qquad \text{(a)}$$

and we also know that

$$u = \sum_{i=1}^{4} \phi_i u_i = \boldsymbol{\phi}^T \mathbf{u}^n \qquad \text{(b)}$$

where

$$\phi_1 = (1 - \xi)(1 - \eta), \qquad \phi_2 = \xi(1 - \eta)$$
$$\phi_3 = \xi\eta \quad , \qquad \phi_4 = (1 - \xi)\eta \qquad \text{(c)}$$

Hence

$$\begin{Bmatrix} \dfrac{\partial u}{\partial \xi} \\ \dfrac{\partial u}{\partial \eta} \end{Bmatrix} = \begin{bmatrix} (\eta - 1) & (1 - \eta) & \eta & -\eta \\ (\xi - 1) & -\xi & \xi & (1 - \xi) \end{bmatrix} \begin{Bmatrix} u_{n_1} \\ u_{n_2} \\ u_{n_3} \\ u_{n_4} \end{Bmatrix} \tag{d}$$

$$= \mathbf{B}^* \mathbf{u}^n$$

The relationship between both systems of coordinates is expressed by the Jacobian (3.48)

$$\mathbf{J} = \begin{bmatrix} \dfrac{\partial x}{\partial \xi} & \dfrac{\partial y}{\partial \xi} \\ \dfrac{\partial x}{\partial \eta} & \dfrac{\partial y}{\partial \eta} \end{bmatrix} = \mathbf{B}^* \begin{bmatrix} x_1 & y_1 \\ x_2 & y_2 \\ x_3 & y_3 \\ x_4 & y_4 \end{bmatrix} \tag{e}$$

Next, we have to invert the elements of **J**. In order to avoid the algebra, let us define the elements of **J** as J_{ij}, which are known after having done the multiplication of equation (e); i.e.

$$\mathbf{J} = \begin{bmatrix} J_{11} & J_{12} \\ J_{21} & J_{22} \end{bmatrix} \tag{f}$$

We can invert (f)

$$\mathbf{J}^{-1} = \frac{1}{|\mathbf{J}|} \begin{bmatrix} J_{22} & -J_{12} \\ -J_{21} & J_{11} \end{bmatrix} \tag{g}$$

where

$$|\mathbf{J}| = J_{11} J_{22} - J_{12} J_{21}$$

One can now form the derivatives $\partial u/\partial x$, $\partial u/\partial y$, i.e.

$$\begin{Bmatrix} \dfrac{\partial u}{\partial x} \\ \dfrac{\partial u}{\partial y} \end{Bmatrix} = \mathbf{J}^{-1} \begin{Bmatrix} \dfrac{\partial u}{\partial \xi} \\ \dfrac{\partial u}{\partial \eta} \end{Bmatrix} = \mathbf{J}^{-1} \mathbf{B}^* \begin{Bmatrix} u_{n_1} \\ u_{n_2} \\ u_{n_3} \\ u_{n_4} \end{Bmatrix} \tag{h}$$

The matrix $\mathbf{J}^{-1}\mathbf{B}^*$ can be easily computed and will be represented as **B**, where

$$\underset{(2 \times 4)}{\mathbf{B}} = \underset{(2 \times 2)}{\mathbf{J}^{-1}} \underset{(2 \times 4)}{\mathbf{B}^*} \tag{i}$$

The derivatives (h) can now be expressed as

$$\begin{Bmatrix} \dfrac{\partial u}{\partial x} \\ \dfrac{\partial u}{\partial y} \end{Bmatrix} = \mathbf{B}\mathbf{u}^n \tag{j}$$

The variational statement corresponding to Laplace's equation is

$$\iint \left(h_x \frac{\partial u}{\partial x} \frac{\partial \delta u}{\partial x} + h_y \frac{\partial u}{\partial y} \frac{\partial \delta u}{\partial y} \right) \mathrm{d}x\, \mathrm{d}y = \int \bar{q}_n\, \delta u\, \mathrm{d}S \tag{k}$$

If we define an element *property* matrix, we can express the left-hand side of (k) as

$$\delta \mathbf{u}^{n,T} \int_0^1 \int_0^1 \mathbf{B}^T \mathbf{H} \mathbf{B}\, (\text{abs. value}\, |\mathbf{J}|)\, \mathrm{d}\xi\, \mathrm{d}\eta\, \mathbf{u}^n \tag{l}$$

where

$$\mathbf{H} = \begin{bmatrix} h_x & \cdot \\ \cdot & h_y \end{bmatrix}$$

To find the corresponding P matrix we need to integrate the right-hand side of (k) on, for instance, side 2–3. $\bar{q}_n$ is the constant flux normal to the boundary, l is the length. Hence

$$\int_2^3 \bar{q}_n\, \delta u\, \mathrm{d}S = \delta \mathbf{u}^{n,T} \bar{q}_n l \int_0^1 \begin{Bmatrix} 0 \\ \eta \\ (1-\eta) \\ 0 \end{Bmatrix} \mathrm{d}\eta = \delta \mathbf{u}^{n,T} \bar{q}_n l \begin{Bmatrix} 0 \\ \frac{1}{2} \\ \frac{1}{2} \\ 0 \end{Bmatrix} \tag{m}$$

In general we have

$$\mathrm{d}S = \left[\left(\frac{\mathrm{d}x}{\mathrm{d}\xi} \right)^2 + \left(\frac{\mathrm{d}y}{\mathrm{d}\xi} \right)^2 \right]^{\frac{1}{2}} \mathrm{d}\xi \quad \text{on side } \eta = \text{constant}$$

and

$$\mathrm{d}S = \left[\left(\frac{\mathrm{d}x}{\mathrm{d}\eta} \right)^2 + \left(\frac{\mathrm{d}y}{\mathrm{d}\eta} \right)^2 \right]^{\frac{1}{2}} \mathrm{d}\eta \quad \text{on side } \xi = \text{constant}$$

3.5 SECOND-ORDER CONTINUITY FUNCTIONS FOR RECTANGULAR ELEMENTS

For the bi-harmonic equation $\nabla^4 u$ (Example 1.10), minimum continuity requirements imply continuity of the function and its derivatives. Some rectangular and triangular element functions which can be used in this case will now be discussed. We will concern ourselves only with those elements which have been extensively tested and have been found to be accurate.

The simplest rectangle has nodes at the four corners and three displacement measures per node[3] (Figure 3.11). We take the nodal displacement listing as

$$\mathbf{u}^n = \{u_1, u_{1,x}, u_{1,y} \ldots u_{4,y}\} \tag{3.58}$$

The expansion consists of the complete quadratic and six additional terms. Inter-element compatibility requires continuity of u and the normal derivative. Since only the first derivatives are included in $\mathbf{u}^n$, the expansion for u along a side must contain four parameters, i.e. it must be cubic, in order to satisfy continuity of u. Also, the normal derivative must vary linearly to satisfy continuity of $\partial u/\partial n$.

The terms of successive degrees of polynomials are conveniently represented by Pascal's triangle. Its form up to a tenth degree polynomial is shown below:

$$\begin{array}{c}
1 \\
\xi \quad \eta \\
\xi^2 \quad \xi\eta \quad \eta^2 \\
\xi^3 \quad \xi^2\eta \quad \xi\eta^2 \quad \eta^3 \qquad \text{bicubic} \\
\xi^4 \quad \xi^3\eta \quad \xi^2\eta^2 \quad \xi\eta^3 \quad \eta^4 \\
\xi^5 \quad \xi^4\eta \quad \xi^3\eta^2 \quad \xi^2\eta^3 \quad \xi\eta^4 \quad \eta^5 \qquad \text{biquintic} \\
\xi^6 \quad \xi^5\eta \quad \xi^4\eta^2 \quad \xi^3\eta^3 \quad \xi^2\eta^4 \quad \xi\eta^5 \quad \eta^6 \\
\xi^7 \quad \xi^6\eta \quad \xi^5\eta^2 \quad \xi^4\eta^3 \quad \xi^3\eta^4 \quad \xi^2\eta^5 \quad \xi\eta^6 \quad \eta^7 \\
\xi^8 \quad \xi^7\eta \quad \xi^6\eta^2 \quad \xi^5\eta^3 \quad \xi^4\eta^4 \quad \xi^3\eta^5 \quad \xi^2\eta^6 \quad \xi\eta^7 \quad \eta^8 \\
\ldots \quad \xi^6\eta^3 \quad \xi^5\eta^4 \quad \xi^4\eta^5 \quad \xi^3\eta^6 \quad \ldots \\
\xi^6\eta^4 \quad \xi^5\eta^5 \quad \xi^4\eta^6
\end{array} \tag{3.59}$$

A complete cubic has ten terms. We need two additional terms which are cubic or less on the boundaries $\xi = 0, 1$ and $\eta = 0, 1$. The restriction is indicated by the broken lines. One can choose the terms arbitrarily but one should preserve the symmetry of the expansion

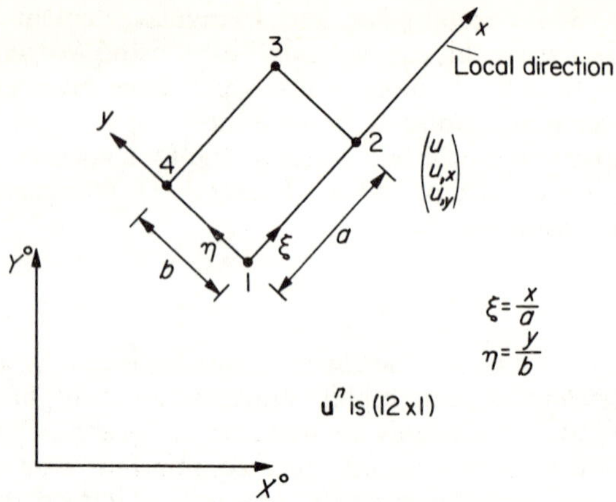

Figure 3.11 Coordinate system—first-order rectangular bending element

to ensure invariance of the element matrices. There are three possible combinations:

$$\begin{array}{llll} & \xi^3\eta & \text{and} & \xi\eta^3 \\ \text{or} & \xi^3\eta^2 & \text{and} & \xi^2\eta^3 \\ \text{or} & \xi^2\eta^2 & \text{and} & \xi^3\eta^3 \end{array}$$

We have succeeded in satisfying continuity of u. Now, continuity of $\partial u/\partial n$ requires $\partial u/\partial n$ to vary linearly on a side since it is specified only at the node points. Irrespective of what combination we choose, we cannot avoid ending up with a cubic variation for the normal derivative. Therefore it is not possible to establish a continuous expansion which satisfies both continuity conditions if the nodal variables are restricted to w and the first derivatives.

A non-conforming element based on the first choice ($\xi^3\eta$ and $\xi\eta^3$) will be completely equivalent to the higher-order expansions (3.44) and has been extensively employed in plate bending analyses. The complete expansion is

$$u = \alpha_1 + \alpha_2\xi + \alpha_3\eta + \alpha_4\xi^2 + \alpha_5\xi\eta + \alpha_6\eta^2 + \alpha_7\xi^3 + \alpha_8\xi^2\eta + \alpha_9\xi\eta^2 + \alpha_{10}\eta^3 + \alpha_{11}\xi^3\eta + \alpha_{12}\xi\eta^3 \quad (3.60)$$

The interpolation functions (3.44) can be specialised for the ξ, η system of coordinates in Figure 3.11 and are shown below for reference:

$$u = \phi_1 u_1 + \phi_2(\partial u/\partial \xi)_1 + \phi_3(\partial u/\partial \eta)_1 + \ldots + \phi_{12}(\partial u/\partial \eta)_4$$

$$\begin{aligned}
\phi_1 &= 2(\eta - 1)(\xi - 1)\{\tfrac{1}{2}(1 + \xi + \eta) - \xi^2 - \eta^2\} \\
\phi_2 &= a(\eta - 1)(\xi - 1)^2\xi \\
\phi_3 &= -b(\eta - 1)^2(\xi - 1)\eta \\
\phi_4 &= 2(\eta - 1)\xi(\eta^2 + \xi^2 - \tfrac{3}{2}\xi - \tfrac{1}{2}\eta) \\
\phi_5 &= a(\eta - 1)\xi^2(\xi - 1) \\
\phi_6 &= b(\eta - 1)^2\xi\eta \\
\phi_7 &= 2\eta\xi\{-\eta^2 - \xi^2 - \tfrac{1}{2} + \tfrac{3}{2}(\xi + \eta)\} \\
\phi_8 &= a\eta\xi^2(\xi - 1) \\
\phi_9 &= b\eta^2\xi(\eta - 1) \\
\phi_{10} &= 2\eta(\xi - 1)(\eta^2 + \xi^2 - \tfrac{1}{2} - \tfrac{3}{2}\eta) \\
\phi_{11} &= a\eta\xi(\xi - 1)^2 \\
\phi_{12} &= -b\eta^2(\xi - 1)(\eta - 1)
\end{aligned} \tag{3.61}$$

The expansion contains all linear terms, which do not produce energy, and the quadratic terms, which are important for completeness as they can reproduce the constant energy density state for the element. The linear terms are necessary to apply the essential boundary conditions. Hence the solution will converge to the exact solution as the mesh size is decreased. Results for this element are generally sufficiently accurate for engineering applications (see Figure 3.17).

We return again to the search for a compatible element. It is clear that we have to allow for a cubic variation of the normal derivative in order to employ a single expansion for the entire element. The expansion consists of all terms above the upper broken line in (3.59)

$$u = \text{equation (3.60)} + \alpha_{13}\xi^2\eta^2 + \alpha_{14}\xi^3\eta^2 + \alpha_{15}\xi^2\eta^3 + \alpha_{16}\xi^3\eta^3$$

Four conditions on $\partial u/\partial n$ are required per side. However, we can add only four additional nodal quantities, which leaves us with just one choice, namely, the cross derivative $u_{,xy}$ at the corners. The interpolation functions for (a) were obtained by Bogner, Fox and Schmidt[6] and are listed below. Figure 3.12 defines the node numbering scheme and nodal displacement listing.

$$u = \phi_1 u_1 + \phi_2 u_{1,x} + \phi_3 u_{1,y} + \phi_4 u_{1,xy} + \ldots + \phi_{16} u_{4,xy} \tag{3.62}$$

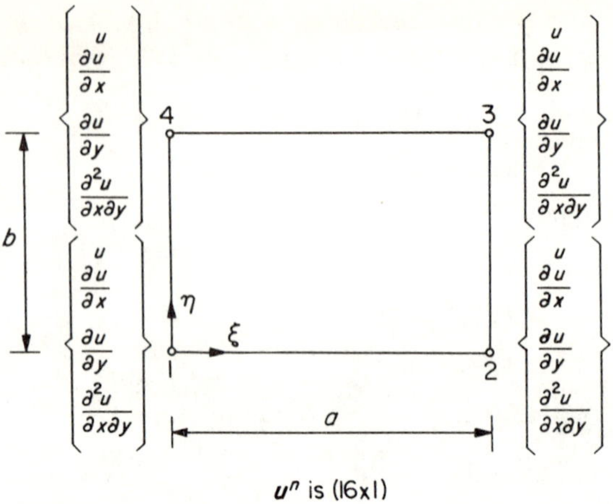

Figure 3.12 Bicubic bending element

where

$$\phi_1 = f_1(\xi)f_1(\eta) \qquad \phi_2 = ag_1(\xi)f_1(\eta)$$
$$\phi_3 = bf_1(\xi)g_1(\eta) \qquad \phi_4 = abg_1(\xi)g_1(\eta)$$
$$\phi_5 = f_2(\xi)f_1(\eta) \qquad \phi_6 = ag_2(\xi)f_1(\eta)$$
$$\phi_7 = bf_2(\xi)g_1(\eta) \qquad \phi_8 = abg_2(\xi)g_1(\eta)$$
$$\phi_9 = f_2(\xi)f_2(\eta) \qquad \phi_{10} = ag_2(\xi)g_2(\eta)$$
$$\phi_{11} = bf_2(\xi)g_2(\eta) \qquad \phi_{12} = abg_1(\xi)f_2(\eta)$$
$$\phi_{13} = f_1(\xi)f_2(\eta) \qquad \phi_{14} = ag_1(\xi)f_2(\eta)$$
$$\phi_{15} = bf_1(\xi)g_2(\eta) \qquad \phi_{16} = abg_1(\xi)g_2(\eta)$$

and

$$\left.\begin{aligned} f_1(S) &= 1 - 3S^2 + 2S^3 \\ f_2(S) &= 3S^2 - 2S^3 \\ g_1(S) &= S - 2S^2 + S^3 \\ g_2(S) &= S^3 - S^2 \end{aligned}\right\} \text{cubic (Hermitian) polynomials} \quad (3{\cdot}63)$$

Results for this element show a considerable improvement in accuracy by comparison with the non-conforming one previously discussed. Nevertheless the cross-derivative, $\partial^2 u/\partial x\, \partial y$ is a 'non-essential' measure and is inconvenient due to its higher order when transformations are required.

Wilson and Brebbia[7] suggested a simple modification to (3.62) which results in an element with 12 degrees of freedom. They replace the cross derivatives with finite difference approximations in terms of the first derivatives at the corners. For example, they take

$$ab\left(\frac{\partial^2 u}{\partial x\,\partial y}\right)_1 = \frac{1}{2}\left[a\left\{\left(\frac{\partial u}{\partial x}\right)_4 - \left(\frac{\partial u}{\partial x}\right)_1\right\} + b\left\{\left(\frac{\partial u}{\partial y}\right)_2 - \left(\frac{\partial u}{\partial y}\right)_1\right\}\right] \tag{3.64}$$

at node 1. The resulting expansion contains all the linear and quadratic terms. Continuity of u is satisfied but the normal slope is discontinuous, i.e. the element is non-conforming.

Results for this element apparently lie between the two previous ones.[7] Both elements with 12 degrees of freedom are non-conforming and therefore may not provide bounds on energy, whereas a conforming element always gives bounded solutions.

Bogner, Fox and Schmidt[6] have also developed a more refined model based on biquintic interpolation (diagonals after ξ^5, η^5 in Pascal's triangle) which has continuity of u, $u_{,n}$ and $u_{,nn}$. Nine nodal displacement measures are required per node:

$$\{u, u_{,x}, u_{,y}, u_{,xx}, u_{,yy}, u_{,xy}, u_{,xyy}, u_{,yyx}, u_{,xxyy}\} \tag{3.65}$$

As would be expected, this element is quite accurate.

Rather than attempt to generate an expansion that is valid over the entire rectangular element, one can subdivide the element into regions and work with a different expansion over each region. The individual expansions must be continuous (up to the first derivatives) across the interior boundaries as well as the exterior boundaries. We describe here the procedure for a rectangle suggested by Deak and Pian.[8] The treatment of the triangle will be discussed later.

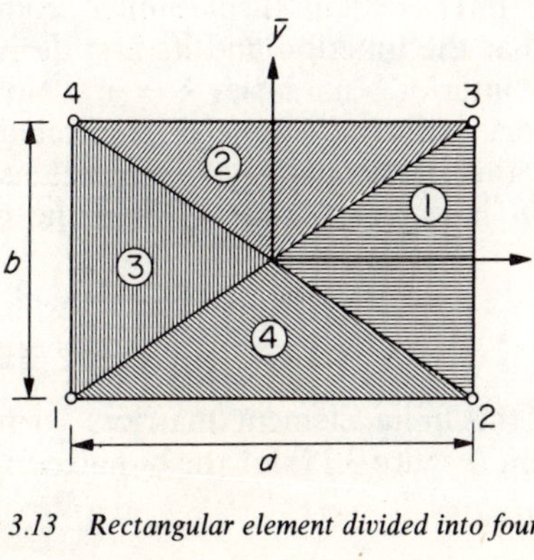

Figure 3.13 Rectangular element divided into four regions

Deak and Pian reduce a rectangle into four triangles as shown in Figure 3.13. They work with dimensionless coordinates

$$\bar{x} = 2x/a \qquad \bar{y} = 2y/b \tag{3.66}$$

and express the function as

$$u = \sum_{i=1}^{12} \alpha_i f_i(\bar{x}, \bar{y}) \tag{3.67}$$

$$\left.\begin{array}{l}
\left.\begin{array}{ll}
f_1 = 1 & f_2 = \bar{x}^2 \\
f_3 = \bar{y}^2 & f_4 = \bar{x} \\
f_5 = \bar{x}^3 & f_7 = \bar{y} \\
f_8 = \bar{y}^3 & f_{10} = \bar{x}\bar{y} \\
\multicolumn{2}{l}{f_{11} = 3\bar{x}^3\bar{y} + 3\bar{y}^3\bar{x} - \bar{x}^3\bar{y}^3 - 5\bar{x}\bar{y}}
\end{array}\right\} \text{for regions } 1, 2, 3, 4 \\
f_6 = \begin{cases}
\bar{x}^2 - 2\bar{x} + \bar{y}^2 & \text{for region 1} \\
2\bar{x}\bar{y} - 2\bar{x} & \text{for region 2} \\
-\bar{x}^2 - 2\bar{x} - \bar{y}^2 & \text{for region 3} \\
-2\bar{x}\bar{y} - 2\bar{x} & \text{for region 4}
\end{cases} \\
f_9 = \begin{cases}
\bar{y}^2 - 2\bar{y} + \bar{x}^2 & \text{for region 2} \\
2\bar{x}\bar{y} - 2\bar{y} & \text{for region 1} \\
-\bar{y}^2 - 2\bar{y} - \bar{x}^2 & \text{for region 4} \\
-2\bar{x}\bar{y} - 2\bar{y} & \text{for region 3}
\end{cases} \\
f_{12} = \begin{cases}
\frac{1}{4}(\bar{x}^3\bar{y}^3 - \bar{y}\bar{x}^5 - 3\bar{x}\bar{y}^3 + 3\bar{y}\bar{x}^3) & \text{for regions 1, 3} \\
\frac{1}{4}(\bar{x}\bar{y}^5 - \bar{x}^3\bar{y}^3 - 3\bar{x}\bar{y}^3 + 3\bar{y}\bar{x}^3) & \text{for regions 2, 4}
\end{cases}
\end{array}\right\} \tag{3.68}$$

The expansion leads to a cubic variation for u and a linear variation for the normal slope on the exterior boundaries and therefore satisfies complete inter-element displacement compatibility. One can easily show that the function and its first derivatives are continuous across the interior boundaries $\bar{x} = \pm\bar{y}$. Note that with this model we still work with only three displacement quantities per node. Results for this element do not converge as rapidly as the bicubic model but it is more accurate than the non-conforming rectangle.

Example 3.5

We will develop the finite element matrices corresponding to a rectangular element (Figure 3.11) for the bi-harmonic equation

$$\nabla^4 u = p \tag{a}$$

The variational statement corresponding to this equation is—see Example 1.10—

$$\iint \left\{ \left(\frac{\partial^2 u}{\partial x^2}\right)\left(\frac{\partial^2 \delta u}{\partial x^2}\right) + 2\left(\frac{\partial^2 u}{\partial x\,\partial y}\right)\left(\frac{\partial^2 \delta u}{\partial x\,\partial y}\right) + \left(\frac{\partial^2 u}{\partial y^2}\right)\left(\frac{\partial^2 \delta u}{\partial y^2}\right)\right\} \mathrm{d}x\,\mathrm{d}y$$

$$= \iint p\,\delta u\,\mathrm{d}x\,\mathrm{d}y \tag{b}$$

Let us assume we apply function (3.62),

$$u = \boldsymbol{\phi}^T \mathbf{u}^n \tag{c}$$

Hence,

$$\begin{Bmatrix} \dfrac{\partial^2 u}{\partial x^2} \\ \dfrac{\partial^2 u}{\partial y^2} \\ \sqrt{2}\dfrac{\partial^2 u}{\partial x\,\partial y} \end{Bmatrix} = \begin{bmatrix} \dfrac{1}{a^2}\left\{\dfrac{\partial^2\phi_1}{\partial\xi^2}, \dfrac{\partial^2\phi_2}{\partial\xi^2} \cdots \dfrac{\partial^2\phi_{16}}{\partial\xi^2}\right\} \\ \dfrac{1}{b^2}\left\{\dfrac{\partial^2\phi_1}{\partial\eta^2}, \dfrac{\partial^2\phi_2}{\partial\eta^2} \cdots \dfrac{\partial^2\phi_{16}}{\partial\eta^2}\right\} \\ \dfrac{\sqrt{2}}{ab}\left\{\dfrac{\partial^2\phi_1}{\partial\xi\,\partial\eta}, \dfrac{\partial^2\phi_2}{\partial\xi\,\partial\eta} \cdots \dfrac{\partial^2\phi_{16}}{\partial\xi\,\partial\eta}\right\} \end{bmatrix} \mathbf{u}^n \tag{d}$$

$$= \mathbf{B}(\xi, \eta)\mathbf{u}^n$$

We can now write (b) as

$$\delta \mathbf{u}^{n,T} \iint \mathbf{B}^T\mathbf{B}\,\mathrm{d}x\,\mathrm{d}y\ \ \mathbf{u}^n = \delta\mathbf{u}^{n,T} \iint \boldsymbol{\phi} p\,\mathrm{d}x\,\mathrm{d}y \tag{e}$$

As the $\delta\mathbf{u}^{n,T}$ are arbitrary, this gives

$$\mathbf{K}\mathbf{u}^n = \mathbf{P} \tag{f}$$

where

$$\mathbf{K} = \frac{1}{ab}\int_0^1\int_0^1 \mathbf{B}^T\mathbf{B}\,\mathrm{d}x\,\mathrm{d}y$$

$$\mathbf{P} = \frac{1}{ab}\int_0^1\int_0^1 \boldsymbol{\phi}^T p\,\mathrm{d}x\,\mathrm{d}y$$

The integrations can be carried out by hand or using the integration formula of the Appendix.

3.6 SECOND-ORDER CONTINUITY FUNCTIONS FOR TRIANGULAR ELEMENTS

The generation of interpolation functions for triangular elements is simplified considerably by working with the dimensionless oblique coordinates discussed previously and shown in Figure 3.14.

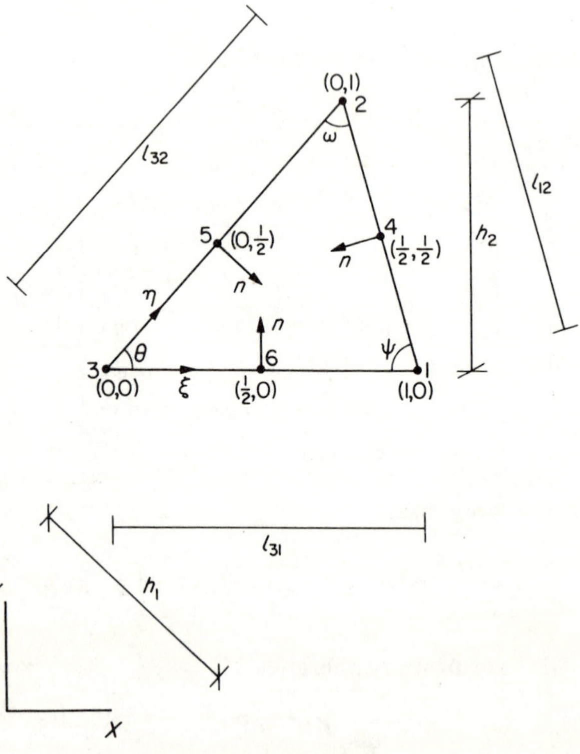

Figure 3.14 Triangular coordinates

Since the minimum choice of nodal variables is u, $u_{,x}$, $u_{,y}$, the expansion for u must be at least of third degree on the boundary. Let us consider the simplest triangle, i.e. nodes at the vertices and three displacement measures per node. A complete cubic polynomial has ten terms:

$$u = \alpha_1 + \alpha_2\xi + \alpha_3\eta + \alpha_4\xi^2 + \alpha_5\xi\eta + \alpha_6\eta^2 + \alpha_7\xi^3 + \alpha_8\xi^2\eta + \alpha_9\xi\eta^2 + \alpha_{10}\eta^3 \tag{3.69}$$

We work initially with derivatives with respect to ξ, η and then transform to Cartesian derivatives. Evaluating (3.69) at the nodes leads to

$$\alpha_1 = u_3 \qquad \alpha_2 = (\partial u/\partial \xi)_3 = u_{3,\xi} \qquad \alpha_3 = u_{3,\eta}$$

$$\begin{aligned} \alpha_4 &= 3u_1 - 3\alpha_1 - 2\alpha_2 - u_{1,\xi} \\ \alpha_6 &= 3u_2 - 3\alpha_1 - 2\alpha_3 - u_{2,\eta} \\ \alpha_7 &= u_{1,\xi} + \alpha_2 + 2\alpha_1 - 2u_1 \\ \alpha_{10} &= u_{2,\eta} + \alpha_3 + 2\alpha_1 - 2u_2 \end{aligned} \tag{3.70}$$

and

$$\begin{aligned} \alpha_5 + \alpha_8 &= u_{1,\eta} - \alpha_3 \\ \alpha_5 + \alpha_9 &= u_{2,\xi} - \alpha_2 \end{aligned} \tag{a}$$

Equation (a) shows that α_5, α_8 and α_9 are not uniquely defined.

It is clear that we must include an additional nodal measure. Since the normal slope varies quadratically, an obvious choice is the normal slope at the mid-point of one side. The general expressions for the normal slopes at the three mid-points are

$$\begin{aligned} h_3(\partial u/\partial n)_4 \equiv h_3 u_{4,n} &= |-C_3 u_{,\xi} - (1 - C_3)u_{,\eta}|_{\xi=\eta=\frac{1}{2}} \\ h_1 u_{5,n} &= |u_{,\xi} - C_1 u_{,\eta}|_{\xi=0,\eta=\frac{1}{2}} \\ h_2 u_{6,n} &= |-C_2 u_{,\xi} + u_{,\eta}|_{\xi=\frac{1}{2},\eta=0} \end{aligned} \tag{3.71}$$

$$C_1 = (l_{31}/l_{32})\cos\theta \qquad C_2 = (l_{32}/l_{31})\cos\theta \qquad C_3 = (l_{32}/l_{12})\cos\omega$$

Taking $u_{4,n}$ as the tenth parameter results in

$$\begin{aligned} \alpha_8 &= u_{1,\xi} - \alpha_3 - \alpha_5 \\ \alpha_9 &= u_{2,\xi} - \alpha_2 - \alpha_5 \\ \alpha_5 &= 4h_3u_{4,n} + 6C_3u_1 - C_3u_{1,\xi} + (1 + C_3)u_{1,\eta} + 6(1 - C_3)u_2 \\ &\quad + (1 - C_3)u_{2,\xi} - (1 - C_3)u_{2,\eta} - 6\alpha_1 - 2\alpha_2 - 2\alpha_3 \end{aligned} \tag{3.72}$$

The final interpolation functions are as follows:

$$\left.\begin{aligned}
u &= \phi_1 u_1 + \phi_2\, \partial u/\partial x|_1 + \phi_3\, \partial u/\partial y|_1 + \ldots + \phi_9\, \partial u/\partial y|_3 \\
&\quad + \phi_{10}\, \partial u/\partial n|_4 \\
\phi_1 &= f_1 + 6C_3 f_{10} \\
\phi_2 &= a_2(f_2 - C_3 f_{10}) - a_1\{f_3 + (1 + C_3)f_{10}\} \\
\phi_3 &= -b_2(f_2 - C_3 f_{10}) + b_1\{f_3 + (1 + C_3)f_{10}\} \\
\phi_4 &= f_4 + 6(1 - C_3)f_{10} \\
\phi_5 &= a_2\{f_5 + (1 - C_3)f_{10}\} - a_1\{f_6 - (1 - C_3)f_{10}\} \\
\phi_6 &= -b_2\{f_5 + (1 - C_3)f_{10}\} + b_1\{f_6 - (1 - C_3)f_{10}\} \\
\phi_7 &= f_7 - 6f_{10} \\
\phi_8 &= a_2(f_8 - 2f_{10}) - a_1(f_9 - 2f_{10}) \\
\phi_9 &= -b_2(f_8 - 2f_{10}) + b_1(f_9 - 2f_{10}) \\
\phi_{10} &= 4h_3 f_{10}
\end{aligned}\right\} \quad (3.73)$$

where

$$f_1 = \xi^2(3 - 2\xi) \qquad f_2 = \xi^2(\xi - 1) \qquad f_3 = \xi^2\eta$$
$$f_4 = \eta^2(3 - 2\eta) \qquad f_5 = \xi\eta^2 \qquad f_6 = \eta^2(\eta - 1)$$
$$f_7 = 1 - f_1 - f_4 \qquad f_8 = \xi\{(\xi - 1)^2 - \eta^2\}$$
$$f_9 = \eta\{(\eta - 1)^2 - \xi^2\} \qquad f_{10} = \xi\eta(1 - \xi - \eta)$$

This expansion satisfies continuity of u on all sides. Expanding (2.86) shows that the normal derivatives at points 5 and 6 involve displacement measures for side 1–2, and therefore continuity of u_n is not satisfied on sides 2–3 and 1–3.

$$\begin{aligned}
4h_1 u_{5,n} &= 6C_3 u_1 + (-a_1 + a_3 C_3)u_{1,x} + (b_1 - b_3 C_3)u_{1,y} \\
&\quad + 6(1 - C_1 - C_3)u_2 + \{-a_3(2 - C_3) - a_1(1 + C_2)\}u_{2,x} \\
&\quad + \{b_3(2 - C_3) + b_1(1 + C_1)\}u_{2,y} - 6(1 - C_1)u_3 \\
&\quad + \{-a_3 + a_1(1 - C_1)\}u_{3,x} + \{b_3 - b_1(1 - C_1)\}u_{3,y} \\
&\quad + 4h_3 u_{4,n}
\end{aligned} \qquad (3.74)$$

$$\begin{aligned}
4h_2 u_{6,n} &= 6(C_3 - C_2)u_1 + \{a_3(1 + C_3) + 2a_2 C_2\}u_{1,x} \\
&\quad + \{-b_3(1 + C_3) - 2b_2 C_2\}u_{1,y} + 6(1 - C_3)u_2 \\
&\quad - a_3(1 - C_2)u_{2,x} + b_3(1 - C_3)u_{2,y} - 6(1 - C_2)u_3 \\
&\quad + \{a_3 - a_2(1 - C_2)\}u_{3,x} + \{-b_3 + b_2(1 - C_2)\}u_{3,y} \\
&\quad + 4h_3 u_{4,n}
\end{aligned}$$

We generate a conforming triangle as follows. First we subdivide the triangle into three triangles as indicated in Figure 3.15. Point C

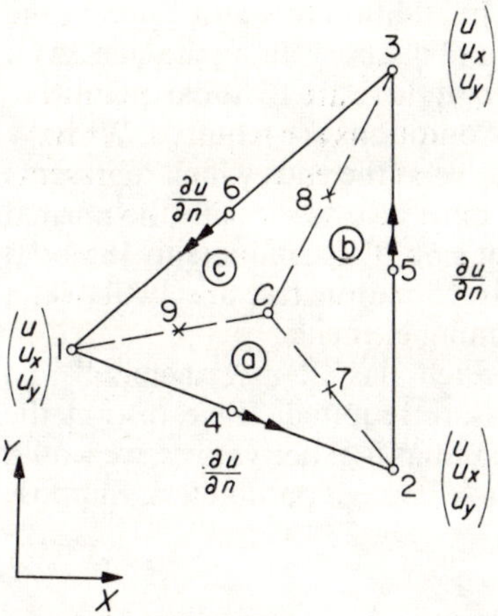

Figure 3.15 Triangular element subdivision—twelve degrees-of-freedom triangle. Note: A nine degrees-of-freedom triangle is obtained by requiring the normal slope to vary linearly on the exterior boundaries:

$$\left(\frac{\partial u}{\partial n}\right)_4 = \frac{1}{2}\left[\left(\frac{\partial u}{\partial n}\right)_1 + \left(\frac{\partial u}{\partial n}\right)_2\right]$$

is located at the centroid for convenience. We take (2.88) as the expansion for each sub-triangle, evaluate the sub-element matrices and then assemble the element matrix according to the following listing:

$$\begin{aligned}\mathbf{u}^n &= \{u_1, u_{1,x}, \ldots, u_{3,y}, u_{4,n}, u_{5,n}, u_{6,n}, u_C, u_{C,x}, u_{C,y}\} \\ &= \{\mathbf{u}_B \mathbf{u}_C\}\end{aligned} \tag{3.75}$$

Continuity of u and the normal slope is satisfied on the exterior boundaries. However, continuity of $u_{,n}$ is not satisfied on the interior boundaries for arbitrary $\mathbf{u}_C$. We enforce continuity of normal slope by equating the 'left' and 'right' normal derivatives at nodes 7, 8 and 9 using (3.74). This leads to a restriction on $\mathbf{u}_C$:

$$\mathbf{u}_C = \mathbf{E}\mathbf{u}_B \tag{3.76}$$

The final element has twelve degrees of freedom (see Figure 3.15).

Finally, a nine-parameter version is obtained by suppressing the mid-side rotations.

This element was developed in References 9 and 10. The results converge reasonably well for the twelve degrees of freedom case. The nine degrees of freedom version is 'too stiff' in comparison with the twelve degrees of freedom model, which is to be expected since we have introduced additional constraints.

We discuss next the second approach, i.e. establishing an interpolation polynomial which is valid for the entire triangle. We have already shown that a cubic polynomial does not have a sufficient number of parameters to satisfy normal slope compatibility. The next choice is a quartic which involves 15 parameters. Five conditions are required to specify u on a side. In addition to u and $\partial u/\partial S$ at the ends, we must work with an additional quantity along the side, say u at the mid-point. Note that at this point there are 12 nodal quantities. The normal slope is cubic and four conditions are required. We have already included the normal derivative at the ends which represents two conditions. It follows that we must also work with the normal derivative at two interior points per side. This amounts to 18 nodal displacement quantities. Since only 15 parameters are available, a quartic is not suitable for a conforming element.

We consider next a quintic which has 21 parameters.[10] To specify u on a side, six conditions are required. If we restrict the choice of nodal quantities to only u and its first derivatives, we would find that a quintic also does not have enough parameters. Suppose

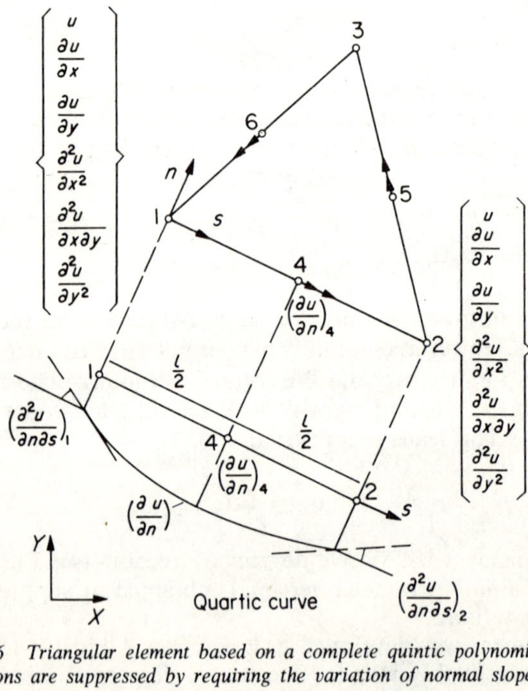

Figure 3.16 Triangular element based on a complete quintic polynomial. The mid-side rotations are suppressed by requiring the variation of normal slope to be cubic rather than quartic.

$$\left(\frac{\partial u}{\partial n}\right)_4 = \frac{1}{2}\left[\left(\frac{\partial u}{\partial n}\right)_1 + \left(\frac{\partial u}{\partial n}\right)_2\right] + \frac{l}{8}\left[\left(\frac{\partial^2 u}{\partial s\,\partial n}\right)_1 - \left(\frac{\partial^2 u}{\partial s\,\partial n}\right)_2\right]$$

we remove this restriction and include the three second derivatives as nodal quantities for the corner nodes. The six conditions on u are u, $\partial u/\partial S$, and $\partial^2 u/\partial S^2$ at the ends. At this point, we have six quantities per corner node and a total of 18 displacements. We can introduce three additional quantities. Now, the normal slope is quartic and requires five conditions. We already have $u_{,n}$ and $u_{,ns}$ at the ends. The additional conditions must involve $u_{,n}$ at an interior point on the side, say, at the mid-point. There are 21 unknowns in all and we see that a quintic polynomial is suitable. The nodal quantities are summarised in Figure 3.16. As expected the results for this element are excellent even for coarse grids (Figure 3.17).

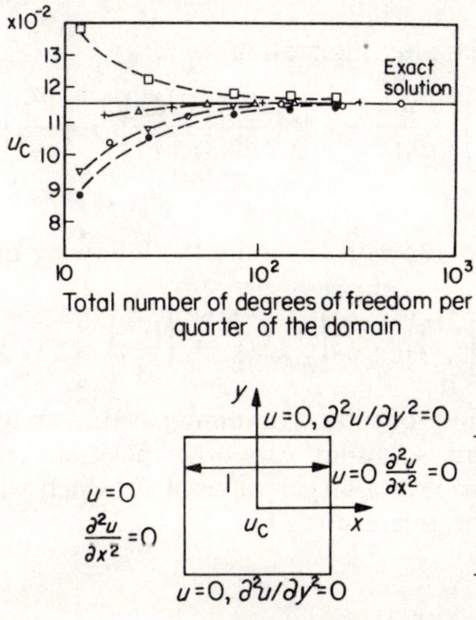

Figure 3.17 Centre point results for bi-harmonic equation. □ *Twelve degrees of freedom (non fully compatible) rectangular element (equation 3.61)*[3]; + *Sixteen degrees of freedom rectangle (equation 3.62)*[6]; ▽ *Twelve degrees of freedom rectangle (equation 3.67)*[8]; ○ *Twelve degrees of freedom triangle (equations 3.69 to 3.76)*[10]; ● *Nine degrees of freedom triangle (similar as previous but with degree of freedom at midsides suppressed)*[8]; △ *Eighteen degrees of freedom triangle*[11]

COMPARISON STUDIES

We will now compare results for some of the second-order continuity functions described above.

The equation under study is the bi-harmonic

$$\nabla^4 u = \Delta_c \qquad \text{(a)}$$

where Δ_c is a delta function equal to 1 at the centre of the 1×1 domain and zero at any other point (see Figure 3.17). The corresponding boundary conditions are

$$u = 0 \quad \text{and} \quad \frac{\partial^2 u}{\partial n^2} = 0 \quad \text{on all boundaries} \tag{b}$$

Note that the first boundary condition is *essential* and the *second* natural. Hence the latter is imposed through the variational statement

$$\iint \left(\frac{\partial^4 u}{\partial x^4} + 2\frac{\partial^4 u}{\partial x^2\, \partial y^2} + \frac{\partial^4 u}{\partial y^4} \right) \delta u \, \mathrm{d}x\, \mathrm{d}y = \Delta_c\, \delta u + \int_S \left(\frac{\partial^2 u}{\partial n^2} \frac{\partial \delta u}{\partial n} \right) \mathrm{d}S \tag{c}$$

Equation (c) gives, after integration,

$$\delta F = \iint \left\{ \left(\frac{\partial^2 u}{\partial x^2}\right)\left(\frac{\partial^2 \delta u}{\partial x^2}\right) + 2\left(\frac{\partial^2 u}{\partial x\, \partial y}\right)\left(\frac{\partial^2 \delta u}{\partial x\, \partial y}\right) + \left(\frac{\partial^2 u}{\partial y^2}\right)\left(\frac{\partial^2 \delta u}{\partial y^2}\right) \right\} \mathrm{d}x\, \mathrm{d}y$$

$$- \Delta_c\, \delta u = 0 \tag{d}$$

This relationship allows us to write the following quadratic functional:

$$F = \tfrac{1}{2} \iint \left[\left(\frac{\partial^2 u}{\partial x^2}\right)^2 + 2\left(\frac{\partial^2 u}{\partial x\, \partial y}\right)^2 + \left(\frac{\partial^2 u}{\partial y^2}\right)^2 \right] \mathrm{d}x\, \mathrm{d}y - 1 \times u_c \tag{e}$$

which (see Section 1.5) has a minimum value at the equilibrium position (i.e. exact solution). Any other position (i.e. approximate solution) will produce a larger value of F, which will be called F_a and which will be greater than F:

$$F_a > F \tag{f}$$

We can use this result to establish bounds on approximate solutions.

The governing equations for a linear discrete system have the following general form:

$$\mathscr{P} = \mathscr{K}\mathscr{U} \tag{g}$$

where $\mathscr{P}$ contains the actions (for instance 1 for the centre point in the case of equation (a)), $\mathscr{U}$ contains the discrete unknowns and $\mathscr{K}$ is the system influence matrix. Noting that these equations follow from (d), we can write

$$F = \tfrac{1}{2}\mathscr{U}^{\mathrm{T}}\mathscr{K}\mathscr{U} - \mathscr{U}^{\mathrm{T}}\mathscr{P} \tag{h}$$

The value of F at the equilibrium position is obtained by substituting for $\mathscr{K}\mathscr{U}$ using (g):

$$F\big|_{\text{eq. position}} = \tfrac{1}{2}\mathscr{U}^{\mathrm{T}}\mathscr{P} - \mathscr{U}^{\mathrm{T}}\mathscr{P} = -\tfrac{1}{2}\mathscr{U}^{\mathrm{T}}\mathscr{P} \tag{i}$$

For the above case there is only one source or action term $\mathscr{P}_c = 1$ acting on the system. Let u_c be the exact value and $u_c^{(a)}$ an approximation. The exact and approximate F functionals are

$$2F = -u_c P_c$$
$$2F_{\text{approx.}} = -u_c^{(a)} P_c \tag{j}$$

Noting (f) we obtain a bound on u_c,

$$u_c^{(a)} \leqslant u_c \tag{k}$$

This result shows that an approximate compatible solution will give a lower bound. This is evident in Figure 3.17 where the results for the non-compatible rectangular element converge to the known exact solution but do not give a bound.

Results for the two fully compatible rectangular elements converge in the right way to the correct solution, although the convergence is faster for the 16 degrees of freedom case. Unfortunately for this case the cross derivative $\partial^2 u/\partial x\, \partial y$ is required as nodal variable, which is generally inconvenient when transformations are involved.

The triangular element with nine degrees of freedom shows very poor convergence, which reflects the fact that there have been several constraints imposed in its development. The same formulation without removal of the mid-side node slopes (12 degrees of freedom triangle) shows better convergence.

Finally, the quintic 18 degrees of freedom triangular element shows excellent results even for the coarsest grid. The element requires continuity of second-order derivatives as well as slopes. The inclusion of second-order derivatives as nodal unknowns implies their continuity at the nodes, which is not valid when there are discontinuities in element thickness or material properties.

It should be mentioned that, although the finite element method gives bounds in the total energy when the elements satisfy completeness and admissibility, these bounds converge *monotonically* only if the discretisations comprise a minimising sequence. This means that by suitably specialising the nodal unknowns for the nth discretisation, we must be able to reproduce the displacement patterns corresponding to the $n-1$ previous discretisations. In order to satisfy this requirement, the nth discretisation must contain all the previous nodes and the element expansions must be invariant, i.e. their form must not depend on the orientation or dimensions of the element. As an illustration, if over a square region the element pattern is 2×2 square elements, the next minimising sequence will be given by 4×4, the other by 8×8, etc. A 3×3 subdivision will not correspond to the same minimising sequence since it contains a different set of nodes.

REFERENCES

1. BREBBIA, C. A. and CONNOR, J. J., *Fundamentals of Finite Element Techniques for Structural Engineers*, Butterworths (1973)
2. FELIPPA, C. A., *Refined Finite Element Analysis of Linearised Non-linear Two Dimensional Structures*, Ph.D. dissertation, California University, Berkeley (1966)
3. MELOSH, R. J., 'Basis for Derivation of Matrices for the Direct Stiffness Method', *AIA aerospace J.*, **1**, 1631 (1963)
4. EMERY, A. F. and CARSON, W. W., 'An Evaluation of the Use of the Finite Element Method in the Computation of Temperature', *Trans. ASME, J. Heat Transfer* (May 1971)
5. ERGATOUDIS, J., IRONS, B. M. and ZIENKIEWICZ, O. C., 'Curved Isoparametric, Quadrilateral Elements for Finite Element Analysis', *Int. J. Solids Structures*, **4**, 31–42 (1968)
6. BOGNER, F. K., FOX, R. L. and SCHMIDT, L. A., 'The Generation of Interelement Compatible Stiffness and Mass Matrices by the use of Interpolation Formulas', Conf. on Matrix Methods in Structural Mechanics, Wright-Patterson Air Force Base, Ohio, October (1965)
7. WILSON, R. R. and BREBBIA, C. A., 'Dynamic Behaviour of Steel Foundations for Turbo-Alternators', *J. Sound. Vibr.*, **18**, 405–416 (1971)
8. DEAK, A. L. and PIAN, T. H., 'Application of the Smooth Surface Interpolation to the Finite Element Analysis', *AIA aerospace J.*, 187–189, January (1967)
9. CLOUGH, R. W. and TOCHER, J. L., 'Finite Element Stiffness Matrix for Analysis of Plate Bending', Conf. on Matrix Methods in Structural Mechanics, Wright-Patterson Air Force Base, Ohio, October (1965)
10. FELIPPA, C. A. and CLOUGH, R. W., 'A Refined Quadrilateral Element for Analysis of Plate Bending', 2nd Conf. on Matrix Methods in Structural Mechanics, Wright-Patterson Air Force Base, Ohio, October (1968)
11. BELL, K. A., 'A Refined Triangular Plate Bending Element', *Int. J. Numerical Meth. Engng.*, **1**, No. 1 (1969)

EXERCISES

3-1 Develop a linear and quadratic one-dimensional element to solve the diffusion equation

$$h\frac{\mathrm{d}^2u}{\mathrm{d}x^2} = p$$

3-2 Consider the axisymmetric triangular element shown in Figure 3.18. If u does not vary with respect to θ, we can write Laplace's equation as

$$\frac{\partial^2 u}{\partial r^2} + \frac{1}{r}\frac{\partial u}{\partial r} + \frac{\partial^2 u}{\partial z^2} = 0 \qquad \text{(a)}$$

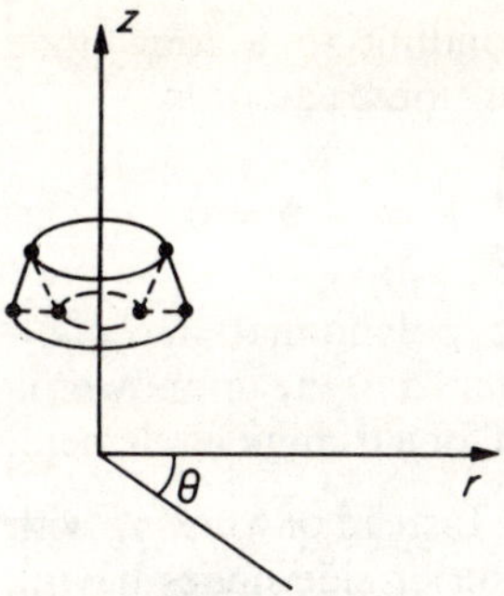

Figure 3.18 Axisymmetric element

Propose a linear *u* function for this element and develop the corresponding matrices.

3-3 Discuss how general asymmetric loading can be taken into account using the axisymmetric element developed in Exercise 3.2.

3-4 Using four three nodal matrices, such as the one developed in Example 3.1, form a square element [Figure 3.19(a)]. Once this is done condense the fifth node. Compare the result with that obtained by joining two triangles [Figure 3.19(b)].

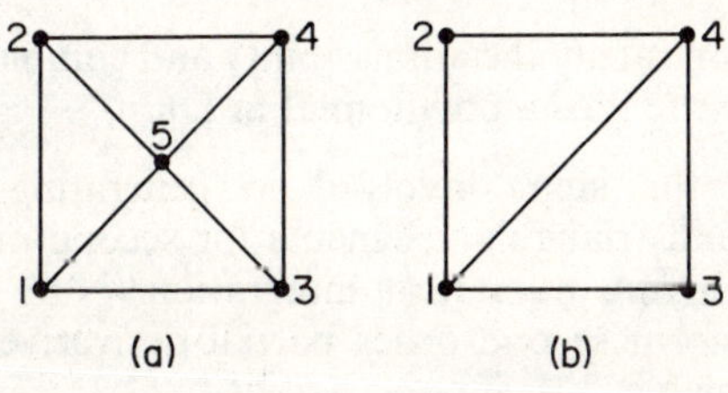

Figure 3.19 Assemblage of triangular elements

3-5 Explain how the isoparametric transformation can be applied for a six node triangle as shown in Figure 3.20.

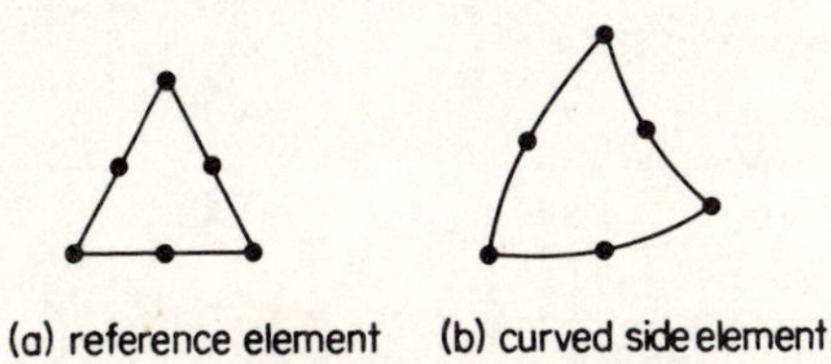

Figure 3.20 Curved sides triangular element

3-6 Find the element matrices corresponding to a four node tetrahedron for the case of the extended harmonic equation

$$\frac{\partial}{\partial x}\left(h_x \frac{\partial u}{\partial x}\right) + \frac{\partial}{\partial y}\left(h_y \frac{\partial u}{\partial y}\right) + \frac{\partial}{\partial z}\left(h_z \frac{\partial u}{\partial z}\right) + \lambda u + p = 0 \qquad \text{(a)}$$

3-7 Starting with a complete quadratic polynomial in ξ_1, ξ_2, generate the equations relating the parameters α to the unknowns u_i and verify that equation (3.28) is the solution for a triangular element.

3-8 Consider the bicubic expansion (3.62). Instead of working with $u_{,\xi\eta}$ at the corners, could one introduce interior side nodes having only the normal slope as a variable? Comment.

3-9 Using the polynomials shown in (3.62), we can write the following function over a rectangular element:

$$\begin{aligned} u = {} & \phi_1 u_1 + \phi_5 u_2 + \phi_9 u_3 + \phi_{13} u_4 + \phi_2 \left(\frac{\partial u}{\partial x}\right)_1 + \phi_6 \left(\frac{\partial u}{\partial x}\right)_2 \\ & + \phi_{10} \left(\frac{\partial u}{\partial x}\right)_3 + \phi_{14} \left(\frac{\partial u}{\partial x}\right)_4 + \phi_3 \left(\frac{\partial u}{\partial y}\right)_1 + \phi_7 \left(\frac{\partial u}{\partial y}\right)_2 \\ & + \phi_{11} \left(\frac{\partial u}{\partial y}\right)_3 + \phi_{15} \left(\frac{\partial u}{\partial y}\right)_4 \end{aligned} \qquad \text{(a)}$$

Will the function satisfy the admissibility and completeness requirements? *Hint*: write it as a polynomial in ξ, η.

3-10 Discuss the steps involved in generating isoparametric quadrilateral and triangular elements for second-order continuity functions employing numerical integration. Note that now one needs to transform second-order partial derivatives and that the Jacobian is variable.

4 Basic Principles and Governing Equations of Fluid Mechanics

4.1 EULERIAN AND LAGRANGIAN FORMULATIONS: MATERIAL DERIVATIVE

One can specify the position of a material point with respect to a 'fixed' orthogonal reference frame as shown in Figure 4.1. The initial coordinates at $t = 0$ are a_i and the coordinates at time t are x_i.* Quantities of interest such as pressure, velocity and temperature are functions of position and time. Since there are two choices as to position variables, the governing equations can have two different forms.

In the Lagrange approach, the independent variables are time t and the initial coordinates a_i. The spatial coordinates and dependent variables are expressed as

$$\begin{aligned} f &= f(a_1, a_2, a_3, t) \\ x_i &= a_i + u_i(a_1, a_2, a_3, t) \end{aligned} \tag{4.1}$$

where u_i is the displacement. This viewpoint is quite reasonable for solids since the deformations (and consequently u) are small.

A fluid generally experiences significant deformation and one is not interested in the initial position. The actual coordinates x_i and

* Note that we are now changing from x, y, z notation to $x_1x_2x_3$ indicial notation for simplicity.

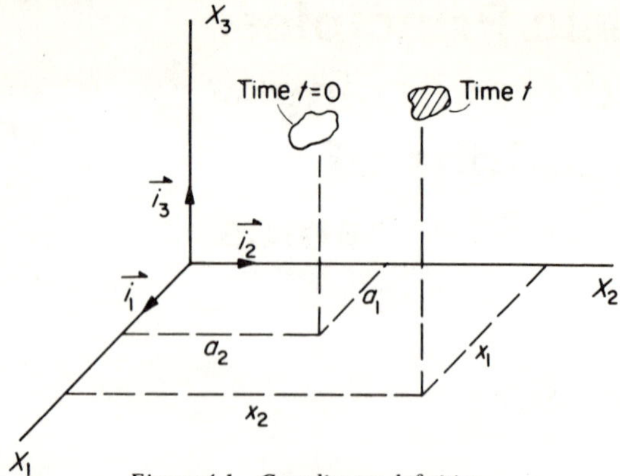

Figure 4.1 Coordinates definition

time are more convenient as independent variables. In the Eulerian approach, one takes

$$a_i = x_i - u_i(x_1, x_2, x_3, t)$$
$$f = f(x_1, x_2, x_3, t) \tag{4.2}$$

We will follow the Eulerian approach throughout this and other chapters unless otherwise stated.

Next we consider the determination of the rate of change of a dependent variable with time. Let f denote a variable associated with the material point (x_i, t) as shown in Figure 4.2. During the time

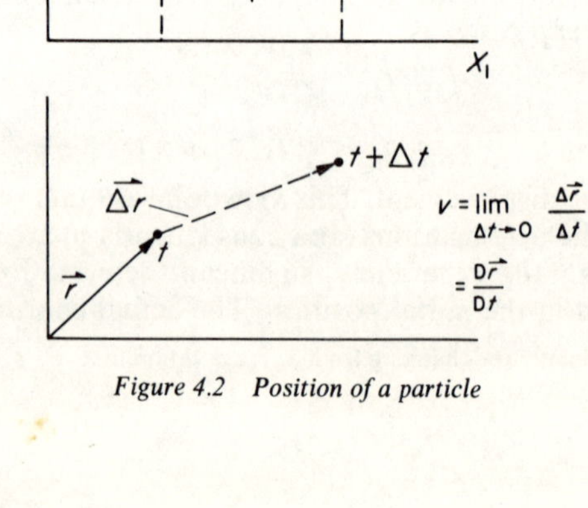

Figure 4.2 Position of a particle

interval Δt, the point moves to $x_i + \Delta x_i$ and the variable changes to $f + \Delta f$. Assuming f is a continuous function, we expand it in a Taylor series about (x_i, t), obtaining

$$\Delta f = \delta f + \tfrac{1}{2}\delta^2 f + \ldots$$

where*

$$\delta f = \frac{\partial f}{\partial x_i}\Delta x_i + \frac{\partial f}{\partial t}\Delta t \tag{4.3}$$

and

$$\delta^2 f = \delta(\delta f)$$

The limit, as $\Delta t \to 0$, of the ratio $\Delta f/\Delta t$ is written as $\mathrm{D}f/\mathrm{D}t$.

$$\frac{\mathrm{D}f}{\mathrm{D}t} = \lim_{\Delta t \to 0} \frac{\Delta f}{\Delta t} \tag{4.4}$$

Since we are actually following the material point (particle), the term 'material' derivative is used for $\mathrm{D}f/\mathrm{D}t$. Other names for (4.4) are 'particle' derivative and 'Stokes' derivative.

To expand (4.4), we note that the velocity vector for a particle is defined as

$$\vec{v} = \text{velocity vector} = \lim_{\Delta t \to 0} \frac{\Delta \vec{r}}{\Delta t} \equiv \frac{\mathrm{D}\vec{r}}{\mathrm{D}t} \tag{4.5}$$

The components of the velocity and position vectors are

$$\vec{v} = v_j \vec{i}_j \qquad \vec{r} = x_j \vec{i}_j \tag{a}$$

Then

$$v_j = \lim_{\Delta t \to 0} \frac{\Delta x_j}{\Delta t} = \frac{\mathrm{D}x_j}{\mathrm{D}t} \tag{4.6}$$

and taking (4.3) into account, equation (4.4) becomes

$$\frac{\mathrm{D}f}{\mathrm{D}t} = \frac{\partial f}{\partial t} + v_i \frac{\partial f}{\partial x_i} \tag{4.7}$$

The first term is the local time derivative, i.e. with the point fixed in position. The remaining terms represent the contribution of movement, and are called the 'convective' terms.

* We are using the indicial convention for summation over repeated subscripts.

In the Lagrange approach, $f = f(a_i, t)$ and δf reduces to

$$\delta f = \frac{\partial f}{\partial t} \Delta t \tag{4.8}$$

since the initial coordinates for the point are constant. The material and local derivatives now coincide:

$$\begin{aligned} \frac{\mathrm{D}f}{\mathrm{D}t} &= \frac{\partial f}{\partial t} \\ v_i &= \frac{\partial u_i}{\partial t} \end{aligned} \tag{4.9}$$

Although (4.9) is simpler than (4.7), the Eulerian approach is generally more convenient for fluids due to the large distortions which may occur.

Later we shall need to evaluate the material derivative of integral expressions. The approach is essentially the same as for a function. As an illustration, we consider a scalar function f, where f is the amount of the variable per unit mass. The total amount of f in the volume is

$$\int_{\text{mass}} f \,\mathrm{d}m = \int_{\text{volume}} f\rho \,\mathrm{d}V \tag{4.10}$$

where ρ is the mass density, m the mass.

The variation of integral (4.10) with respect to time can be written

$$\frac{\mathrm{D}}{\mathrm{D}t} \iiint f\rho \,\mathrm{d}V = \lim_{\Delta t \to 0} \left\{ \frac{\Delta(\int \rho f \,\mathrm{d}V)}{\Delta t} \right\} \tag{4.11}$$

The increase given by (4.11) should be equal to the rate of increase of the total amount of f inside a control surface plus the net rate of outward flux of f carried by transport through the control surface. For a velocity vector $\vec{v}$ acting on a part $\mathrm{d}S$ of the control surface (Figure 4.3), the amount of flux through $\mathrm{d}S$ per unit time is

$$(f\rho)v_n \,\mathrm{d}S = (f\rho)v_i \alpha_{ni} \,\mathrm{d}S \tag{4.12}$$

where the α_{ni} terms are the direction cosines of the velocity components with respect to the normal.

We can now write

$$\frac{\mathrm{D}}{\mathrm{D}t} \iiint (f\rho) \,\mathrm{d}V = \iiint \frac{\partial (f\rho)}{\partial t} \,\mathrm{d}V + \iint_S (f\rho) v_i \alpha_{ni} \,\mathrm{d}S \tag{4.13}$$

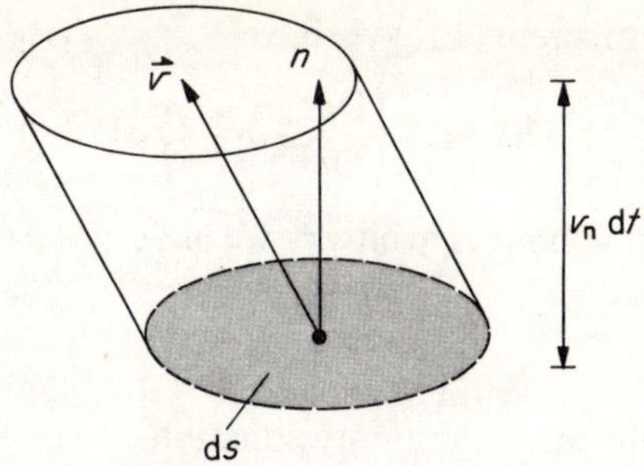

Figure 4.3 Flux through ds

Finally, it is convenient to transform the surface integral in (4.13) to a volume integral using Gauss's formula, which in general can be written as

$$\iiint_V g\frac{\partial h}{\partial x_i}\,\mathrm{d}V = \iint_S \alpha_{ni} gh\,\mathrm{d}S - \iiint_V h\frac{\partial g}{\partial x_i}\,\mathrm{d}V \qquad (4.14)$$

This gives for equation (4.13) the following formula:

$$\frac{\mathrm{D}}{\mathrm{D}t}\iiint_V (f\rho)\,\mathrm{d}V = \iiint_V \left\{\frac{\partial}{\partial t}(f\rho) + \frac{\partial}{\partial x_i}(\rho f v_i)\right\}\mathrm{d}V \qquad (4.15)$$

Formula (4.15) is called Reynolds transport theorem and gives the material derivative of a volume integral for the case where the integrand and the volume over which the integral is taken vary with time.

CONSERVATION OF MASS

Let us apply (4.15) to the case of mass conservation. The total amount of mass in an element $\mathrm{d}V$ is $\rho\,\mathrm{d}V$. Hence the total mass is

$$M = \iiint_V \rho\,\mathrm{d}V \qquad (4.16)$$

where ρ is a function of space and time.

If no mass is created or destroyed inside V, the net rate of mass inflow throughout the control surface plus the rate of change of mass within the volume is equal to zero:

$$\frac{\mathrm{D}M}{\mathrm{D}t} = \frac{\mathrm{D}}{\mathrm{D}t}\iiint_V \rho\,\mathrm{d}V = 0 \qquad (4.17)$$

From Reynolds theorem (4.15) we have

$$\frac{D}{Dt}\int \rho \, dV = \int \left(\frac{D\rho}{Dt} + \rho \frac{\partial v_i}{\partial x_i} \right) dV = 0 \tag{4.18}$$

Since the control volume is arbitrary we have

$$\frac{D\rho}{Dt} + \rho \frac{\partial v_i}{\partial x_i} = 0 \tag{4.19}$$

which is called the mass continuity equation.

Now (4.15) reduces to

$$\frac{D}{Dt}\int (f\rho) \, dV = \int \frac{Df}{Dt} \rho \, dV = \int \left(\frac{\partial f}{\partial t} + v_i \frac{\partial f}{\partial x_i} \right) \rho \, dV \tag{4.20}$$

4.2 DEFORMATION RATE MEASURES

The internal resistance of a fluid to motion depends on the time rate of change of the measures which define the deformation. These measures are called strains and consist of two types:

1 Extensional strain—the relative change in length of a differential line element
2 Shearing strain—the decrease in the angle between two orthogonal line elements

Once the extensional and shearing strains for a triply orthogonal set of line elements at a point are specified, the strains for an arbitrary set of directions through the point can be determined by applying strain transformation laws.

Our objective here is to establish expressions for the deformation rates. We consider initially the two-dimensional case and then generalise to three dimensions.

Figure 4.4 shows the initial (time t) and deformed (time $t + \Delta t$) position of two differential line elements. The movement of a line element consists of rigid body translation and rotation, and relative incremental extension $\Delta\varepsilon$. We define the extensional strain rate as the material derivative of the length dL of a differential line element,

$$\begin{gathered} \frac{D}{Dt}(dL) = \dot{\varepsilon} \, dL \\ \Downarrow \\ \dot{\varepsilon} = \lim_{\Delta t \to 0} \left\{ \frac{\Delta(dL)}{(dL)\,\Delta t} \right\} = \lim_{\Delta t \to 0} \frac{\Delta\varepsilon}{\Delta t} \end{gathered} \tag{4.21}$$

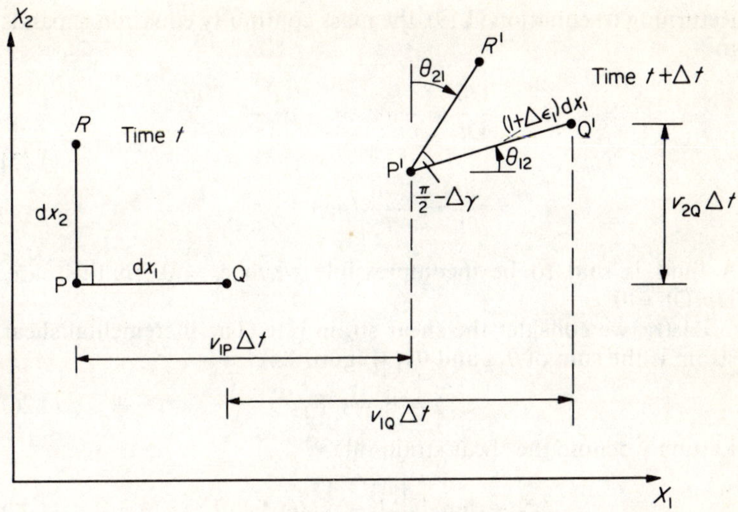

Figure 4.4 Deformation measures

Applying (4.21) to the line element parallel to X_1, we obtain, up to the first order,

$$\Delta v_1 = \frac{\partial v_1}{\partial x_1}\Delta x_1 + \frac{\partial v_1}{\partial x_2}\Delta x_2$$

which gives

$$\dot{\varepsilon}_1 = \lim_{\Delta t \to 0}\left(\frac{\Delta \varepsilon_1}{\Delta t}\right) = \frac{\partial v_1}{\partial x_1} \tag{4.22}$$

and similarly

$$\dot{\varepsilon}_2 = \frac{\partial v_2}{\partial x_2}$$

The volumetric strain rate is defined in a similar way:

$$\frac{\mathrm{D}}{\mathrm{D}t}(\mathrm{d}V) = \dot{\varepsilon}_v\,\mathrm{d}V \tag{4.23}$$

where $\mathrm{d}V$ is the differential volume. One obtains for the three-dimensional case

$$\dot{\varepsilon}_v = \lim_{\Delta t \to 0}\left(\frac{\Delta \varepsilon_v}{\Delta t}\right) = \frac{\partial v_1}{\partial x_1} + \frac{\partial v_2}{\partial x_2} + \frac{\partial v_3}{\partial x_3} = \frac{\partial v_i}{\partial x_i} \tag{4.24}$$

$$= \text{divergence}\ (\vec{v}) = \vec{\nabla} \cdot \vec{v}$$

Returning to equation (4.19), the mass continuity equation expands to

$$\begin{aligned} \frac{\mathrm{D}\rho}{\mathrm{D}t} &= -\rho\dot{\varepsilon}_v = -\rho\frac{\partial v_i}{\partial x_i} \\ &\Downarrow \\ \frac{\partial \rho}{\partial t} &= -\frac{\partial}{\partial x_i}(\rho v_i) \end{aligned} \tag{4.25}$$

A fluid is said to be incompressible when $\dot{\varepsilon}_v = 0$. In this case, $\mathrm{D}\rho/\mathrm{D}t = 0$.

Lastly, we consider the shear strain rate. The incremental shear strain is the sum of θ_{12} and θ_{21} (Figure 4.4):

$$\gamma = \theta_{12} + \theta_{21} \tag{4.26}$$

Letting $\dot{\gamma}$ denote the shear strain rate,

$$\dot{\gamma}_{12} = \lim_{\Delta t \to 0}\left(\frac{\Delta\gamma}{\Delta t}\right) = \frac{\mathrm{D}}{\mathrm{D}t}(\theta_{12} + \theta_{21}) \tag{4.27}$$

and evaluating the θ derivatives results in

$$\dot{\gamma}_{12} = \frac{\partial v_1}{\partial x_2} + \frac{\partial v_2}{\partial x_1} \tag{4.28}$$

At this point, it is convenient to introduce the strain rate terms e_{ij}:

$$e_{ij} = \frac{1}{2}\left(\frac{\partial v_i}{\partial x_j} + \frac{\partial v_j}{\partial x_i}\right) \tag{4.29}$$

The strain rate measures are related to e_{ij} by

$$\begin{aligned} \dot{\varepsilon}_i &= e_{ii} \qquad \text{(no sum)} \\ \dot{\gamma}_{ij} &= e_{ij} + e_{ji} = 2e_{ij} \\ \dot{\varepsilon}_v &= e_{11} + e_{22} + e_{33} \equiv e_v \end{aligned} \tag{4.30}$$

When $\dot{\gamma} = 0$, the line elements remain orthogonal and

$$\dot{\theta}_{12} = \omega_3 \qquad \dot{\theta}_{21} = -\omega_3 \tag{4.31}$$

where ω is the angular velocity about the X_3 axis. This suggests that the difference between $\dot{\theta}_{12}$ and $\dot{\theta}_{21}$ can be taken as a measure of the 'average' angular velocity about X_3. We introduce a shear symmetric tensor ω_{ij}

$$\omega_{ij} = \frac{1}{2}\left(\frac{\partial v_j}{\partial x_i} - \frac{\partial v_i}{\partial x_j}\right) \qquad i \neq j;\, i, j = 1, 2, 3 \tag{4.32}$$

Cyclic permutation of the subscripts gives the average angular velocities about the three axes:

$$\omega_{12} \to \omega_3 \qquad \omega_{23} \to \omega_1 \qquad \omega_{31} \to \omega_2 \tag{4.33}$$

It is not difficult to show that ω_{ij} is invariant for certain transformations of axes. For example,

$$\omega'_{12} = \omega_{12} \quad \text{for } (X_1, X_2, X_3) \text{ and } (X'_1, X'_2, X_3) \tag{4.34}$$

This is why we can interpret ω_{12} as the 'average' rotation about X_3. The set ω_{ij} is called the 'vorticity' tensor, and the contracted components ω_i define the vorticity vector $\vec{\omega}$.

$$\begin{aligned} \vec{\omega} &= \omega_1\vec{i}_1 + \omega_2\vec{i}_2 + \omega_3\vec{i}_3 \\ &= \tfrac{1}{2}\,\text{curl}\,(\vec{v}) = \tfrac{1}{2}\vec{\nabla} \times \vec{v} \end{aligned} \tag{4.35}$$

A flow is *irrotational* when the average angular velocities are zero:

$$\begin{aligned} &\omega_{ij} = 0 \\ \therefore\quad &\frac{\partial v_i}{\partial x_j} = \frac{\partial v_j}{\partial x_i} \qquad \begin{matrix} i = 1, 2, 3 \\ j = 2, 3, 1 \end{matrix} \end{aligned} \tag{4.36}$$

Equation (4.36) requires the velocity vector to be the gradient of a continuous function Φ, such that

$$\vec{\omega} = \vec{0} \Rightarrow \vec{v} = \vec{\nabla}\Phi = \frac{\partial \Phi}{\partial x_j}\vec{i}_j \tag{4.37}$$

One can interpret Φ as a velocity potential. Assuming irrotational flow simplifies the governing equations considerably. We will comment further on this point later on.

4.3 EQUILIBRIUM EQUATIONS

We consider a specific volume at time t as shown in Figure 4.5. The external actions are represented by a distributed surface loading ($\vec{p}$) per unit surface area and a distributed body force ($\vec{b}$) per unit mass.

For equilibrium, the time rate of change of total momentum equals the sum of all external forces. The change rate of momentum is

$$\frac{\mathrm{D}}{\mathrm{D}t}\int \rho\vec{v}\,\mathrm{d}V \tag{4.38}$$

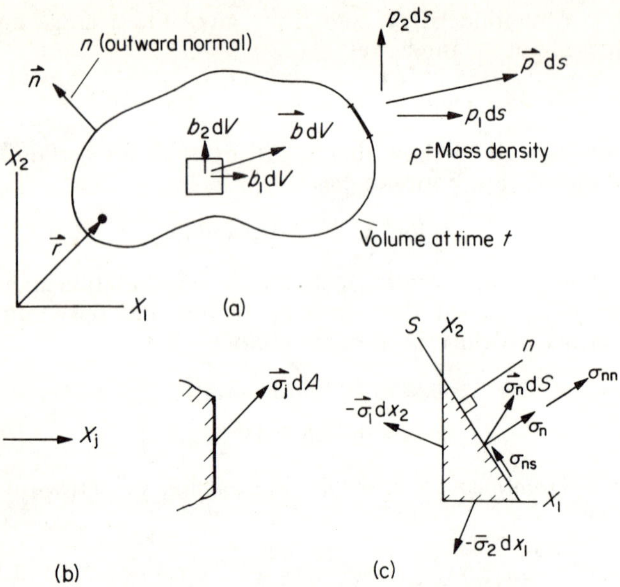

Figure 4.5 Two-dimensional body definitions

which, taking into account conservation of mass [equation (4.30)], can be written

$$\int \rho \frac{D\vec{v}}{Dt} dV \tag{4.39}$$

Hence Newton's law can be expressed as

$$\iiint \rho\vec{b}\, dV + \iint \vec{p}\, dS = \iiint \rho \frac{D\vec{v}}{Dt} dV \tag{4.40}$$

Similarly, the time rate of change of the total moment of momentum is equal to the vector sum of the moments of the external forces; i.e.

$$\iiint \rho(\vec{r} \times \vec{b})\, dV + \iint (\vec{r} \times \vec{p})\, dS = \iiint \rho\left(\vec{r} \times \frac{D\vec{v}}{Dt}\right) dV \tag{4.41}$$

where the integrals apply to the position at time t, and $D\vec{v}/Dt$ is the acceleration vector. The differential equilibrium equations are obtained by expanding the surface integral involving $\vec{p}$ in terms of stress vectors and then applying Gauss's integration by parts formula.

We define $\vec{\sigma}_j$ as the stress vector (force per unit area) acting on the $+j$ face, i.e. the face whose outward normal points in the $+X_j$ direction, as shown in Figure 4.5(b). The Cartesian component representation is

$$\vec{\sigma}_j = \sigma_{jk}\vec{i}_k \tag{4.42}$$

We establish stress transformation laws by enforcing force equilibrium for the differential element of Figure 4.5(c). This yields

$$\begin{aligned} \vec{\sigma}_n &= \alpha_{nj}\vec{\sigma}_j \\ \alpha_{nj} &= \cos(n, X_j) \end{aligned} \tag{4.43}$$

and

$$\begin{aligned} \sigma_{nn} &= \vec{i}_n \cdot \vec{\sigma}_n = \alpha_{nj}\alpha_{nk}\sigma_{jk} \\ \sigma_{ns} &= \vec{i}_s \cdot \vec{\sigma}_n = \alpha_{nj}\alpha_{sk}\sigma_{jk} \end{aligned} \tag{4.44}$$

Now, $\vec{\sigma}_n = \vec{p}$ on the boundary surface. Then, using (4.43) and Gauss's formula, (4.40) expands to the following scalar equations:

$$\frac{\partial \sigma_{jk}}{\partial x_j} + \rho b_k = \rho \frac{\mathrm{D}v_k}{\mathrm{D}t} \quad \text{in } V \tag{4.45}$$

with

$$p_k = \alpha_{nj}\sigma_{jk} = \bar{p}_k \quad \text{on } S$$

Similarly, we could expand (4.41); after applying (4.45) we would find that

$$\sigma_{jk} = \sigma_{kj} \quad \text{in } V \tag{4.46}$$

The force equilibrium equation can be transformed further by substituting for the material derivative and utilising the mass continuity equation (4.27). One obtains

$$\iint_S \vec{p}\,\mathrm{d}S + \iiint_V \rho\vec{b}\,\mathrm{d}V = \iint_S (\rho v_n)\vec{v}\,\mathrm{d}S + \iiint_V \frac{\partial}{\partial t}(\rho\vec{v})\,\mathrm{d}V \tag{4.47}$$

and the corresponding scalar equations

$$\frac{\partial \sigma_{jk}}{\partial x_j} + \rho b_k = \frac{\partial}{\partial x_j}(\rho v_j v_k) + \frac{\partial}{\partial t}(\rho v_k) \tag{4.48}$$

Equation (4.48) is generally called the momentum equation. When applying (4.47), one considers the volume as shown in Figure 4.6. The first term on the right-hand side is the outward momentum flux and the second is the local rate of change of momentum.

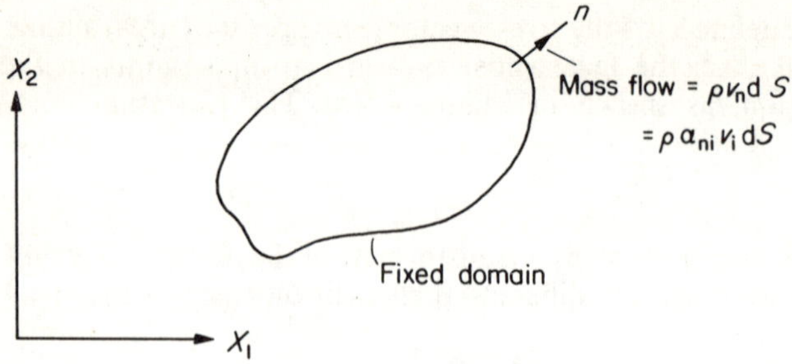

Figure 4.6 Volume definition

The stress components consist of pressure and friction terms. We write

$$\sigma_{ij} = -p\delta_{ij} + \tau_{ij} \tag{4.49}$$

where τ_{ij} denotes the viscous components. Later, we shall include additional dissipative terms due to turbulence in τ_{ij}. The momentum equations are now

$$-\frac{\partial p}{\partial x_k} + \frac{\partial \tau_{jk}}{\partial x_j} + \rho b_k = \frac{\partial}{\partial x_j}(\rho v_j v_k) + \frac{\partial}{\partial t}(\rho v_k) \tag{4.50}$$

with

$$p_n = -p + \alpha_{nj}\alpha_{nk}\tau_{jk}$$

$$p_s = \alpha_{nj}\alpha_{sk}\tau_{jk}$$

If the τ_{ij} terms are neglected, the fluid is said to be frictionless or *inviscid*. For this case,

$$\sigma_{ij} \Rightarrow -p\delta_{ij} \tag{4.51}$$

and the stress transformation relations reduce to

$$\begin{aligned} \sigma_{nn} &= -p \\ \sigma_{ns} &= 0 \end{aligned} \tag{4.52}$$

According to (4.52), an inviscid fluid cannot resist a tangential boundary surface force, i.e. one can apply only a normal boundary surface force.

4.4 THE ENERGY EQUATION

The first law of thermodynamics expresses, at an equilibrium position, the energy balance requirement for a system. This law involves the introduction of the concept of *internal* energy of the body, which we call $\mathscr{U}$. The internal energy is related to other variables through the *equations of state*. In a solid, for instance, the internal energy may be a function of the strains (and sometimes also of the temperature), in a fluid it may be a function of the pressure and density, etc. The absolute value of the total internal energy of the system is impossible to know in practice as it includes such variables as the kinetic energy of the atoms for instance. In what follows we will consider only changes in the internal energy.

We will assume that the system receives only work produced by the external forces and heat flux. Other types of energy will not be considered here.

$\mathscr{W}$, the work done by the external forces, plus the heat energy $\mathscr{Q}$ has to be equal to the increase of the total energy of the system, which is the sum of the internal energy $\mathscr{U}$, and the kinetic energy increment $\mathscr{T}$.

$$\mathscr{W} + \mathscr{Q} = \mathscr{U} + \mathscr{T} \tag{4.53}$$

We can now define the power input as the external work per unit time, and the heat input as heat per unit time:

$$\mathrm{d}\mathscr{W} + \mathrm{d}\mathscr{Q} = \left(\frac{\mathrm{D}\mathscr{W}}{\mathrm{D}t} + \frac{\mathrm{D}\mathscr{Q}}{\mathrm{D}t}\right)\mathrm{d}t \tag{4.54}$$

One can write

$$\mathscr{U} + \mathscr{T} = \int_{t_1}^{t_2}\left(\frac{\mathrm{D}\mathscr{W}}{\mathrm{D}t} + \frac{\mathrm{D}\mathscr{Q}}{\mathrm{D}t}\right)\mathrm{d}t \tag{4.55}$$

or simply

$$\frac{\mathrm{D}\mathscr{W}}{\mathrm{D}t} + \frac{\mathrm{D}\mathscr{Q}}{\mathrm{D}t} = \frac{\mathrm{D}}{\mathrm{D}t}(\mathscr{U} + \mathscr{T}) \tag{4.56}$$

The expanded forms (with our notation) are

$$\begin{aligned}\frac{\mathrm{D}\mathscr{W}}{\mathrm{D}t} &= \iint_S p_i v_i\,\mathrm{d}S + \iiint_V \rho b_i v_i\,\mathrm{d}V\\ \frac{\mathrm{D}\mathscr{Q}}{\mathrm{D}t} &= \iint_S -q_n\,\mathrm{d}S + \iiint f\rho\,\mathrm{d}V\end{aligned} \tag{4.57}$$

where q_n is the outward heat flux, f is the distributed heat source per unit mass.

We can write the right-hand side terms in (4.56) as

$$\frac{\mathrm{D}}{\mathrm{D}t}\int(\rho u + t)\,\mathrm{d}V \tag{4.58}$$

where u is the internal energy per unit volume and per unit mass, and $t = \frac{1}{2}\rho v_i^2$ is the kinetic energy per unit volume.

Using equation (4.20) to expand the material derivatives in (4.58), we obtain

$$\begin{aligned} \frac{D\mathcal{T}}{Dt} &= \frac{D}{Dt}\iiint_V \tfrac{1}{2}\rho(v_i)^2\,dV = \iiint \rho v_i \frac{Dv_i}{Dt}\,dV \\ \frac{D(\mathcal{U})}{Dt} &= \iiint \rho\frac{Du}{Dt}\,dV \end{aligned} \tag{4.59}$$

Requiring the forces to satisfy the equilibrium equations (4.50), and expressing the normal heat flux in terms of its orthogonal components,

$$\begin{aligned} q_n &= \alpha_{nj}q_j \\ q_j &= \text{heat flux across } X_j \text{ face} \end{aligned} \tag{4.60}$$

we obtain from equation (4.56) the following expressions for the internal energy, after applying Gauss's theorem:

$$\rho\frac{Du}{Dt} = -\frac{\partial q_i}{\partial x_i} + f\rho - pe_v + \tau_{ij}e_{ij} \tag{4.61}$$

The first two terms on the right-hand side are due to external heat input;* the last term represents the rate of dissipation of mechanical energy resulting from friction. Mechanical energy dissipation is irreversible and positive definite, i.e. it is positive for arbitrary e_{ij}. We will utilise this property later on to establish stress–strain relations.

The mechanical energy dissipation and volumetric expansion are generally neglected for a fluid, and the internal energy can be written

$$du = C_p\,dT \tag{4.62}$$

where C_p is the coefficient of specific heat at constant pressure. Equation (4.61) can sometimes be approximated by

$$\rho_0 C_p\frac{DT}{Dt} = -\frac{\partial q_i}{\partial x_i} + f\rho_0 \tag{4.63}$$

where ρ_0 is constant.

Equation (4.63) implies the assumption that no thermo-mechanical coupling exists.

* It should be understood that the heat terms are multiplied by a constant which is the mechanical equivalent of heat.

Example 4.1

One can think of a 'continuum' system as being composed of many particles interconnected by springs as shown in Figure 4.7(a). Forces like F_1 are the external forces and the forces in the springs are internal.

Consider for instance just one particle and one spring [Figure 4.7(b)] under a u displacement of the particle which is subjected to an

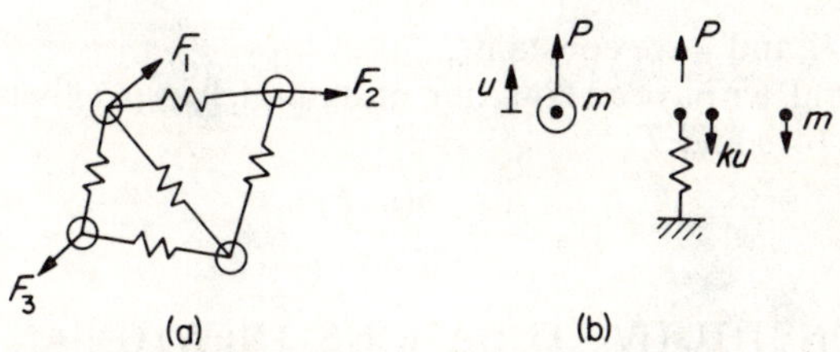

Figure 4.7 Particles and spring systems. (a) *Several degrees of freedom system;* (b) *Single degree of freedom system*

external force P; ku is the internal force (where k is the spring constant), v is the velocity. We have the following total work for the particle:

$$\mathscr{W}_T = \int_{t_1}^{t_2} Fv\,\mathrm{d}t = \int_{t_1}^{t_2} (P - ku)v\,\mathrm{d}t$$
$$= \int_{t_1}^{t_2} (P - ku)\,\mathrm{d}u = Pu - \frac{ku^2}{2} = \mathscr{W} - \mathscr{U} \tag{a}$$

Next we compute the increment of kinetic energy:

$$\mathscr{T} = \int_{t_1}^{t_2} m\frac{\mathrm{d}v}{\mathrm{d}t}v\,\mathrm{d}t = \frac{mv^2}{2} \tag{b}$$

Equations (a) and (b) give

$$Pu - \frac{ku^2}{2} = \frac{mv^2}{2} \quad \text{or} \quad \mathscr{W} - \mathscr{U} = \mathscr{T} \tag{c}$$

EQUATIONS OF STATE

The pressure can be written, for instance, in terms of density and temperature:

$$p = p(\text{density, temperature}) = p(\rho, T) \tag{4.64}$$

In hydraulic applications we generally assume that the mechanical and thermal behaviour are uncoupled. Hence the flow is assumed to satisfy a relationship $f(p, \rho) = 0$, called a barotropic equation of state.

A particular barotropic relationship for adiabatic changes is

$$p = C\rho^{\gamma}$$

or

$$\rho = Kp^{1/\gamma}$$

where C, K and γ are constants.

In general, we have an equation of state such as (4.64) plus another relating u to ρ and T

$$u = u(\rho, T)$$

4.5 CONSTITUTIVE EQUATIONS—NEWTONIAN FLUID

Let us now consider the relationship between stress and strain rates. The term 'Newtonian' fluid refers to a fluid for which the viscous stress–strain rate relations are linear. It is convenient to shift from index to matrix notation at this point. We define $\boldsymbol{\tau}$ and $\mathbf{e}$ as,

$$\begin{aligned} \boldsymbol{\tau} &= \{\tau_{11}\tau_{22}\tau_{33}\tau_{12}\tau_{23}\tau_{31}\} \\ \mathbf{e} &= \{e_{11}e_{22}e_{33}2e_{12}2e_{23}2e_{31}\} \end{aligned} \tag{4.65}$$

and write

$$\boldsymbol{\tau} = \mathbf{De} \tag{4.66}$$

Letting ψ denote the mechanical energy dissipation function, we have

$$\psi = \boldsymbol{\tau}^T\mathbf{e} = \mathbf{e}^T\mathbf{D}^T\mathbf{e} \tag{4.67}$$

where ψ must be a continuous positive definite function of the strain rate measures. Therefore, it follows that $\mathbf{D}$ is *symmetrical* and *positive definite.*

If we require the stress–strain relations to be invariant for all directions (i.e. the fluid is isotropic), $\mathbf{D}$ has only two independent elements and (4.66) reduces to

$$\tau_{ij} = \lambda e_v\delta_{ij} + 2\mu e_{ij} \tag{4.68}$$

where λ, μ are material coefficients. The mean stress σ is defined as the average of the normal stresses

$$\begin{aligned} \sigma &= \tfrac{1}{3}(\sigma_{11} + \sigma_{22} + \sigma_{33}) \\ &= -p + \tfrac{1}{3}(\tau_{11} + \tau_{22} + \tau_{33}) = -p + \lambda e_v + \tfrac{2}{3}\mu e_v \end{aligned} \tag{4.69}$$

Taking $\sigma = -p$ is known as the 'Stokes' condition and leads to

$$\lambda = -\tfrac{2}{3}\mu \tag{4.70}$$

The volumetric deformation for many fluids is small in comparison to the shearing deformations, and it is generally quite reasonable to assume the fluid is incompressible. For this case,

$$\tau_{ij} \approx 2\mu e_{ij} = \mu\left(\frac{\partial v_i}{\partial x_j} + \frac{\partial v_j}{\partial x_i}\right) \tag{4.71}$$

and the mass continuity equation (4.27) simply becomes

$$e_v = \frac{\partial v_i}{\partial x_i} \approx 0 \tag{4.72}$$

As a consequence of assuming incompressibility, the pressure has to be determined from the equilibrium equations.

4.6 NAVIER–STOKES EQUATIONS—INCOMPRESSIBLE NEWTONIAN FLUID

It is necessary, at this point, to summarise the equations. We restrict this summary to incompressible flow, no mechanical–thermal coupling, and a Newtonian fluid. In Section 4.8 we discuss how the equations are modified to account for turbulence.

Figure 4.8 shows the classification of the boundary zones. The zone S_v represents the surface area on which the velocities are prescribed. If S_v is a fixed physical boundary (e.g. a wall), the velocity components are zero (no slip). Symmetry and inflow velocity constraints are also included in S_v.

We define S_p as the surface zone on which prescribed boundary forces are acting. Sometimes S_p is a 'free' surface and the boundary forces are due to atmospheric pressure and wind shear.

Lastly, S_T denotes the boundary segment on which the temperature is prescribed. The outward normal heat flux q_n is prescribed on the remaining segment S_f.

Continuity

$$e_v = \frac{\partial v_i}{\partial x_i} = 0 \quad \text{in } V \tag{4.73}$$

$$\left.\begin{aligned} v_n &= \bar{v}_n \\ v_s &= \bar{v}_s \end{aligned}\right\} \quad \text{on } S_v \tag{4.74}$$

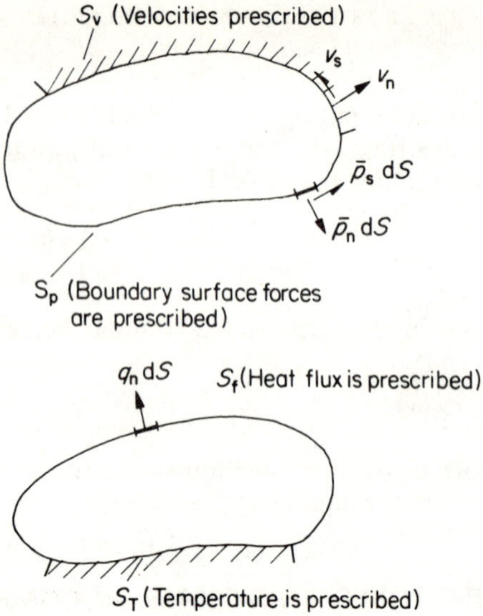

Figure 4.8 Classification of boundary zones

Equilibrium

$$-\frac{\partial}{\partial x_k}\left(\frac{p}{\rho_0}\right) + b_k + \nu\nabla^2 v_k = \frac{\partial}{\partial x_j}(v_j v_k) + \frac{\partial}{\partial t}v_k \quad \text{in } V \qquad (4.75)$$

where $\nu = \mu/\rho_0$ is called the kinematic viscosity. The Laplacian in (4.75) has been obtained by substituting (4.71) in the equilibrium equations:

$$\left.\begin{aligned} -p + 2\mu\frac{\partial v_n}{\partial n} &= \bar{p}_n \\ \mu\left(\frac{\partial v_s}{\partial n} + \frac{\partial v_n}{\partial s}\right) &= \bar{p}_s \end{aligned}\right\} \quad \text{on } S_p \qquad (4.76)$$

Heat rate balance

$$\left.\begin{aligned} C_p\left(\frac{\partial T}{\partial t} + \frac{\partial}{\partial x_i}(v_i T)\right) &= f - \frac{1}{\rho_0}\frac{\partial q_i}{\partial x_i} \qquad (4.77) \\ q_i &= -K^m \rho_0 C_p \frac{\partial T}{\partial x_i} \qquad (4.78) \end{aligned}\right\} \quad \text{in } V$$

$$\left.\begin{aligned} T &= \bar{T} \quad \text{on } S_T \\ q_n &= \bar{q}_n \quad \text{on } S_f \end{aligned}\right\} \tag{4.79}$$

Note: we have assumed that Fourier's law governs the heat diffusion [equation (4.78)].

The variables are the velocity components v_i, pressure and temperature. They have to satisfy the governing differential equations (4.73, 4.75 and 4.77) and prescribed boundary conditions. The formulation is complete in the sense that there are a sufficient number of equations. However, since the equations are nonlinear, one has to resort to a numerical solution scheme except for relatively simple problems.

Note that for this case the flow is *barotropic* (i.e. the mechanical and thermal behaviour are uncoupled) and we have ten equations (three equations of motion, one equation of continuity, six constitutive equations), and ten unknowns (six stress components, three velocity components and one pressure). For the case of compressible flow the pressure and density are coupled through an equation of state.

Friction is neglected by setting the viscosity μ equal to zero. As a consequence, we cannot prescribe the tangential boundary force p_s, or the tangential velocity v_s.

Stokes flow refers to the case where the nonlinear convective acceleration terms are negligible in comparison to the local acceleration. One sets

$$\frac{\partial}{\partial x_j}(v_j v_k) \approx 0 \tag{4.80}$$

in (4.75).

Alternative forms of (4.75) can be generated by differentiation. For instance the pressure term can be eliminated by cross differentiation and requiring p to be continuous. This leads to equilibrium equations relating the vorticity (angular velocity)

$$\tfrac{1}{2}F_i + \nu\nabla^2\omega_i = \omega_l \frac{\partial}{\partial x_l}\, v_i + \frac{D\omega_i}{Dt} \tag{4.81}$$

where

$$F_i = \frac{\partial}{\partial x_j} b_k - \frac{\partial}{\partial x_k}\, b_j$$

(i, j, k in cyclic order 1, 2, 3).

For two-dimensional flow (X_1, X_2 plane), only ω_3 is finite, and in the absence of body forces (4.81) reduces to

$$\nu \nabla^2 \omega = \frac{D\omega}{Dt} \tag{4.82}$$

We note that the frictional stress contribution vanishes when the flow is frictional.

4.7 THE PRINCIPLE OF VIRTUAL POWER

We derive the Principle of Virtual Power by operating in the stress equilibrium equations and stress boundary relations [equations (4.45)]:

$$\left.\begin{aligned} \frac{\partial}{\partial x_j}\sigma_{jk} + \rho b_k &= \rho \frac{Dv_k}{Dt} \\ \sigma_{jk} &= \sigma_{kj} \end{aligned}\right\} \text{ in } V \tag{4.83}$$
$$\bar{p}_k = \alpha_{nj}\sigma_{jk} \quad \text{on } S$$

Let us multiply (4.83) by a piecewise continuous function δv_k, and integrate over the domain

$$\iiint \left\{ \frac{\partial}{\partial x_j}\sigma_{jk} + \rho b_k - \rho \frac{Dv_k}{Dt} \right\} \delta v_k \, dV + \iint (\bar{p}_k - \alpha_{nj}\sigma_{jk})\, \delta v_k \, dS = 0 \tag{4.84}$$

This expression must be satisfied for arbitrary δv_k if the stress field is an equilibrium field. Integrating the internal stress term with Gauss's formula, and substituting for σ_{jk} in terms of p and τ, we obtain

$$\iint \bar{p}_k \, \delta v_k \, dS + \iiint \rho b_k \, \delta v_k \, dV = \iiint \left(-p\, \delta e_v + \tau_{jk}\, \delta e_{jk} + \rho \frac{Dv_k}{Dt} \delta v_k \right) dV \tag{4.85}$$

where

$$\begin{aligned} \delta e_v &= \frac{\partial}{\partial x_i}(\delta v_i) \\ \delta e_{jk} &= \frac{1}{2}\left(\frac{\partial}{\partial x_j}\delta v_k + \frac{\partial}{\partial x_k}\delta v_j \right) \end{aligned} \tag{4.86}$$

If we consider δv_k as a 'virtual' velocity, the left-hand side of (4.85) can be interpreted as external virtual power. The right-hand side defines the virtual mechanical work rate of internal stresses and the virtual work rate of the inertia force. With this interpretation, it seems appropriate to call (4.85) the Principle of Virtual Power. We emphasise that it applies for arbitrary piecewise continuous δv_k and arbitrary t. It is an alternative statement of the equilibrium equations.

The principle of virtual power is the basis for finite element models in fluid mechanics. Its role corresponds to that of the principle of virtual displacements in solid mechanics. If we take $\delta v_k = v_k$, the actual velocity, the principle coincides with the first law of thermodynamics specialised for no heat input [see (4.56) to (4.50)].

Example 4.2

We can use the principle of virtual power to establish *consistent* boundary conditions. For example, if one assumes the fluid to be inviscid ($v = 0$),

$$\iiint \sigma_{jk}\,\delta e_{jk}\,\mathrm{d}V \Rightarrow \iiint -p\frac{\partial}{\partial x_k}\,\delta v_k\,\mathrm{d}V \tag{a}$$

Integrating the last expression by parts, we obtain

$$\iiint -p\frac{\partial \delta v_k}{\partial x_k}\,\mathrm{d}V = \iint -p\,\delta v_n\,\mathrm{d}S + \iiint \frac{\partial p}{\partial x_k}\,\delta v_k\,\mathrm{d}S \tag{b}$$

Thus, expanding the surface force term in the principle of virtual power, we have

$$\iint \bar{p}_k\,\delta v_k\,\mathrm{d}S = \iint (\bar{p}_n\,\delta v_n + \bar{p}_s\,\delta v_s)\,\mathrm{d}S \tag{c}$$

where n is the normal and s the tangential direction.

Hence the *consistent* stress surface force boundary conditions [compare (b) and (c)] are

$$\bar{p}_n = -p, \qquad \bar{p}_s = 0 \tag{d}$$

and the equilibrium equations

$$b_k = \frac{\partial}{\partial x_k}(p) + \rho\frac{\mathrm{D}v_k}{\mathrm{D}t} \tag{e}$$

Note that in this case we cannot prescribe tangential velocity.

Example 4.3

We can incorporate incompressibility by weighting the continuity equation with respect to δp and integrating over the volume (see Lagrangian multipliers, Chapter 1),

$$e_v = 0$$
$$\Downarrow \qquad \text{(a)}$$
$$\iiint e_v \, \delta p \, \mathrm{d}V = 0 \quad \text{for arbitrary } \delta p$$

Combining (a) and (4.85) leads to

$$\iint (\bar{p}_n \, \delta v_n + \bar{p}_s \, \delta v_s) \, \mathrm{d}S + \iiint \rho b_k \, \delta v_k \, \mathrm{d}V$$
$$= \iiint \left(-p \, \delta e_v - e_v \, \delta p + \tau_{jk} \, \delta e_{jk} + \rho \frac{\mathrm{D}v_k}{\mathrm{D}t} \, \delta v_k \right) \mathrm{d}V \qquad \text{(b)}$$

for arbitrary t, δv_k, δp.

Additional variational statements are generated later on in the chapters on finite element applications.

4.8 TURBULENCE

At this stage we will divide the flow into two general categories, *laminar* and *turbulent* flow. If the flow particles follow streamlines which do not intersect each other, the flow is laminar. Viscous flow, for which the inertial forces are small or can be neglected by comparison with the viscous flow, are laminar flows. The relationship between inertial and viscous forces is given by the Reynolds number (Re)

$$Re = \frac{\text{inertial forces}}{\text{viscous forces}} = \frac{\text{density} \times \text{speed} \times \text{size}}{\text{viscous forces}}$$

For very small Reynolds numbers the inertia forces can be neglected by comparison with the viscous ones (Re between 0 and 10^{-2}). For Reynolds numbers between 10^{-2} and 10^3, we need to consider both forces, and for $Re > 10^3$ the viscous forces are neglected.

If the Reynolds number is large the flow is no longer laminar as the particles tend to move in a random manner. These are called *turbulent* flows and, although random in nature, they can be analysed in terms of average velocities and pressure using statistical concepts.

Turbulent flows are accounted for by interpreting the instantaneous flow variables as the sum of an average and a random deviation. We write,

$$v = v' + v''$$
$$p = p' + p'' \qquad (4.87)$$
$$T = T' + T''$$

where ( )$'$ are the ensemble averages and ( )$''$ the deviations. Introducing (4.87) in the instantaneous momentum equations results in additional terms relating the deviations.

We consider first the equilibrium equations (4.50):

$$-\frac{\partial}{\partial x_k}p + \rho b_k + \frac{\partial}{\partial x_j}\tau^f_{jk} = \rho_0\frac{\partial}{\partial x_j}(v_j v_k) + \rho_0\frac{\partial}{\partial t}v_k \qquad (4.88)$$

where τ^f denotes the frictional (viscous) stresses. Substituting for v and p and averaging over the total ensemble results in

$$-\frac{\partial}{\partial x_k}p' + \langle \rho b_k \rangle + \frac{\partial}{\partial x_j}(\langle \tau^f_{jk} \rangle - \rho_0 \langle v''_j v''_k \rangle)$$
$$= \rho_0\frac{\partial}{\partial x_j}(v'_j v'_k) + \rho_0\frac{\partial}{\partial t}v'_k \qquad (4.89)$$

where $\langle\ \rangle$ now represents the ensemble average. By definition

$$\langle f'g \rangle = f'\langle g \rangle$$
$$\langle f'' \rangle = 0 \qquad (4.90)$$

We interpret the second-order velocity deviation terms as equivalent stresses (they are generally called Reynolds stresses), and write

$$\tau = \tau^f + \tau^t$$
$$\langle \tau^t_{jk} \rangle = -\rho_0 \langle v''_j v''_k \rangle \qquad (4.91)$$

If we take the body force terms according to (4.53) and assume a Newtonian fluid, the equations have the same form as (4.88) with τ^f replaced with τ, and the instantaneous variables replaced with ensemble average measures. Finally, the modified forms of (4.73) to (4.76) are

$$\frac{\partial v_i}{\partial x_i} = 0$$
$$-\frac{\partial}{\partial x_k}\left(\frac{p}{\rho_0}\right) + b_k + \nu\nabla^2 v_k + \frac{1}{\rho_0}\tau^t_{jk,j} = \frac{\partial}{\partial x_j}(v_j v_k) + \frac{\partial}{\partial t}v_k \qquad (4.92)$$

and

$$p_n = -p + \alpha_{nj}\alpha_{nk}\tau^t_{jk} + 2\mu\frac{\partial v_n}{\partial n}$$
$$p_s = \alpha_{nj}\alpha_{sk}\tau^t_{jk} + \mu\left(\frac{\partial v_s}{\partial n} + \frac{\partial v_n}{\partial s}\right) \tag{4.93}$$

where all variables are now ensemble averages. We have dropped the prime superscript to simplify the notation.

The essential difficulty lies in evaluating the turbulent stresses. In addition, our consideration of internal energy in Section 4.4 led to the following energy balance equation [see (4.63)]:

$$\rho_0 C_p \frac{\mathrm{D}T}{\mathrm{D}t} = -\frac{\partial q_i}{\partial x_i} + f\rho_0 \tag{4.94}$$

where the mechanical energy dissipation, $\psi = \tau_{ij}e_{ij}$, has been neglected with respect to the flux and distributed source terms.

The fact that ψ is positive definite was used to establish the viscous stress–strain relations. Extending this argument to include turbulence, we write

$$\tau_{ij}e_{ij} = (\tau^f_{ij} + \tau^t_{ij})e_{ij} = \psi^f + \psi^t \tag{4.95}$$

and require ψ^t to be a positive definite continuous quadratic function of e_{ij}, the ensemble averaged strain rate measures. Using the notation of (4.67), this implies

$$\boldsymbol{\tau}^t = \mathbf{D}^t\,\mathbf{e} \tag{4.96}$$

($\mathbf{D}^t$ is symmetrical and positive definite.) Since the Reynolds stresses are flow properties (whereas the frictional stresses are material properties), $\mathbf{D}^t$ is dependent on the velocity field. The problem is not yet resolved and various forms for $\mathbf{D}^t$ are used. The most frequent choice is the 'isotropic' form

$$\tau^t_{ij} = 2\rho_0\eta e_{ij}$$
$$\eta = \text{'eddy' viscosity} \tag{4.97}$$

Recently, there has been a shift over to the orthotropic form (no coupling between shear and extension) for circulation analysis of lakes and coastal waters, where the vertical flow differs appreciably from the lateral flow.

$$\left.\begin{aligned}\tau^t_{11} &= D_{11}e_{11} + D_{12}e_{22} + D_{13}e_{33}\\ \tau^t_{22} &= D_{13}e_{11} + D_{22}e_{22} + D_{23}e_{33}\\ \tau^t_{33} &= D_{13}e_{11} + D_{23}e_{22} + D_{33}e_{33}\\ \tau^t_{12} &= 2D_{44}e_{12}\\ \tau_{23} &= 2D_{55}e_{23}\\ \tau^t_{23} &= 2D_{66}e_{31}\end{aligned}\right\} \qquad (4.98)$$

The fully anisotropic form ($\mathbf{D}^t$ is fully populated) is more appropriate but accurate values of the coefficients are not known, even for the isotropic case.

We return now to the approximate rate equation (with the mechanical dissipation neglected). Taking the ensemble average leads to

$$\rho_0 C_p \frac{\mathrm{D}T'}{\mathrm{D}t} = \rho_0 f - \frac{\partial}{\partial x_i}(q'_i + \rho_0 C_p \langle v''_i T'' \rangle) \qquad (4.99)$$

We expand the flux using Fourier's law

$$q'_i = -K^m \rho_0 C_p \frac{\partial T'}{\partial x_i} \qquad (4.100)$$

where K^m is a molecular diffusion coefficient (units of length^2/time). One can interpret (4.100) as an application of Fick's law to the diffusion of heat. The deviation term is treated in a similar way; i.e. we write

$$\langle v''_i T'' \rangle = -K^t_{ij} \frac{\partial T'}{\partial x_j} \qquad (4.101)$$

and interpret K^t_{ij} as an equivalent diffusion coefficient due to turbulence. Combining the two terms

$$K^m + K^t_{ij} = K_{ij} \qquad (4.102)$$

allows us to write the equations as (we drop the prime superscript for convenience)

$$\begin{aligned}\frac{\partial T}{\partial t} &= \frac{f}{C_p} + \frac{\partial}{\partial x_i}\left(K_{ij}\frac{\partial T}{\partial x_j}\right) \quad \text{in } V\\ T &= \bar{T} \quad \text{on } S_T\\ q_n &= -\rho_0 C_p \alpha_{ni}\left(K_{ij}\frac{\partial T}{\partial x_j}\right) \quad \text{on } S_f\end{aligned} \qquad (4.103)$$

Again, one is faced with the problem of estimating the turbulent coefficients. In turbulent flow, K^m is negligible with respect to K^t. Although K^t is a flow property, it is in many cases assumed to be constant and isotropic ($K^t_{ij} = K\delta_{ij}$).

BIBLIOGRAPHY

BATCHELOR, G. K., *An Introduction to Fluid Dynamics*, Cambridge University Press (1967)

CARSLAW, H. S. and JAEGER, J. C., *Conduction of Heat in Solids*, Oxford (1959)

LANDAU, L. D. and LIFSHITZ, E. M., *Fluid Mechanics*, Pergamon Press (1959)

LANGHAAR, H. L., *Energy Methods in Applied Mechanics*, John Wiley (1962)

LONGWELL, P. A., *Mechanics of Fluid Flow*, McGraw-Hill (1966)

MALVERN, L. E., *Introduction to the Mechanics of a Continuous Medium*, Prentice-Hall (1969)

PRANDTL, L., *The Essentials of Fluid Dynamics*, Blackie, London (1969)

ROHSENOW, W. M. and CHOI, H., *Heat, Mass and Momentum Transfer*, Prentice-Hall (1961)

SCHLIGHTING, H., *Boundary Layer Theory*, McGraw-Hill (1960)

SHELLAND, A. H. P., *Non-Newtonian Flow and Heat Transfer*, McGraw-Hill (1967)

EXERCISES

4-1 Carry out the details of deriving $\sigma_{ij} = \sigma_{ji}$ from equation (4.28).

4-2 Write the continuity equation for the case of incompressible irrotational flow as defined in (4.37) as a function of the velocity potential Φ.

4-3 Consider the two-dimensional flow as described by the velocity vector

$$\vec{v} = (-2x_1^2 + 3x_2)\vec{i}_1 + (3x_1 + 2x_2)\vec{i}_2 \qquad \text{(a)}$$

Determine whether the flow is irrotational or incompressible or both.

4-4 Show that for an inviscid fluid ($\sigma_{ij} = -p\delta_{ij}$) the momentum equations (4.49) can be written

$$-\frac{1}{\rho}\nabla p + \vec{b} = \frac{\partial \vec{v}}{\partial t} + \vec{v} \,.\, \nabla \vec{v}$$

4-5 Show that the energy equation for no internal heat sources and heat governed by Fourier's law can be written for an inviscid fluid as

$$\frac{\mathrm{D}T}{\mathrm{D}t} = \vec{\nabla} \,.\, (K\vec{\nabla}T) - p\vec{\nabla} \,.\, \vec{v} \qquad \text{(a)}$$

where $K = K^m \rho_0 C_p$.

4-6 Consider the case of incompressible flow ($\vec{\nabla} . \vec{v} = 0$); the heat transfer formula (a) of Exercise 4.5 now simplifies to

$$\frac{DT}{Dt} = \vec{\nabla} . (K\vec{\nabla}T) \qquad \text{(a)}$$

If convective terms can be neglected (e.g. case of heat transfer in solids), we have

$$\frac{\partial T}{\partial t} = \vec{\nabla} . (K\vec{\nabla}T) \qquad \text{(b)}$$

The right-hand side of formula (b) represents a Laplace's equation. Indicate how this equation, with the corresponding boundary conditions,

$$T = \bar{T} \quad \text{on } S_T \qquad \text{(c)}$$

and

$$K\frac{\partial T}{\partial n} = g \quad \text{on } S_f$$

can be written in Galerkin's form.

4-7 Apply a finite element three node triangular element for the above variational statement. Discuss ways in which the final matrices may be integrated.

4-8 For a linearly elastic isotropic solid, the constitutive equations can be written

$$\sigma_{ij} = \bar{\lambda} e_v \delta_{ij} + 2\bar{\mu} e_{ij} \qquad \text{(a)}$$

where $\bar{\lambda}$ and $\bar{\mu}$ are Lame's constants. e_v, e_{ij} are strain-displacement relationships similar to those shown in equations (4.29) and (4.30) but substituting v_i by u_i. For instance,

$$e_v = \frac{\partial u_1}{\partial x_1} + \frac{\partial u_2}{\partial x_2} + \frac{\partial u_3}{\partial x_3} \qquad \text{(b)}$$

Write the equilibrium equations, stress boundary conditions and principle of virtual power for the solid in terms of displacements.

4-9 We can write for a linearly isotropic solid (Exercise 4.8)

$$\sigma_{ij} = \bar{\lambda} e_v \delta_{ij} + 2\bar{\mu} e_{ij} \qquad \text{(a)}$$

where the Lame constants $\bar{\lambda}$ and $\bar{\mu}$ are related to the modulus of elasticity shear modulus G and Poisson's ratio v by

$$\bar{\mu} = G = \frac{E}{2(1+v)}, \qquad \bar{\lambda} = \frac{vE}{(1+v)(1-2v)} \qquad \text{(b)}$$

Verify that the constitutive equations (a) can now be written

$$\sigma_{11} = \frac{E}{(1+\nu)(1-2\nu)}[(1-\nu)e_{11} + \nu e_{22} + \nu e_{33}]$$

$$\sigma_{22} = \frac{E}{(1+\nu)(1-2\nu)}[\nu e_{11} + (1-\nu)e_{22} + \nu e_{33}]$$

$$\sigma_{33} = \frac{E}{(1+\nu)(1-2\nu)}[\nu e_{11} + \nu e_{22} + (1-\nu)e_{33}]$$

$$\sigma_{12} = \frac{E}{2(1+\nu)}(2e_{12})$$

$$\sigma_{23} = \frac{E}{2(1+\nu)}(2e_{23}), \qquad \sigma_{13} = \frac{E}{2(1+\nu)}(2e_{13})$$

Relate the mean stress $\sigma = \frac{1}{3}(\sigma_{11} + \sigma_{22} + \sigma_{33})$ to e_v and discuss the case $\nu = \frac{1}{2}$.

4-10 Find the solution for incompressible viscous flow between parallel plates (Figure 4.9), one of which is moving with a velocity V. Note that the flow is two-dimensional and consider gravity.

Figure 4.9 Flow between parallel plates

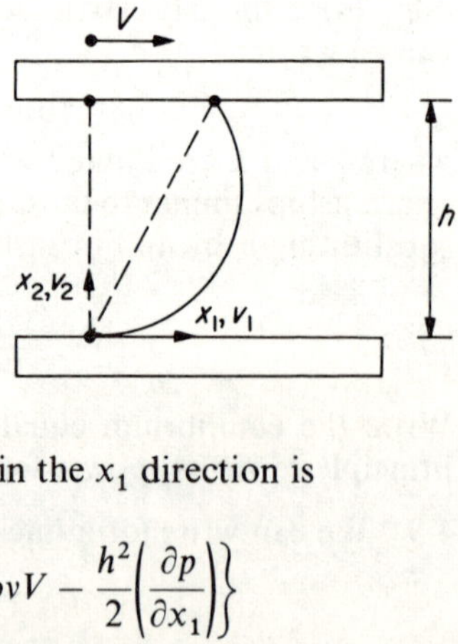

Show that the solution for velocity in the x_1 direction is

$$\rho v v_1 = \frac{x_2^2}{2}\left(\frac{\partial p}{\partial x_1}\right) + \frac{x_2}{h}\left\{\rho v V - \frac{h^2}{2}\left(\frac{\partial p}{\partial x_1}\right)\right\}$$

4-11 Show that the rate of change of the internal energy for an inviscid incompressible fluid is equal to the rate of change of heat input and that the rate of external power input is equal to the rate of change of the kinetic energy of the system.

4-12 Consider a two-dimensional problem for which the velocity potential is

$$\Phi = 4x_1x_2 + \text{constant}$$

Verify that for this case the flow is irrotational and incompressible. Plot the path of some of the particles.

5 Inviscid Fluids

5.1 BASIC PRINCIPLES

In this chapter we will specialise the governing equations of fluid flow for the case of inviscid flow.

We will define a *steady* flow as the case for which the velocity at any point is independent of time (i.e. $\partial\bar{v}/\partial t = 0$). A steady flow problem requires spatial boundary conditions but not initial conditions.

Let us summarise the governing equations for a fluid, previously deduced:

Equation of motion

$$\frac{\partial \sigma_{ij}}{\partial x_j} + \rho b_j = \rho \frac{\mathrm{D} v_j}{\mathrm{D} t} \tag{5.1}$$

Continuity

$$\frac{\partial \rho}{\partial t} + \rho \frac{\partial v_k}{\partial x_k} = 0 \tag{5.2}$$

Constitutive equations (Stokes type flow)

$$\sigma_{ij} = -p\delta_{ij} + 2\mu(e_{ij} - \tfrac{1}{3}e_{ij}\delta_{ij}) \tag{5.3}$$

Equation of state (barotropic flow)

$$f(p, \rho) = 0 \tag{5.4}$$

Strain rate–velocity relationships

$$e_{ij} = \frac{1}{2}\left\{\frac{\partial v_i}{\partial x_j} + \frac{\partial v_j}{\partial x_i}\right\} \tag{5.5}$$

For a liquid with zero viscosity the constitutive equations reduce to

$$\sigma_{ii} = -p\delta_{ii} \tag{5.6}$$

or simply

$$\sigma = -p \tag{5.7}$$

The state of stress is then defined by a single variable p.

The equilibrium equations are now

$$-\frac{\partial p}{\partial x_j} + \rho b_j = \rho \frac{\mathrm{D}v_j}{\mathrm{D}t} \tag{5.8}$$

with boundary conditions

$$p_n = -p \quad \text{or } v_n \text{ given} \tag{5.9}$$

but

$$p_s = 0, \qquad v_s \text{ not necessary}$$

which implies that we cannot apply a tangential boundary force to an inviscid fluid.

Incompressibility (ρ = constant = ρ_0) implies that the continuity equation gives

$$\frac{\partial v_k}{\partial x_k} = 0 \tag{5.10}$$

RELATIONSHIP BETWEEN VISCOSITY AND VORTICITY

Consider a sphere of incompressible, inviscid fluid initially at rest. When the forces are incremented in the fluid they will produce certain surface forces on this sphere. These surface forces must act normally to the surface since the fluid is frictionless (i.e. it would not resist tangential forces). Thus, their component is applied at the centre of the sphere and the body forces at the mass centre. The sphere then cannot rotate as no couple can be applied to it.

In order to define this *irrotationality* condition mathematically, let us consider the equation of motion in terms of velocities. We have

$$-\frac{\partial}{\partial x_j}\left(\frac{p}{\rho_0}\right) + v\frac{\partial^2 v_j}{\partial x_k} + b_j = \frac{\mathrm{D}v_j}{\mathrm{D}t} \tag{5.11}$$

or

$$-\vec{\nabla}\,.\left(\frac{p}{\rho_0}\right) + v\nabla^2\vec{v} + \vec{b} = \frac{\mathrm{D}\vec{v}}{\mathrm{D}t}$$

where

$$\nabla^2(\ ) = \frac{\partial^2(\ )}{\partial x_1^2} + \frac{\partial^2(\ )}{\partial x_2^2} + \frac{\partial^2(\ )}{\partial x_3^2} = \vec{\nabla} . \{\vec{\nabla}(\ )\}$$

$$\vec{\nabla}(\ ) = \left\{ \frac{\partial(\ )}{\partial x_1}, \frac{\partial(\ )}{\partial x_2}, \frac{\partial(\ )}{\partial x_3} \right\}$$

and $\vec{\nabla} . f$ is the gradient of the differentiable scalar function f, and $\vec{\nabla} . \vec{v}$ is the gradient of a vector and equal to

$$\vec{\nabla} . \vec{v} = \left\{ \frac{\partial(\ )}{\partial x_1}, \frac{\partial(\ )}{\partial x_2}, \frac{\partial(\ )}{\partial x_3} \right\} . (v_1, v_2, v_3)$$

The term with the kinematic viscosity $\nu(\nu = \mu/\rho_0)$ can be expanded, taking into account the incompressibility condition, to

$$\sum_{k=1}^{3} \frac{\partial^2 v_j}{\partial x_k^2} = 2\left\{ \frac{\partial}{\partial x_m}(\omega_l) - \frac{\partial}{\partial x_l}(\omega_m) \right\} \tag{5.12}$$

$j = 1, 2, 3$; $m = 3, 1, 2$; $l = 2, 3, 1$

or simply

$$\nabla^2 \vec{v} = 2 \operatorname{curl}(\vec{\omega}) = 2\vec{\nabla} \times \vec{\omega} \tag{5.13}$$

where

$$\vec{\omega} = \tfrac{1}{2}\vec{\nabla} \times \vec{v}$$

Hence if the flow is irrotational, the $\vec{\omega}$ vector disappears, which implies that the viscosity term also disappears and (5.11) reduces to

$$-\frac{1}{\rho_0}\frac{\partial p}{\partial x_j} + b_j = \frac{\mathrm{D}v_j}{\mathrm{D}t} \tag{5.14}$$

or

$$-\vec{\nabla}\left(\frac{p}{\rho_0}\right) + \vec{b} = \frac{\mathrm{D}\vec{v}}{\mathrm{D}t}$$

5.2 BERNOULLI'S PRINCIPLE

Let us consider a single mass particle acted upon by a force vector $\vec{F}$ (Figure 5.1). The force equilibrium equation is

$$\vec{F} = m\frac{D\vec{v}}{\mathrm{D}t} \tag{5.15}$$

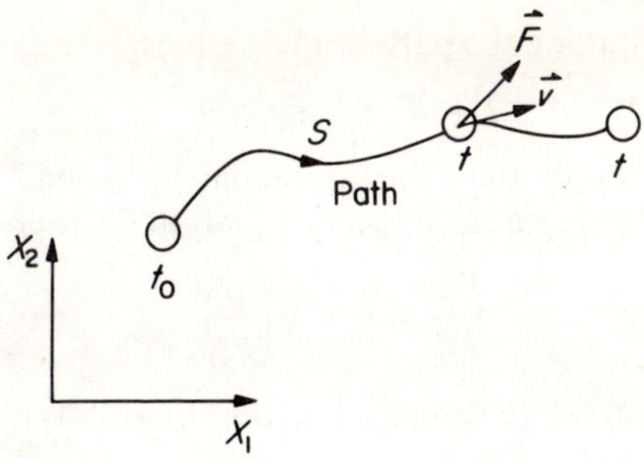

Figure 5.1 Path of a particle

In moving from position t_0 to position t, $\vec{F}$ does work on the particle and consequently the kinetic energy changes. Conservation of energy requires that

$$\text{work} = \text{increment in kinetic energy}$$
$$\Downarrow \tag{5.16}$$
$$\int_{t_0}^{t} \vec{F} \cdot \vec{v}\, \mathrm{d}t = m \int_{t_0}^{t} \vec{v} \cdot \frac{\mathrm{D}\vec{v}}{\mathrm{D}t}\, \mathrm{d}t = m \int_{t_0}^{t} \frac{\mathrm{D}}{\mathrm{D}t}(\tfrac{1}{2}\vec{v} \cdot \vec{v})\, \mathrm{d}t$$

This law applies for arbitrary $\vec{F}$ (i.e. dissipative as well as conservative). If $\vec{F}$ is conservative, it is derivable from a force potential

$$\vec{F} = -\vec{\nabla}\Omega \tag{5.17}$$

where $\Omega = \Omega(x)$ is the potential. The necessary and sufficient condition for a single-valued field with continuous first derivatives to be conservative is

$$\vec{\nabla} \times \vec{F} = \vec{0}$$

Equation (5.16) now becomes

$$\int_{t_0}^{t} \left\{ \vec{\nabla}\Omega \cdot \vec{v} + \frac{\mathrm{D}}{\mathrm{D}t}(\mathscr{T}) \right\} \mathrm{d}t = 0 \tag{5.18}$$

where $\mathscr{T} = \frac{1}{2}m\vec{v} \cdot \vec{v}$ is the kinetic energy of the particle.

Equation (5.18) can also be written as a function of S, a parameter along the path of the particle

$$\int_{S_0}^{S} \frac{\partial \Omega}{\partial s}\, \mathrm{d}S + \int_{t_0}^{t} \frac{\mathrm{D}(\mathscr{T})}{\mathrm{D}t}\, \mathrm{d}t = 0 \tag{5.19}$$

This means that along the path of the particle,

$$\Omega + \mathscr{T} = \text{constant} \tag{5.20}$$

We can now apply the same argument for a fluid particle. The vector form of the force equilibrium equation for frictionless flow is

$$\frac{\mathrm{D}\vec{v}}{\mathrm{D}t} = \vec{b} - \vec{\nabla}\left(\frac{p}{\rho_0}\right) \tag{5.21}$$

When $\vec{b}$ is conservative (usually $\vec{b}$ is due to gravity), we can write

$$b_k = -\frac{\partial \Omega}{\partial x_k} \Rightarrow \vec{b} = -\vec{\nabla}\Omega \tag{5.22}$$

We can now note the identity

$$\begin{aligned} \frac{\mathrm{D}\vec{v}}{\mathrm{D}t} &= \frac{\partial \vec{v}}{\partial t} + \tfrac{1}{2}\vec{\nabla}(\vec{v}\,.\,\vec{v}) - 2(\vec{v} \times \vec{\varpi}) \\ \vec{\varpi} &= \tfrac{1}{2}\vec{\nabla} \times \vec{v} \end{aligned} \tag{5.23}$$

Hence the equilibrium equation (5.21) can be written

$$\frac{\partial \vec{v}}{\partial t} + \vec{\nabla}H - 2(\vec{v} \times \vec{\varpi}) = \vec{0} \tag{5.24}$$

where H is called the total head and is defined by

$$\begin{aligned} H &= \frac{p}{\rho_0} + \Omega + \tfrac{1}{2}V^2 \\ V^2 &= \vec{v}\,.\,\vec{v} = \sum v_i v_i \end{aligned} \tag{5.25}$$

We can now integrate (5.24) multiplied by $\vec{v}$ along the path of the particle (streamline) and obtain the following work statement:

$$\int_{t_0}^{t} \left(\frac{\partial \vec{v}}{\partial t} - 2(\vec{v} \times \vec{\varpi}) + \vec{\nabla}H\right).\,\vec{v}\,\mathrm{d}t = 0 \tag{5.26}$$

The second term drops since $\vec{v} \times \vec{\varpi}$ is orthogonal to $\vec{v}$. Thus,

$$\int_{t_0}^{t} \left(\frac{\partial \vec{v}}{\partial t} + \vec{\nabla}H\right).\,\vec{v}\,\mathrm{d}t = 0 \tag{5.27}$$

For *steady* flow, $\vec{v}$ and p are independent of t, and it follows that

$$H = \frac{p}{\rho_0} + \Omega + \tfrac{1}{2}V^2 = \text{constant along a streamline for steady flow} \tag{5.28}$$

This result is due to Bernoulli. When the flow is steady and irrotational, (5.24) simply becomes

$$\vec{\nabla} H = \vec{0} \tag{5.29}$$

and Bernoulli's principle reduces to

$$H = \text{constant throughout the domain for steady inviscid irrotational flow} \tag{5.30}$$

i.e. the same H constant applies throughout all the field.

Example 5.1

Let us consider the two-dimensional steady flow of an inviscid and incompressible fluid. Navier–Stokes equations can be written

$$\left.\begin{aligned} -\frac{1}{\rho}\frac{\partial p}{\partial x_1} + b_1 - \frac{\partial}{\partial x_1}\{\tfrac{1}{2}(v_1^2 + v_2^2)\} &= 0 \\ -\frac{1}{\rho}\frac{\partial p}{\partial x_2} + b_2 - \frac{\partial}{\partial x_2}\{\tfrac{1}{2}(v_1^2 + v_2^2)\} &= 0 \end{aligned}\right\} \tag{a}$$

If a body force potential exists such as the gravity one, i.e.

$$\Omega = -hg \tag{b}$$

where h is the elevation and g the acceleration due to gravity, we can write

$$\mathrm{d}\left\{\frac{p}{\gamma} + h + \left(\frac{v_1^2 + v_2^2}{2g}\right)\right\} = 0 \tag{c}$$

$$\therefore \qquad \frac{p}{\gamma} + h + \frac{V^2}{2g} = \text{constant} \tag{d}$$

where $\gamma = \rho g$ and $V^2 = v_1^2 + v_2^2$.

Equation (d) is Bernoulli's equation for two-dimensional, inviscid, incompressible and steady flow.

We define the following:

$$\left.\begin{aligned} \frac{p}{\gamma} &= \text{pressure head} \\ h &= \text{elevation head} \\ \frac{V^2}{2g} &= \text{velocity head} \end{aligned}\right. \tag{e}$$

Let us finally consider how equation (d) can be interpreted for the pipe and channel shown in Figure 5.2. Consider two sections with elevations h_1 and h_2. The pressure at one or the other end is known by some manometer reading (pipe), or simply by knowing the position of the free surface (channel). The velocity heads are $V_1^2/2g$ and $V_2^2/2g$ for the two sections.

In practice, the 'constant energy' principle represented by Bernoulli's equation is not valid as there is always a 'loss of energy' due to friction, turbulence and other causes. This loss is represented in Figures 5.2(a) and (b) by the difference between the horizontal broken line and the inclined broken line representing the sum of energies and called the hydraulic gradient. The loss is called the head loss.

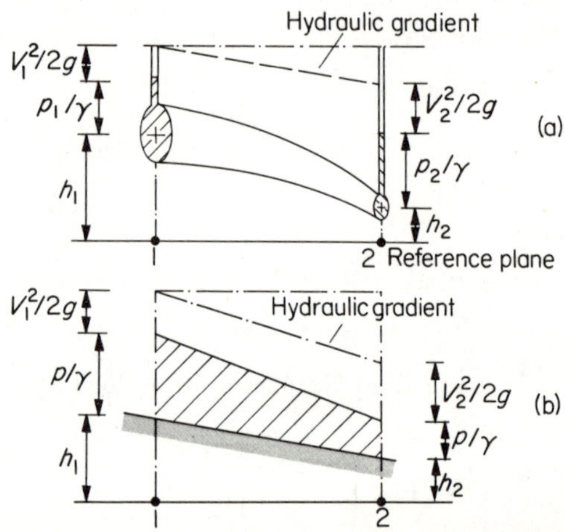

Figure 5.2 Flow in a pipe and channel. (a) *Pipe*: (b) *Channel*

POTENTIAL FLOWS

If the flow of an inviscid fluid is at any time irrotational, it will remain so. These irrotational flows are also called potential flows because the condition

$$2\vec{\omega} = \vec{\nabla} \times \vec{v} = 0 \tag{5.31}$$

is a necessary and sufficient condition for the existence of a scalar potential Φ, such that the following vector *identity* is satisfied:

$$\text{curl}\,(\text{grad}\,\Phi) = \vec{\nabla} \times (\vec{\nabla}\Phi) = 0$$

where the scalar function Φ has continuous first and second derivatives. Hence

$$\vec{v} = \vec{\nabla}\Phi \tag{5.32}$$

or

$$\{v_1, v_2, v_3\} = \left\{\frac{\partial \Phi}{\partial x_1}, \frac{\partial \Phi}{\partial x_2}, \frac{\partial \Phi}{\partial x_3}\right\}$$

The use of the potential greatly simplifies the problem as a single Φ function now replaces the three velocities. The incompressibility equation (5.10) for instance now becomes

$$\nabla^2 \Phi = 0$$

5.3 THE WAVE EQUATION

In many types of inviscid flow problem we can obtain a governing equation of the type

$$\frac{\partial^2 \Phi}{\partial t^2} = C^2 \nabla^2 \Phi \tag{5.33}$$

where Φ is a potential type function.

This equation is called the wave equation and is generally three-dimensional. Equation (5.33) is of the hyperbolic type and we need to know its initial as well as its boundary conditions.

FINITE ELEMENT FORMULATION

The finite element method can now be applied to the wave equation. The variational statement corresponding to equation (5.33) in two dimensions is

$$\iint \left\{ c^2 \left(\frac{\partial^2 \Phi}{\partial x_1^2} + \frac{\partial^2 \Phi}{\partial x_2^2} \right) - \frac{\partial^2 \Phi}{\partial t^2} \right\} \delta\Phi \, \mathrm{d}x_1 \, \mathrm{d}x_2 = \int_{S_2} \left\{ c^2 \frac{\partial \Phi}{\partial n} - f \right\} \delta\Phi \, \mathrm{d}S \tag{5.34}$$

The $\Phi = \bar{\Phi}$ conditions on S_1 are assumed to be identically satisfied.

After integrating (5.34) by parts we obtain

$$\iint \left\{ c^2 \left(\frac{\partial \Phi}{\partial x_1} \frac{\partial \delta\Phi}{\partial x_1} + \frac{\partial \Phi}{\partial x_2} \frac{\partial \delta\Phi}{\partial x_2} \right) + \frac{\partial^2 \Phi}{\partial t^2} \delta\Phi \right\} \mathrm{d}x_1 \, \mathrm{d}x_2 = \int_{S_2} f \, \delta\Phi \, \mathrm{d}S \tag{5.35}$$

The Φ potential can now be approximated on each element by

$$\Phi = \boldsymbol{\phi}^T \boldsymbol{\Phi}^n \quad \text{and} \quad \ddot{\Phi} = \boldsymbol{\phi}^T \ddot{\boldsymbol{\Phi}}^n \tag{5.36}$$

where $\boldsymbol{\phi}$ is an interpolation function and $\boldsymbol{\Phi}^n, \ddot{\boldsymbol{\Phi}}^n$ are the nodal unknowns.

We can substitute (5.36) into (5.35) and obtain for an element

$$\mathbf{K}\boldsymbol{\Phi}^n + \mathbf{M}\,\ddot{\boldsymbol{\Phi}}^n = \mathbf{P} \tag{5.37}$$

where

$$\mathbf{K} = \iint C^2(\boldsymbol{\phi}_{,1}\boldsymbol{\phi}_{,1}^T + \boldsymbol{\phi}_{,2}\boldsymbol{\phi}_{,2}^T)\,\mathrm{d}x_1\,\mathrm{d}x_2$$

$$\mathbf{M} = \iint \boldsymbol{\phi}\boldsymbol{\phi}^T \mathrm{d}x_1\,\mathrm{d}x_2$$

$$\mathbf{P} = \int \boldsymbol{\phi} f\,\mathrm{d}S$$

where

$$\boldsymbol{\phi}_{,1} = \frac{\partial \boldsymbol{\phi}}{\partial x_1}, \qquad \boldsymbol{\phi}_{,2} = \frac{\partial \boldsymbol{\phi}}{\partial x_2}$$

One obtains after assembling all the elements

$$\mathscr{K}\boldsymbol{\Phi} + \mathscr{M}\ddot{\boldsymbol{\Phi}} = \mathscr{P} \tag{5.38}$$

We have already discussed (Chapter 3) that equation (5.38) can be easily solved for the case of harmonic motion such that the excitation and the response are in phase; i.e.

$$\boldsymbol{\Phi} = \mathscr{H}\,\mathrm{e}^{\mathrm{i}\omega t}, \qquad \mathscr{P} = \mathscr{P}_0\,\mathrm{e}^{\mathrm{i}\omega t} \tag{5.39}$$

where $i = \sqrt{-1}$ and ω is the circular frequency.

This case is called a dynamic steady state case and it applies when sufficient time has elapsed after the start of the excitation. Substituting (5.39) into (5.38) we obtain

$$(\mathscr{K} - \omega^2\mathscr{M})\mathscr{H} = \mathscr{P}_0 \tag{5.40}$$

If the frequency ω is known we can solve the system of equations (5.40) for a given value of $\mathscr{P}_0$. In particular, if $\mathscr{P}_0 = \mathbf{0}$, (5.40) reduces to finding the eigenvalues (ω_i) and eigenvectors ($\mathscr{H}_i$) which give a solution to the system.

It is important to point out that for harmonic steady state problems the accuracy of the finite element solution is strongly dependent on the fineness of the mesh. This dependence can be established as a function of

$$r = \frac{L_{\max}}{\lambda_{\min}}$$

where $L_{\max}$ is the maximum size of the largest element and $\lambda_{\min}$ is the minimum wavelength.

The maximum characteristic size should be about 0·1 as a rule.

The above method of solution applies for steady state problems. If $\mathscr{P}$ is an arbitrary function of time, say $\mathscr{P}(t)$, we can solve the problem in two ways; by *modal superposition* or *direct integration.* We will now discuss the modal superposition technique, direct integration methods will be studied in subsequent chapters.

MODAL SUPERPOSITION

Consider equations (5.38) which are now written

$$\mathscr{K}\mathbf{\Phi} + \mathscr{M}\ddot{\mathbf{\Phi}} = \mathscr{P}(t) \tag{5.41}$$

As always we assume that a minimum of essential boundary conditions have been applied. Hence $\mathscr{K}$ is positive definite; as well as $\mathscr{M}$ which is always so. Note that if certain values of $\mathbf{\Phi}$ are known and different from zero, say $\bar{\mathbf{\Phi}}$, they will produce a $\bar{\mathscr{P}}$ vector on the right-hand side once they are multiplied by the corresponding column of $\mathscr{K}$. We here assume that the boundary accelerations corresponding to these Φ values are zero. Hence (5.40) can now be written

$$\mathscr{K}\mathbf{\Phi} + \mathscr{M}\ddot{\mathbf{\Phi}} = \mathscr{P}(t) + \bar{\mathscr{P}} \tag{5.42}$$

where $\mathscr{K}$, $\mathscr{M}$ and $\mathscr{P}$ are the new matrices conveniently reduced, although we keep the same notation for simplicity.

In the modal superposition method one solves first the free vibrations case, equation (5.40). This produces a series of eigenvalues and eigenvectors which can be written as

$$\omega_i, \mathbf{\Phi}_i \quad \text{for } i = 1, 2 \ldots n$$

where the ω_i are all positive for positive definite $\mathscr{K}$ and $\mathscr{M}$ matrices.

By definition,

$$\mathscr{K}\mathbf{\Phi}_i = \omega_i^2 \mathscr{M}\mathbf{\Phi}_i \tag{5.43}$$

The solution vectors $\mathbf{\Phi}_i$ are orthogonal with respect to $\mathscr{K}$ and $\mathscr{M}$; i.e.

$$\left.\begin{aligned}\mathbf{\Phi}_i^T \mathscr{K} \mathbf{\Phi}_j &= \mathbf{0}\\ \mathbf{\Phi}_i^T \mathscr{M} \mathbf{\Phi}_j &= \mathbf{0}\end{aligned}\right\} \quad \text{for } i \neq j \tag{5.44}$$

and only $\neq \mathbf{0}$ when $i = j$. Each eigenvector contains an arbitrary constant which is generally evaluated by normalising $\mathbf{\Phi}_i$ with respect to $\mathscr{M}$; i.e.

$$\mathbf{\Phi}_i^T \mathscr{M} \mathbf{\Phi}_i = 1 \tag{5.45}$$

This leads to

$$\mathbf{\Phi}_i^T \mathscr{K} \mathbf{\Phi}_i = \omega_i^2 \tag{5.46}$$

The solution vector $\mathbf{\Phi}$ can be written as a combination of s different eigenvectors and a vector with the known $\bar{\mathbf{\Phi}}$ values

$$\mathbf{\Phi} = \sum_{i=1}^{s} q_i \mathbf{\Phi}_i + \bar{\mathbf{\Phi}} \tag{5.47}$$

where $s \leqslant n$ and $q_i = q_i(t)$ are interpreted as generalised coordinates.

Substituting $\mathbf{\Phi}$ into equation (5.42) leads to

$$\sum_{i=1}^{s} (\ddot{q}_1 \mathscr{M} \mathbf{\Phi}_i + q_i \mathscr{K} \mathbf{\Phi}_i) = \mathscr{P}(t) \tag{5.48}$$

Premultiplying by $\mathbf{\Phi}_i^T$ and noting the orthogonality relationships we obtain s uncoupled differential equations:

$$\frac{\mathrm{d}^2}{\mathrm{d}t^2}(q_j) + \omega_j^2 q_j = p_j(t) \tag{5.49}$$

where

$$p_j = \mathbf{\Phi}_j^T \mathscr{P}(t)$$

The solution of equation (5.49) is

$$q_j = q_j|_{t=0} \cos \omega_j t + \frac{1}{\omega_j} \left|\frac{\mathrm{d}q_j}{\mathrm{d}t}\right|_{t=0} \sin \omega_j t + \frac{1}{\omega_j} \int_0^t \sin \omega_j(t - \xi) p_j(\xi)\, \mathrm{d}\xi \tag{5.50}$$

where ξ is a dummy integration variable. The first terms, which vanish if the system is initially at rest, contain the initial conditions.

By summing the contributions of all the s nodes, we have

$$\mathbf{\Phi} = \sum_{i=1}^{s} q_i \mathbf{\Phi}_i + \bar{\mathbf{\Phi}} \tag{5.51}$$

The essential difficulty is the choice of s, i.e. the decision as to how many generalised coordinates one should take. The solution is 'exact' only when $s = n$ (exact here refers to the solution of the discretised equations).

5.4 HARMONIC RESPONSE OF COASTAL WATERS

We will now consider a special case of the wave equation. This is the problem of predicting the harmonic response of coastal waters.

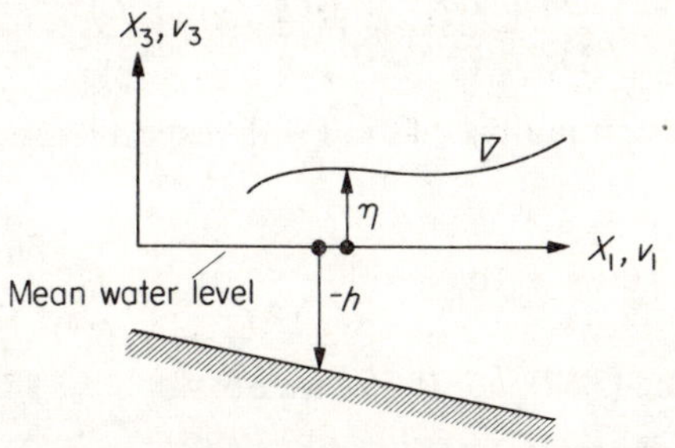

Figure 5.3 Harmonic response—shallow waters

Figure 5.3 defines the notation for this case. We take $\eta(x_1 x_2 t)$ as the elevation above sea level (hydrostatic position) and work with vertically integrated flow measures q_1 and q_2 which are functions of x_1, x_2 and t.

$$q_1 = \int_{-h}^{\eta} v_1 \, dx_3, \qquad q_2 = \int_{-h}^{\eta} v_2 \, dx_3 \tag{5.52}$$

We also assume the amplitude η is small by comparison with the initial depth h, and neglect Coriolis and horizontal friction effects. We will also accept that the pressure is equal to the hydrostatic pressure:

$$p = g\rho(\eta - x_3) \tag{5.53}$$

This equation replaces the third momentum equation, and the other two become

$$\frac{Dv_1}{Dt} = -g\frac{\partial \eta}{\partial x_1}, \qquad \frac{Dv_2}{Dt} = -g\frac{\partial \eta}{\partial x_2} \tag{5.54}$$

Continuity gives

$$\frac{\partial v_1}{\partial x_1} + \frac{\partial v_2}{\partial x_2} + \frac{\partial v_3}{\partial x_3} = 0 \tag{5.55}$$

One can now apply the 'kinematic condition' which says that the vertical velocity at $x_3 = \eta$ is simply the time variation of the wave height (Figure 5.3); i.e.

$$v_3\big|_\eta = \lim_{\Delta t \to 0} \frac{\Delta \eta}{\Delta t} = \frac{\mathrm{D}\eta}{\mathrm{D}t} \tag{5.56}$$

where

$$\frac{\mathrm{D}\eta}{\mathrm{D}t} = \frac{\partial \eta}{\partial t} + v_1\big|_\eta \frac{\partial \eta}{\partial x_1} + v_2\big|_\eta \frac{\partial \eta}{\partial x_2} \tag{5.57}$$

Integrating the continuity equation with respect to x_3 using Liebnitz' rule, i.e.

$$\frac{\partial}{\partial x_1} \int_{h_1(x_1x_2)}^{h_2(x_1x_2)} f(x_1x_2x_3)\,\mathrm{d}x_3 = \int_{h_1}^{h_2} \frac{\partial f}{\partial x_1}\,\mathrm{d}x_3 + f\bigg|_{h_2} \frac{\partial h_2}{\partial x_1} - f\bigg|_{h_1} \frac{\partial h_1}{\partial x_1}$$

and substituting (5.57) into (5.55), we obtain

$$\frac{\partial q_1}{\partial x_1} + \frac{\partial q_2}{\partial x_2} + \frac{\partial \eta}{\partial t} = 0 \tag{5.58}$$

Next one integrates the momentum equations with respect to x_3. Using Liebnitz' rule and the kinematic condition, this gives

$$\frac{\partial}{\partial t}\int v_1\,\mathrm{d}x_3 + \frac{\partial}{\partial x_1}\int v_1 v_1\,\mathrm{d}x_3 + \frac{\partial}{\partial x_2}\int v_2 v_1\,\mathrm{d}x_3 = -g(h+\eta)\frac{\partial \eta}{\partial x_1} \tag{5.59}$$

and similarly for the one in the x_2 direction.

Neglecting the convective terms in the momentum equations and remembering that the amplitude η is small by comparison with h, we can write

$$\frac{\partial q_1}{\partial t} = -gh\frac{\partial \eta}{\partial x_1}, \qquad \frac{\partial q_2}{\partial t} = -gh\frac{\partial \eta}{\partial x_2} \tag{5.60}$$

On a *land* boundary, we prescribe the integrated normal velocity:

$$q_n = \alpha_{n1} q_1 + \alpha_{n2} q_2 = \bar{q}_n \quad \text{on } S_2$$

On an *ocean* boundary, the elevation is specified:

$$\eta = \bar{\eta}(x_1 x_2 t) \quad \text{on } S_1$$

The differential equations (5.58) and (5.60) can be transformed to a single differential equation of second order by differentiating (5.60) with respect to x_1 and x_2 respectively and substituting them into the continuity equation (5.58) differentiated with respect to t. This gives

$$\frac{\partial}{\partial x_1}\left(h\frac{\partial \eta}{\partial x_1}\right) + \frac{\partial}{\partial x_2}\left(h\frac{\partial \eta}{\partial x_2}\right) - \frac{1}{g}\frac{\partial^2 \eta}{\partial t^2} = 0 \tag{5.61}$$

where

$$\begin{aligned} \eta &= \bar{\eta} \quad \text{on } S_1 \\ h\frac{\partial \eta}{\partial n} &= -\frac{\partial}{\partial t}\left(\frac{1}{g}\bar{q}_n\right) \quad \text{on } S_2 \end{aligned} \tag{5.62}$$

The normal velocity boundary condition has been replaced with the equilibrium equation for the normal direction, which is consistent as the order of the differential equation governing the problem has increased.

Resonance and harmonic response due to surge excitation can be investigated by expressing η as

$$\eta(x_1 x_2 t) = H(x_1 x_2)\,\mathrm{e}^{\mathrm{i}\omega t} \tag{5.63}$$

where ω is the circular frequency. Then (5.61) becomes

$$\frac{\partial}{\partial x_1}\left(h\frac{\partial H}{\partial x_1}\right) + \frac{\partial}{\partial x_2}\left(h\frac{\partial H}{\partial x_2}\right) + \frac{\omega^2}{g}H = 0 \tag{5.64}$$

For surge excitation,

$$\begin{aligned} H &= \bar{H} \quad \text{on } S_1 \\ \frac{\partial H}{\partial n} &= f \quad \text{on } S_2 \end{aligned} \tag{5.65}$$

For harmonic resonance, we prescribe different sets of homogeneous boundary conditions:

$$H = 0 \quad \text{on } S_1, \qquad \frac{\partial H}{\partial n} = 0 \quad \text{on } S_2 \tag{5.66}$$

and determine the frequencies and modal shapes.

Equations (5.54) become, for harmonic motion,

$$\frac{\partial^2 u_1}{\partial t^2} \approx -g\frac{\partial H}{\partial x_1}e^{i\omega t}$$
$$\frac{\partial^2 u_2}{\partial t^2} \approx -g\frac{\partial H}{\partial x_2}e^{i\omega t} \tag{5.67}$$

where u_1 and u_2 are horizontal displacements. Equations (5.67) give, after integration over a half-period,

$$|u_1| = \frac{g}{\omega^2}\frac{\partial H}{\partial x_1} \tag{5.68}$$

$$|u_2| = \frac{g}{\omega^2}\frac{\partial H}{\partial x_2} \tag{5.69}$$

which are the maximum horizontal displacements.

FINITE ELEMENT FORMULATION

From (5.64) and the boundary conditions, we can write the following weighted residual expression:

$$\iint\left[\frac{\partial}{\partial x_1}\left(h\frac{\partial H}{\partial x_1}\right) + \frac{\partial}{\partial x_2}\left(h\frac{\partial H}{\partial x_2}\right) + \frac{\omega^2}{g}H\right]\delta H\,\mathrm{d}x_1\,\mathrm{d}x_2 = \int_{S_2}\left(h\frac{\partial H}{\partial n} - f\right)\delta H\,\mathrm{d}S \tag{5.70}$$

Integrating by parts, we have

$$\iint\left(h\frac{\partial H}{\partial x_1}\frac{\partial \delta H}{\partial x_1} + h\frac{\partial H}{\partial x_2}\frac{\partial \delta H}{\partial x_2} - \frac{\omega^2}{g}H\,\delta H\right)\mathrm{d}x_1\,\mathrm{d}x_2 = \int_{S_2} f\,\delta H\,\mathrm{d}S \tag{5.71}$$

If $f = 0$, the right-hand side of equation (5.71) disappears. Assume that the H variable can be approximated on each element by

$$H = \boldsymbol{\phi}^T\mathbf{H}^n \tag{5.72}$$

where $\boldsymbol{\phi}$ is an interpolation function; $\mathbf{H}^n$ are the nodal unknowns; for an element, we have

$$\delta\mathbf{H}^{n,T}\iint\left[h(\boldsymbol{\phi}_{,1}\boldsymbol{\phi}_{,1}^T + \boldsymbol{\phi}_{,2}\boldsymbol{\phi}_{,2}^T) - \frac{\omega^2}{g}\boldsymbol{\phi}\boldsymbol{\phi}^T\right]\mathrm{d}x_1\,\mathrm{d}x_2\,\mathbf{H}^n = \delta\mathbf{H}^{n,T}\int\boldsymbol{\phi}f\,\mathrm{d}S \tag{5.73}$$

We can write (5.73) as

$$\mathbf{K}\mathbf{H}^n - \omega^2\mathbf{M}\mathbf{H}^n = \mathbf{P} \tag{5.74}$$

where

$$\mathbf{K} = \iint h(\boldsymbol{\phi}_{,1}\boldsymbol{\phi}_{,1}^T + \boldsymbol{\phi}_{,2}\boldsymbol{\phi}_{,2}^T)\,\mathrm{d}x_1\,\mathrm{d}x_2$$

$$\mathbf{M} = \frac{1}{g}\iint \boldsymbol{\phi}\boldsymbol{\phi}^T\,\mathrm{d}x_1\,\mathrm{d}x_2$$

$$\mathbf{P} = \int \boldsymbol{\phi} f\,\mathrm{d}S$$

which are similar to matrices (5.36). For the whole continuum we have

$$\mathscr{K}\mathscr{H} - \omega^2\mathscr{M}\mathscr{H} = \mathscr{P} \tag{5.75}$$

where $\mathscr{K}$ and $\mathscr{M}$ are both symmetric.

Example 5.2

If we know the frequency of the waves and the elevation $\bar{\mathscr{H}}$ at a given (ocean type) boundary, equation (5.75) can be written

$$(\mathscr{K} - \omega^2\mathscr{M})\mathscr{H} = \bar{\mathscr{P}} \tag{a}$$

where the elements of the $\bar{\mathscr{P}}$ vector are due to multiplying the known $\bar{\mathscr{H}}$ elevations by the elements of $\mathscr{K}$ and $\omega^2\mathscr{M}$ and passing the result to the right-hand side. One can now solve (a) and obtain a series, the values for the elevations.

As an illustration the case of the Duncan basin built during the Second World War in Table Bay Harbour, South Africa was analysed. This basin has been extensively studied[1] as the features of the Bay are such that some of the seiche frequencies are greatly amplified (Figure 5.4). This fact has been demonstrated by model experiment, harmonic analyses of seichograms and simple theoretical analysis, which can give reasonable results here as the shape of the basin is approximately rectangular.

The experimental values for amplitudes at berth E and O are depicted in Figure 5.4(b). Note that the first significant period is around $T = 11{\cdot}45$ minutes (theoretical value) and that this period clearly shows in the experimental curve, although it is very much damped. This damping is expected because the period $T = 11{\cdot}45$ minutes corresponds to the water flowing freely in and out of the basin which in practice does not happen.

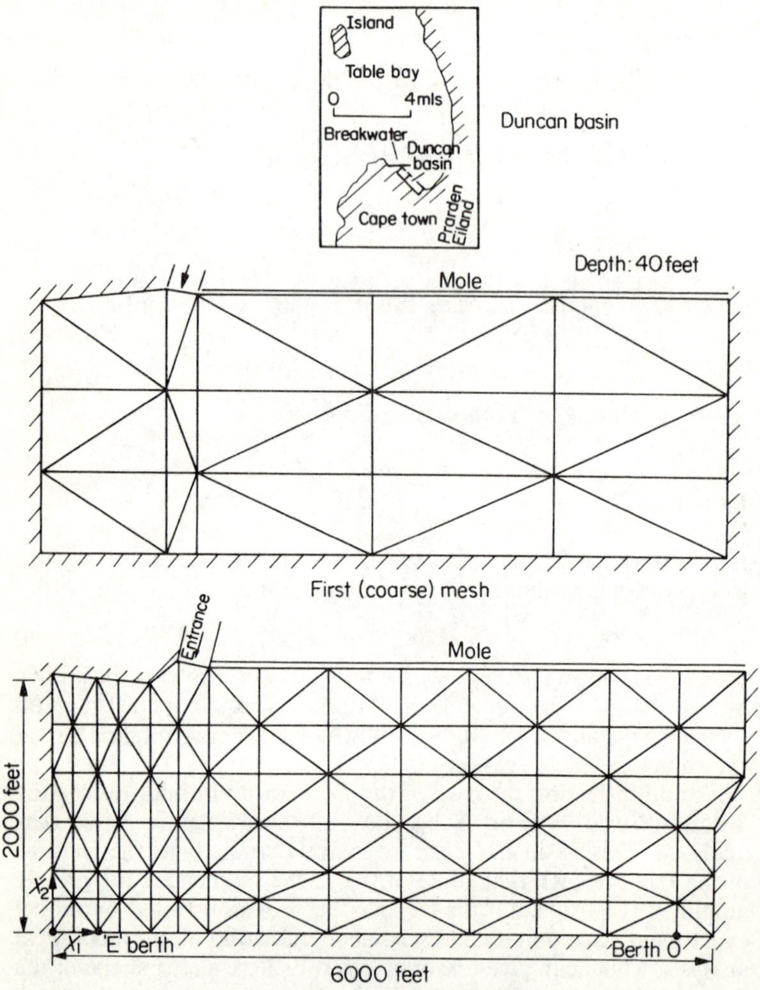

Figure 5.4(a)   Harmonic response analysis of Duncan basin

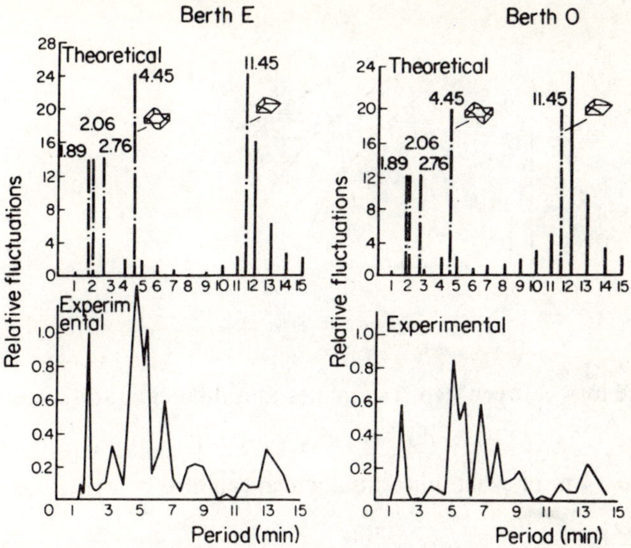

Figure 5.4(b) Harmonic response analysis of Duncan basin

The finite element analysis was carried out by dividing the basin into 168 six node triangular elements, which give 377 nodal unknowns (second mesh). Periods T of 1, 2, 3, ... etc. minutes were chosen ($\omega = 2\pi/T$ = circular frequency) and an elevation $\overline{H} = 1$ assumed for the nodes at the entrance. The resulting elevations at berths E and O are shown in Figure 5.4(b) as solid lines.

Finally, the eigenvalues of the system were found employing the first mesh in order to use less computing time. These values were compared against those obtained by solving the equations for $T = 1, 2, 3, \ldots$ etc. minutes, and good agreement was found (see Figure 5.4(b), broken lines).

5.5 STREAM FUNCTION FORMULATION

When we assume that the flow is laminar the particles follow a series of lines called streamlines, and their velocities are tangential to them (Figure 5.5).* Here we consider only the time independent two-dimensional incompressible case.

* For steady flow these streamlines do not change with time.

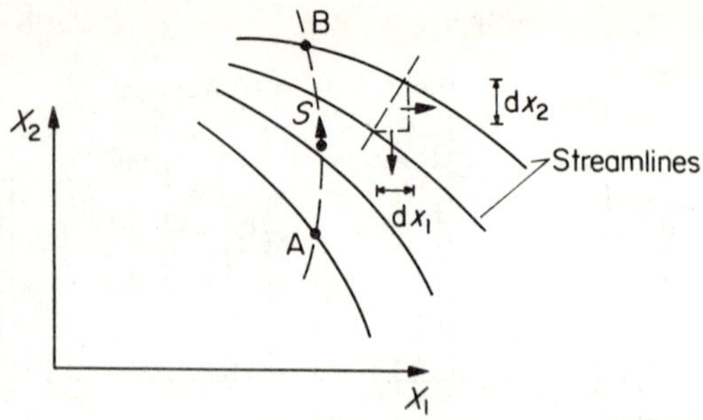

Figure 5.5 Streamlines

The loss between two streamlines at a differential distance is

$$d\psi = v_1\,dx_2 - v_2\,dx_1 \tag{5.76}$$

where $d\psi$ is an exact differential which can also be written as

$$d\psi = \frac{\partial\psi}{\partial x_2}\,dx_2 + \frac{\partial\psi}{\partial x_1}\,dx_1 \tag{5.77}$$

Thus

$$v_1 = \frac{\partial\psi}{\partial x_2}, \qquad v_2 = -\frac{\partial\psi}{\partial x_1} \tag{5.78}$$

ψ is called the stream function and is such that the incompressibility condition is identically satisfied. For a streamline, $d\psi = 0$, hence we have ψ = constant. Consider now a series of streamlines, as shown in Figure 5.5, and a curve AB which intercepts them. In what follows we assume unit thickness. The flux through AB is given by

$$Q = \int_A^B v_n\,dS = \int_A^B (v_1\alpha_{n1} + v_2\alpha_{n2})\,dS \tag{5.79}$$

where α_{n1} and α_{n2} are the direction cosines of the normal to AB with respect to axes x_1 and x_2. Thus

$$\alpha_{n1} = \frac{dx_2}{dS}, \qquad \alpha_{n2} = -\frac{dx_1}{dS} \tag{5.80}$$

Hence equation (5.79) becomes

$$Q = \int_A^B (v_1\,dx_2 - v_2\,dx_1)$$

Substituting (5.78) we have

$$Q = \int_A^B \left(\frac{\partial \psi}{\partial x_2} \mathrm{d}x_2 + \frac{\partial \psi}{\partial x_1} \mathrm{d}x_1 \right) = \int_A^B \mathrm{d}\psi$$
$$Q = \psi(B) - \psi(A) \tag{5.81}$$

The total flux is then equal to the difference between the stream functions from A to B.

Note that by definition streamlines and equipotentials are orthogonal. To show this, consider an increment of ψ and Φ functions,

$$\mathrm{d}\psi = \frac{\partial \psi}{\partial x_1} \mathrm{d}x_1 + \frac{\partial \psi}{\partial x_2} \mathrm{d}x_2$$
$$\mathrm{d}\Phi = \frac{\partial \Phi}{\partial x_1} \mathrm{d}x_1 + \frac{\partial \Phi}{\partial x_2} \mathrm{d}x_2 \tag{5.82}$$

We can then define two vectors:

$$\vec{e}_\psi = \left(\frac{\partial \psi}{\partial x_1}, \frac{\partial \psi}{\partial x_2} \right), \qquad \vec{e}_\Phi = \left(\frac{\partial \Phi}{\partial x_1}, \frac{\partial \Phi}{\partial x_2} \right) \tag{5.83}$$

which using equations (5.32) and (5.78) can be proved to be orthogonal; i.e.

$$\vec{e}_\psi \cdot \vec{e}_\Phi = 0 \tag{5.84}$$

Hence streamlines and equipotentials are orthogonal.

VARIATIONAL FORMULATION

For the case of *incompressible* flow we have

$$\vec{\nabla} \cdot \vec{v} = 0 \tag{5.85}$$

and for *irrotational* flow

$$\vec{\nabla} \times \vec{v} = \vec{0} \tag{5.86}$$

We also know that the velocities can be expressed as a function of a potential function Φ

$$\vec{v} = \vec{\nabla} \Phi \tag{5.87}$$

The continuity equation becomes

$$\nabla^2 \Phi = 0 \tag{5.88}$$

plus boundary conditions $\Phi = \bar{\Phi}$ on S_1 and $\bar{v}_n = \partial \Phi / \partial n$ on S_2. We are interested in finding Φ and the velocities. Afterwards the pressure can be found through Bernoulli's equation.

We can write the following variational statement:

$$\int (\nabla^2 \Phi)\, \delta\Phi \, \mathrm{d}V = \int_{S_2} \left(\frac{\partial \Phi}{\partial n} - \bar{v}_n \right) \delta\Phi \, \mathrm{d}S \tag{5.89}$$

where the Φ functions are assumed to satisfy the boundary conditions on S_1.

Integrating by parts we have

$$-\int \vec{\nabla}\Phi \,.\, \vec{\nabla}\, \delta\Phi \, \mathrm{d}V + \int_{S_2} \bar{v}_n \, \delta\Phi \, \mathrm{d}S = 0 \tag{5.90}$$

The functional corresponding to (5.90) can be written

$$F(\Phi) = \tfrac{1}{2} \int \vec{\nabla}\Phi \,.\, \vec{\nabla}\Phi \, \mathrm{d}V - \int_{S_2} v_n \Phi \, \mathrm{d}S \tag{5.91}$$

Note that the first term in equation (5.91) is the kinetic energy of the system divided by the mass density. The second term can be related to the work done by the impulsive pressure which starts the system from rest.

Another approach used in two-dimensional problems consists in working in terms of the stream function ψ.

Now the incompressibility condition is identically satisfied, and the irrotationality equation gives

$$\nabla^2 \psi = 0 \tag{5.92}$$

plus $\psi = \bar{\psi}$ on S_1 and $\bar{v}_s = -\partial\psi/\partial n$ on S_2. We can propose the weighted expression

$$\int (\nabla^2 \psi)\, \delta\psi \, \mathrm{d}A = \int_{S_2} \left(\frac{\partial \psi}{\partial n} + \bar{v}_s \right) \delta\psi \, \mathrm{d}S \tag{5.93}$$

Integrating by parts, we obtain

$$-\int \vec{\nabla}\psi \,.\, \vec{\nabla}\, \delta\psi \, \mathrm{d}A = \int_{S_2} \bar{v}_s \, \delta\psi \, \mathrm{d}S \tag{5.94}$$

which gives the functional

$$F(\psi) = \tfrac{1}{2} \int \vec{\nabla}\psi \,.\, \vec{\nabla}\psi \, \mathrm{d}A + \int \bar{v}_s \psi \, \mathrm{d}S$$

Example 5.3

Let us consider the case of *confined* flow around a cylinder (Figure 5.6). We assume that the flow is laminar upstream and downstream.

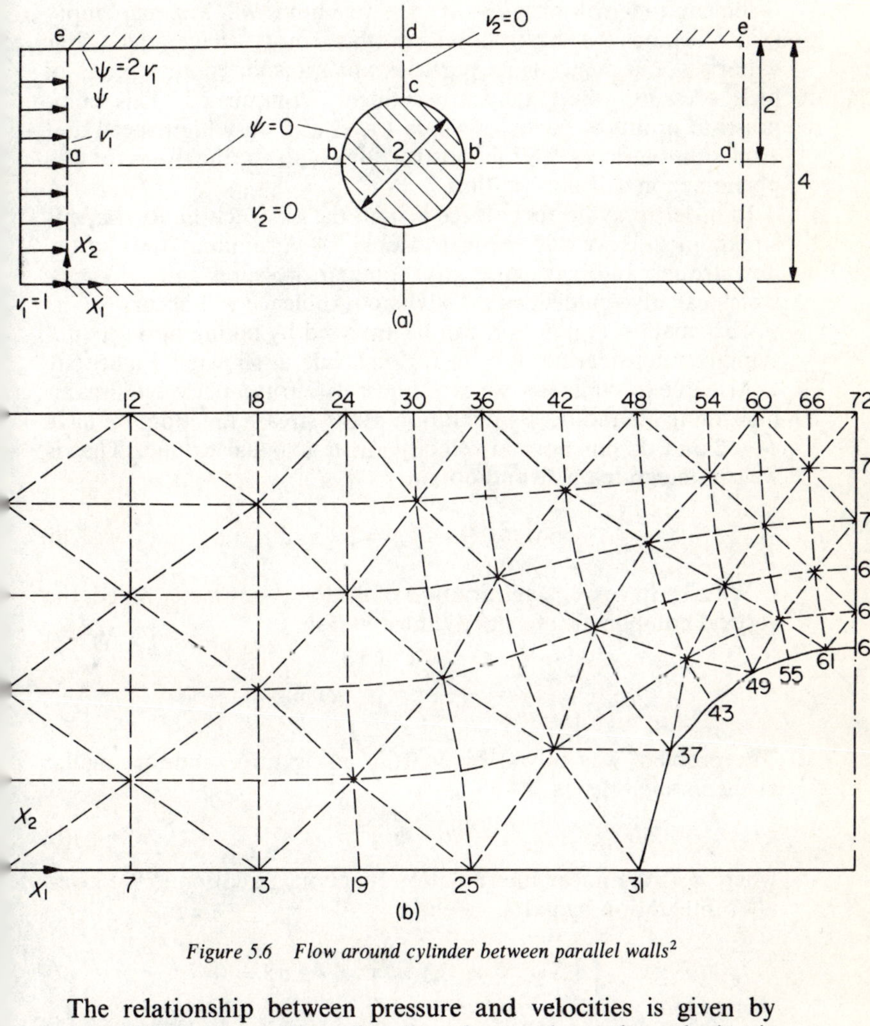

Figure 5.6 Flow around cylinder between parallel walls[2]

The relationship between pressure and velocities is given by Bernoulli's formula. Thus, for the places where the velocity is large, the pressure will be small, and vice versa. The fluid particles will tend to travel from regions of larger pressures to those with smaller ones. Consider the points b, c and b′. The fluid goes from b to c and c to b′. At b, the velocities are small (pressure large); at c, the velocities are large (pressure small) and at b′, the velocities are small again (pressure large). Then aside from the main current the liquid will tend to travel from b′ to c and because of this the turbulence behind the object increases. If the velocity increases the turbulent zone will progress as the streamlines will leave an empty space behind the obstruction. In this empty space, the water vaporises and some of its particles are transported to regions of high pressure where they are suddenly compressed. This compression produces particles which travel at a very high speed and erode the surfaces of the obstruction. This is the basis of the phenomenon called cavitation.

In order to avoid turbulence behind the obstruction, its shape is streamlined as we will see in Example 5.4. A laminar study of the flow around an obstruction gives laminar pressures and velocities, which can give guidelines as to where turbulence will occur.

The smallest region that can be analysed by taking into account symmetry considerations is the region abcde, as shown in Figure 5.6.

Also due to symmetry we can choose $\psi = 0$ as a reference streamline, along ab and bc. By definition of the stream function, we have $\psi = 2$ and dc (for normal velocity v_1 on ae equal to one). That is we can integrate on ae and obtain

$$\psi_e = Q - \psi_a = 2v_1 = 2 \tag{a}$$

We also have a linear variation of ψ along ae. Finally on cd, the vertical component of velocity must vanish

$$v_2 = -\frac{\partial \psi}{\partial x_1} = 0, \quad \text{on } S_2$$

The problem was solved by Martin[2] using three node triangular elements such that

$$\psi = \boldsymbol{\phi}^T \boldsymbol{\psi}^n \tag{b}$$

where $\boldsymbol{\phi}$ is the linear interpolation function. Equation (5.93) gives, after integration by parts,

$$\iint \vec{\nabla}\psi \,.\, \vec{\nabla}\,\delta\psi \,\mathrm{d}A = \int_c^d \bar{v}_s\,\delta\psi\,\mathrm{d}S = 0 \tag{c}$$

Substituting (b) into (c) we have

$$\delta\boldsymbol{\psi}^{n,T}\iint \boldsymbol{\phi}\boldsymbol{\phi}^T\,\mathrm{d}A\,\boldsymbol{\psi}^n = \delta\boldsymbol{\psi}^{n,T}\int_c^d \boldsymbol{\phi}\bar{v}_s\,\mathrm{d}S \tag{d}$$

or

$$\mathbf{K}\boldsymbol{\psi}^n = \mathbf{P} \tag{e}$$

For the whole continuum we have, after applying boundary conditions $\psi_i = \bar{\psi}_i$

$$\mathscr{K}\boldsymbol{\psi} = \mathscr{P} + \bar{\mathscr{P}} \tag{f}$$

Note that $\bar{\mathscr{P}}$ is due to the known values of stream function S_1. This system of equations can be solved and the nodal values of the stream functions found.

Martin[2] has solved the example shown in Figure 5.6, but using, instead of the abcde domain, the half-symmetry region abcb'a'e'e, on which the ψ values are known at every point; hence the only boundaries are of type S_1. This is called a Dirichlet type problem. His results are shown in Table 5.1.

Table 5.1

Node	*X_2 coordinate*	ψ
67	1·00	0
68	1·16	0·3419
69	1·33	0·6880
70	1·54	1·1072
71	1·76	1·5364
72	2·00	2·0000

The velocities at the vertical centre line can then be calculated. They are plotted in Figure 5.7.

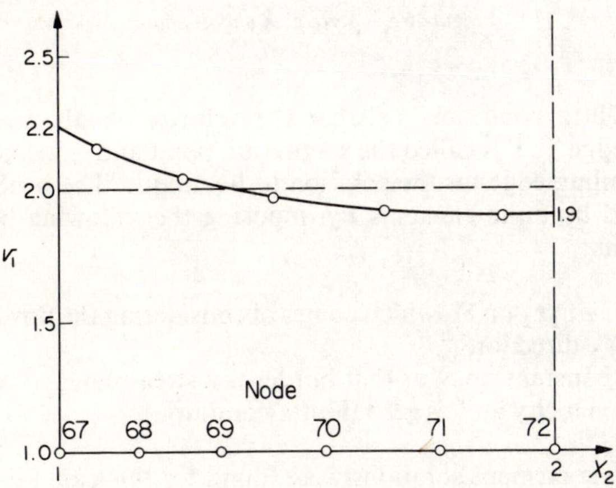

Figure 5.7 Velocity at vertical centreline (c–d)[2]

Martin also proposed applying exactly the $\partial\psi/\partial n$ condition on S_2, which is called a Neumann type condition, as a constaint equation relating the nodal unknowns. We usually prefer to apply them in a weighted residual fashion, as indicated in equation (5.93). In this way the conditions are not generally identically satisfied but are more easily applied.

Example 5.4

This example will show how finite elements can simplify to a great extent the solution of a complex problem.

Let us consider an aerofoil in a uniform flow shown in Figure 5.8 as solved by de Vries and Norrie.[3] This problem requires satisfaction

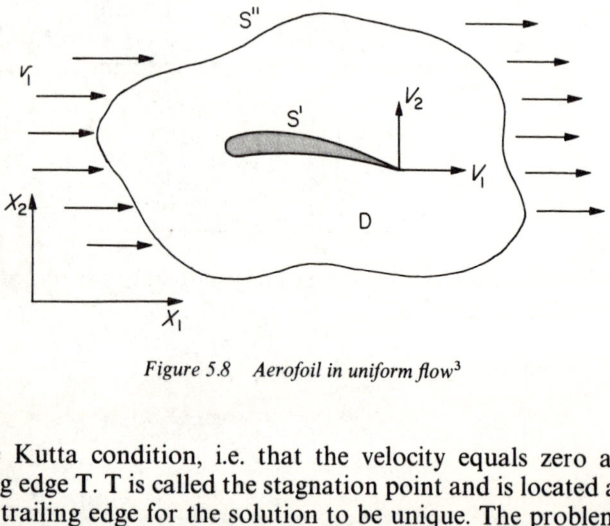

Figure 5.8 Aerofoil in uniform flow[3]

of the Kutta condition, i.e. that the velocity equals zero at the trailing edge T. T is called the stagnation point and is located at the sharp trailing edge for the solution to be unique. The problem can be solved by finite elements by imposing the following boundary conditions:

1 $\psi = -v_1 x_2$ on S'', which comes of considering the flow uniform on X_1 direction
2 ψ is constant on S' as that border is a streamline
3 The velocity at T is zero (Kutta condition)

Three finite element solutions were found for this case (all of them using three node elements).

1 $\nabla^2\psi_1 = 0$ in the domain with boundary conditions

$$\psi_1 = 0 \quad \text{on } S'$$

$$\psi_1 = -v_1 x_2 \quad \text{on } S''$$

2 $\nabla^2\psi_2 = 0$ in the domain with boundary conditions

$$\psi_2 = 0 \quad \text{on } S'$$

$$\psi_2 = 1 \quad \text{on } S''$$

3 $\nabla^2\psi_3 = 0$ in the domain with boundary conditions

$$\psi_3 = 1 \quad \text{on } S'$$

$$\psi_3 = 0 \quad \text{on } S''$$

The final solution for ψ will be given by the following sum:

$$\psi = \psi_1 + \alpha\psi_2 + \beta\psi_3 \tag{a}$$

where the parameters α and β may now be determined by imposing the Kutta condition $v_1 = v_2 = 0$ at T. Note that v_1 and v_2 are lineal functions of α and β.

Figure 5.9 shows the flow pattern around an NACA 4412 aerofoil in a uniform flow field with an 8° angle calculated in the way described above.

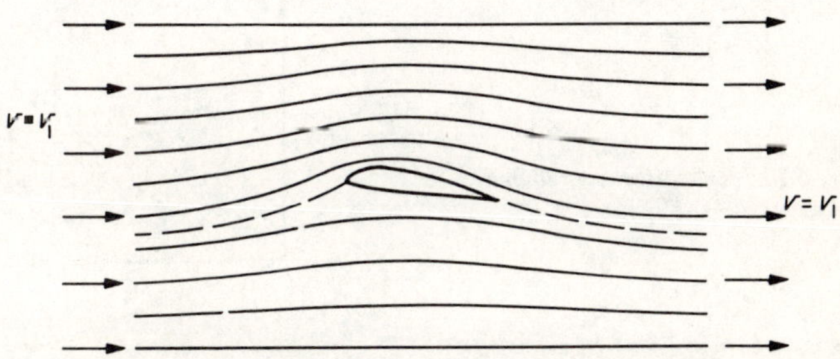

Figure 5.9 Stream function lines for an NACA 4412 aerofoil (Kutta condition prescribed)[3]

A 'static' check of the solution may be obtained by evaluating the *circulation* which is the integral of the tangent velocity around a closed curve, i.e.

$$\oint v_s \, dS \tag{b}$$

Thus, applying Gauss's theorem,

$$\oint v_s \, \mathrm{d}S = \oint -\frac{\partial \psi}{\partial n} \mathrm{d}S = -\oint \left(\frac{\partial \psi}{\partial x_1} \alpha_{n1} + \frac{\partial \psi}{\partial x_2} \alpha_{n2} \right) \mathrm{d}S$$

$$= \iint \left(\frac{\partial^2 \psi}{\partial x_1^2} + \frac{\partial^2 \psi}{\partial x_2^2} \right) \mathrm{d}A$$

$$= -\iint \left(\frac{\partial v_2}{\partial x_1} - \frac{\partial v_1}{\partial x_2} \right) \mathrm{d}A$$

$$= -\iint 2\omega \, \mathrm{d}A \qquad \text{(c)}$$

Equation (c) shows that, if the flow is irrotational, the circulation is zero around the closed curve bounding a surface within the fluid.

5.6 CYLINDRICAL COORDINATES

A body whose shape is generated by rotating a plane area through a full revolution around an axis lying in the plane of the area (see Figure 5.10) is called a solid of revolution. It is convenient to work with a cylindrical system of coordinates, the polar angle θ and two coordinates in the r, z plane, as shown in the figure.

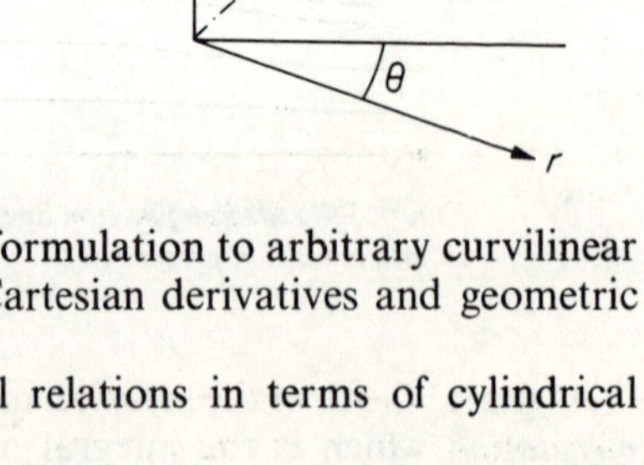

Figure 5.10 Cylindrical coordinates

One can extend the previous formulation to arbitrary curvilinear coordinates by modifying the Cartesian derivatives and geometric parameters.

We will list the fundamental relations in terms of cylindrical coordinates.

Velocities in terms of potential

$$v_r = \frac{\partial \Phi}{\partial r}, \qquad v_\theta = \frac{1}{r}\frac{\partial \Phi}{\partial \theta}, \qquad v_z = \frac{\partial \Phi}{\partial z} \qquad (5.95)$$

The *gradient* of the scalar potential is

$$\vec{\nabla}\Phi = \left(\frac{\partial \Phi}{\partial r}, \frac{1}{r}\frac{\partial \Phi}{\partial \theta}, \frac{\partial \Phi}{\partial z}\right) \tag{5.96}$$

The *divergence* of $\vec{v}$ becomes

$$\vec{\nabla} \cdot \vec{v} = \frac{1}{r}\frac{\partial}{\partial r}(rv_r) + \frac{1}{r}\frac{\partial v_\theta}{\partial \theta} + \frac{\partial v_z}{\partial z} \tag{5.97}$$

The *curl* is given by the following cross product:

$$\vec{\nabla} \times \vec{v} = \left\{\frac{1}{r}\frac{\partial v_z}{\partial \theta} - \frac{\partial v_\theta}{\partial z}, \frac{\partial v_r}{\partial z} - \frac{\partial v_z}{\partial r}, \frac{1}{r}\left(\frac{\partial (rv_\theta)}{\partial r} - \frac{\partial v_r}{\partial \theta}\right)\right\}$$

Finally the *Laplacian* operator can be obtained by evaluating $\vec{\nabla} \cdot \vec{\nabla}$ and is equal to

$$\nabla^2\Phi = \frac{1}{r}\frac{\partial}{\partial r}\left(r\frac{\partial \Phi}{\partial r}\right) + \frac{1}{r^2}\frac{\partial^2 \Phi}{\partial \theta^2} + \frac{\partial^2 \Phi}{\partial z^2} \tag{5.98}$$

For the two-dimensional case we can define a stream function ψ such that

$$v_r = \frac{1}{r}\frac{\partial \psi}{\partial \theta}, \qquad v_\theta = -\frac{\partial \psi}{\partial r} \tag{5.99}$$

The Laplacian for this case is

$$\frac{1}{r}\frac{\partial}{\partial r}\left(r\frac{\partial \psi}{\partial r}\right) + \frac{1}{r^2}\frac{\partial^2 \psi}{\partial \theta^2} = 0 \tag{5.100}$$

A possible finite element mesh in cylindrical coordinates is shown in Figure 5.11, where the Φ is a function of (r, θ, z). In many cases a

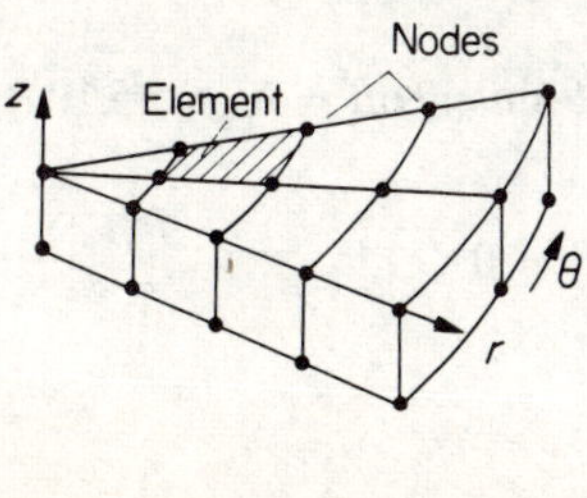

Figure 5.11 Cylindrical mesh (without z variation)

three-dimensional cylindrical problem may be studied as a series of axisymmetric ones if the potential can be decomposed in a Fourier series such as

$$\Phi(r, z, \theta) = \Phi_0(r, z) + \sum_{k=1}^{n} \Phi_k^a(r, z) \sin k\pi\theta + \sum_{k=1}^{n} \Phi_k^s(r, z) \cos k\pi\theta \tag{5.101}$$

where s stands for 'symmetric', and a for 'axisymmetric' terms. The possibility of writing (5.101) will depend on one being able to express the applied boundary potentials and flux in the same way; e.g. for the applied flux only

$$f(r, z, \theta) = f_0(r, z) + \sum_{k=1}^{n} f_k^a(r, z) \sin k\pi\theta + \sum_{k=1}^{n} f_k^s(r, z) \cos k\pi\theta \tag{5.102}$$

We can write the following weighted residual expression, after integration by parts:

$$\int_z \int_r \left(\int_0^{2\pi} \vec{\nabla}\Phi \, . \, \vec{\nabla}\, \delta\Phi \, \mathrm{d}\theta \right) r \, \mathrm{d}r \, \mathrm{d}z = \int_{S_2} \left(\int_0^{2\pi} f \, \delta\Phi \, \mathrm{d}\theta \right) r \, \mathrm{d}S \tag{5.103}$$

where S is the arc length of the boundary curve in the r, z plane. We can expand (5.103) to

$$\int_z \int_r \int_0^{2\pi} \left(\frac{\partial \Phi}{\partial r} \frac{\partial \delta\Phi}{\partial r} + \frac{1}{r^2} \frac{\partial \Phi}{\partial \theta} \frac{\partial \delta\Phi}{\partial \theta} + \frac{\partial \Phi}{\partial z} \frac{\partial \delta\Phi}{\partial z} \right) r \, \mathrm{d}\theta \, \mathrm{d}r \, \mathrm{d}z$$

$$= \int_{S_2} \left(\int_0^{2\pi} f \, \delta\Phi \, \mathrm{d}\theta \right) r \, \mathrm{d}S$$

One can now substitute (5.101) and (5.102) into this last expression and integrate with respect to θ taking into account the following orthogonality conditions:

$$\int_0^{2\pi} \sin n\theta \, \mathrm{d}\theta = \int_0^{2\pi} \cos n\theta \, \mathrm{d}\theta = 0$$

$$\int_0^{2\pi} \cos n\theta \cos m\theta \, \mathrm{d}\theta = \int_0^{2\pi} \sin n\theta \sin m\theta \, \mathrm{d}\theta = \delta_{mn}\pi \tag{5.104}$$

$$\int_0^{2\pi} \sin n\theta \cos m\theta \, \mathrm{d}\theta = 0$$

(δ_{mn} is the Kronecker delta.)

We will finally obtain a system of uncoupled equations of the form

$$\iint \left\{ \frac{\partial \Phi_0}{\partial r} \frac{\partial \delta\Phi_0}{\partial r} + \frac{\partial \Phi_0}{\partial z} \frac{\partial \delta\Phi_0}{\partial z} \right\} r \, \mathrm{d}r \, \mathrm{d}z = \int f_0 \, \delta\Phi_0 \, \mathrm{d}S \tag{5.105}$$

$$\iint \left\{ \left(\frac{\partial \Phi_k^s}{\partial r}\right)\left(\frac{\partial \delta\Phi_k^s}{\partial r}\right) + \frac{(k\pi)^2}{r^2}(\Phi_k^s \, \delta\Phi_k^s) + \left(\frac{\partial \Phi_k^s}{\partial z}\right)\left(\frac{\partial \delta\Phi_k^s}{\partial z}\right) \right\} r \, \mathrm{d}r \, \mathrm{d}z = \int f_k^s(\delta\Phi_k^s) \, \mathrm{d}S \tag{5.106}$$

$$\iint \left\{ \left(\frac{\partial \Phi_k^a}{\partial r}\right)\left(\frac{\partial \delta\Phi_k^a}{\partial r}\right) + \frac{(k\pi)^2}{r^2}(\Phi_k^a \, \delta\Phi_k^a) + \left(\frac{\partial \Phi_k^a}{\partial z}\right)\left(\frac{\partial \delta\Phi_k^a}{\partial z}\right) \right\} r \, \mathrm{d}r \, \mathrm{d}z = \int f_k^a(\delta\Phi_k^a) \, \mathrm{d}S \tag{5.107}$$

The remaining steps are the same as for $k = 0$, i.e. the discretisation and expansions apply for all the modes. We evaluate the symmetric and antisymmetric element matrices, assemble the system equations and solve for the nodal values. This is repeated for successive values of k until convergence is obtained.

We have, from (5.105),

$$\mathscr{K}^0 \mathbf{\Phi}^0 = \mathscr{P}^0 \tag{5.108}$$

and for the symmetric and antisymmetric nodes ($k = 1, 2 \ldots$)

$$\begin{aligned} \mathscr{K}_k^s \mathbf{\Phi}_k^s &= \mathscr{P}_k^s \\ \mathscr{K}_k^a \mathbf{\Phi}_k^a &= \mathscr{P}_k^a \end{aligned} \tag{5.109}$$

Note that the symmetric and antisymmetric forms of $\mathscr{K}_k$ are identical for the same value of k.

Finally a small but practical observation; when one is integrating to obtain the element matrices it is generally convenient to work with the average value of r. This is done in order to avoid singularities when the nodes are on the Z axis, i.e. when $r = 0$ (Figure 5.12).

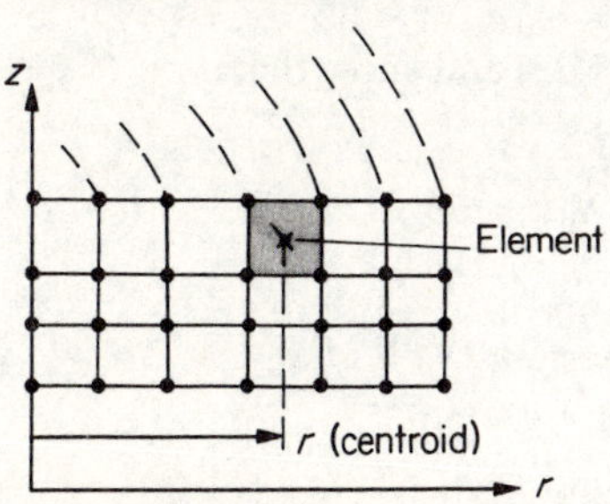

Figure 5.12 Axisymmetric case

Another way of avoiding this singularity is by using numerical integration formulae (see Appendix) with integration points only inside the element.

REFERENCES

1. WILSON, B. W., 'The Mechanism of Seiches in Table Bay Harbour, Cape Town', Proc. 4th Conf. on Coastal Engineering, Chicago, Council for Wave Research, Berkeley, California (1954).
2. MARTIN, H. C., 'Finite Element Analysis of Fluid Flows', Proc. 2nd Conf. Matrix Meth. Str. Mech., AFFDL TR 68-150, Wright-Patterson Air Force Base, Ohio, USA (1969)
3. DE VRIES, G. and NORRIE, D. G., 'The Application of The Finite Element Technique to Potential Flow Problems', *J. Appl. Mech. ASME* (Dec. 1971)

BIBLIOGRAPHY

BATEMAN, H., 'Two Dimensional Motion of a Compressible Fluid and its Associated Variational Principles', *Proc. Roy. Soc.*, **125,** 598–618 (1929)

CHAN, S. T., LAROCK, B. E. and HERRMANN, L. R., 'Free Surface Ideal Fluid Flows by Finite Elements', *J. Hyd. Div., ASCE*, **99,** HY6, June (1973)

DOCTORS, L. J., 'An Application of the Finite Element Technique to Boundary Value Problems of Potential Flows', *Int. J. Num. Meth. Eng.*, **2,** 243 (1970)

LAROCK, B. E., 'Jets from Two-dimensional Symmetric Nozzles of Arbitrary Shape', *J. Fluid Mech.*, **37**, Part 3, 479–489 (1969)

LUKE, J. C., 'A Variational Principle for a Fluid with a Free Surface', *J. Fluid Mech.*, **27,** 395–397 (1967)

MCCORQUODALE, J. A. and LI, C. Y., 'Finite Element Analysis of Sluice Gate Flow', *Trans. Engng. Inst. Canada*, **14,** No. C-2, March (1971)

VOOREN, J. V. D. and LABRUJERE, TH. E., 'Finite Element Solution of the Incompressible Flow over an Airfoil in a Nonuniform Stream', in *Numerical Methods in Fluid Dynamics*, C. Brebbia and J. J. Connor (Eds.), Pentech Press (1974).

EXERCISES

5-1 Expand the term in equation (5.23) and show that

$$\frac{D\vec{v}}{Dt} = \frac{\partial \vec{v}}{\partial t} + \tfrac{1}{2}\vec{\nabla}(\vec{v}\,.\,\vec{v}) - 2(\vec{v} \times \vec{\omega})$$

$$= \frac{\partial \vec{v}}{\partial t} + \tfrac{1}{2}\vec{\nabla}(\vec{v}\,.\,\vec{v}) = \frac{\partial \vec{v}}{\partial t} + (\vec{\nabla}\,.\,\vec{v})\,.\,\vec{v}$$

for an irrotational incompressible fluid.

5-2 Find a variational statement for the two-dimensional Laplacian operator

$$\frac{\partial^2 \Phi}{\partial x^2} + \frac{\partial^2 \Phi}{\partial y^2} = 0 \quad \text{in } A$$

plus

$$\frac{\partial \Phi}{\partial n} = g \quad \text{on } S$$

using the least squares principle. Compare it with (5.93). Comment.

5-3 Outline how to obtain a three-dimensional, ten node tetrahedral element to solve the Laplace equation using homogeneous coordinates $\xi_i (i = 1, 2, 3, 4)$. Write the corresponding matrices and indicate how the element matrices can be computed.

5-4 If the potential Φ for a two-dimensional flow is given by

$$\Phi = \alpha_1 + \alpha_2 x_1 + \alpha_3 x_2 + \alpha_4 x_1^2 + \alpha_5 x_1 x_2 + \alpha_6 x_2^2$$

Determine the expression for the stream function.

5-5 Assume two linear velocity functions on a triangular element such as

$$v_i = \xi_1 v_i^1 + \xi_2 v_i^2 + \xi_3 v_i^3 \quad \text{where } i = 1, 2$$

(the superscripts 1, 2, 3 indicate the nodes). Determine the relationship between the v_i^j values which are necessary for the flow to be incompressible.

5-6 Outline the development of a six node axisymmetric triangular element. Consider both symmetric and antisymmetric nodes.

5-7 Derive the continuity equation for a two-dimensional element in cylindrical coordinates (assuming the fluid is incompressible and there is no variation of velocities with respect to z) by equating flux 'in' to flux 'out' in Figure 5.13. Note that the continuity equation should be the same as the one we can obtain by specialising (5.97).

5-8 Assuming that the exciting wave at the entrance of a harbour can be described by a Fourier series with several components, indicate how the response of the harbour can be found.

5-9 Indicate how one could develop a quadrilateral body of revolution element employing numerical integration. Comment briefly on the extension to higher-order isoparametric elements.

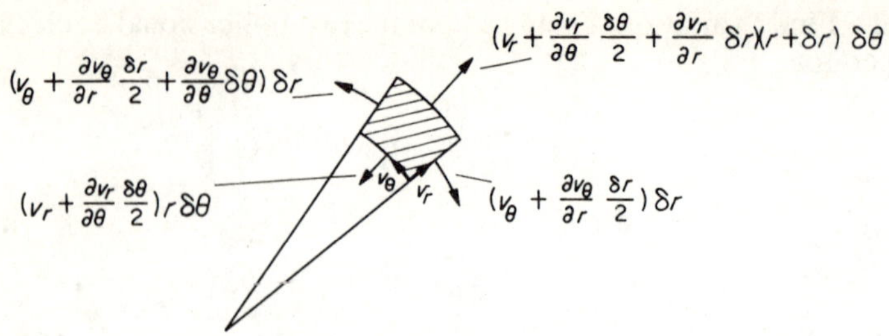

Figure 5.13 Fluid element referred to θ, r coordinates

5-10 Consider the equation

$$\mathscr{K}\mathbf{\Phi} + \mathbf{C}\dot{\mathbf{\Phi}} + \mathscr{M}\ddot{\mathbf{\Phi}} = \mathbf{P}(t) \tag{a}$$

If the matrix $\mathbf{C}$ has a particular form it is possible to solve (a) using the modal superposition method with the eigenvalues and eigenvectors corresponding to

$$(\mathscr{K} - \omega^2\mathscr{M})\mathbf{\Phi} = \mathbf{0} \tag{b}$$

Find the form of $\mathbf{C}$ that makes the modal superposition valid.

6 Flow Through Porous Media

6.1 PRINCIPLES OF GROUNDWATER FLOW

In this chapter we accept that groundwater flow only occurs in the saturated region of a porous soil, and we assume that the pores in a medium such as sand are interconnected and that the fluid can flow through them. The ratio of these voids to the total volume is called porosity and is denoted by

$$n = \frac{\text{voids volume}}{\text{total volume}} \tag{6.1}$$

The relationship governing groundwater flow is known as Darcy's law. Its form can be deduced from the general Navier–Stokes equations under the assumption that the flow is laminar and that inertial forces are negligible when compared with the viscous forces. Given the Reynolds number

$$Re = \frac{\text{inertia forces}}{\text{viscous forces}} = \frac{V d\rho}{\mu} \tag{6.2}$$

where d is the average pore diameter, μ the fluid viscosity, ρ the density and V the velocity, the law applies for $Re < 1$.

The true velocity in the pores is $V = v/n$, where n is the porosity and v an average solid-plus-fluid velocity. Consider the momentum equations for steady flow [see equation (4.50)]; i.e.

$$-\frac{1}{\rho}\frac{\partial p}{\partial x_i} + \frac{1}{\rho}\frac{\partial}{\partial x_k}\tau_{ik} + b_i = 0 \tag{6.3}$$

Assume now that the fluid moves in the pores as a Poiseuille's flow (Example 1.18). If we consider a tube of radius R, the distribution of velocity is proportional to $(R^2 - r^2)$, where r is the radial variable, and inversely proportional to viscosity μ. The friction terms in (6.3) are associated with the second-order derivatives of velocities that have different signs than the velocities themselves (i.e. friction forces act in the opposite direction to velocity).

It follows from dimensional analysis that the friction terms can be written in the following statistical average form:

For a term in the 'i' direction the friction is equal to $\Rightarrow -\frac{1}{c}\frac{v_i}{d^2}\left(\frac{\mu}{\rho n}\right)$

$$\tag{6.4}$$

where d is the average diameter of the pores, c is a dimensionless parameter and v_i is the average velocity component in the x_i direction. We have to consider only the isotropic case for simplicity.

Next we define a potential Ω for the body forces in (6.3) such that

$$b_i = \frac{\partial \Omega}{\partial x_i} \tag{6.5}$$

For instance if the only body force is the gravity acting in the opposite sense to the x_2 direction, we have

$$\Omega = -x_2 g \tag{6.6}$$

where x_2 is called the elevation and g is the acceleration due to gravity.

Equation (6.3) can now be written

$$-\frac{1}{\rho}\frac{\partial p}{\partial x_i} - \frac{\mu}{\rho n c d^2} v_i + \frac{\partial \Omega}{\partial x_i} = 0 \tag{6.7}$$

From equation (6.7) we can obtain Darcy's law:

$$v_i = -\frac{ncd^2}{\mu}\left\{\frac{\partial p}{\partial x_i} - \rho\frac{\partial \Omega}{\partial x_i}\right\} \tag{6.8}$$

or, in vector form,

$$\vec{v} = -\frac{ncd^2}{\mu}\{\vec{\nabla} p - \rho\vec{\nabla}\Omega\} \tag{6.9}$$

The term ncd^2 is a property of the soil only called the intrinsic permeability and will be represented by k, i.e.

$$\vec{v} = -\frac{k}{\mu}\{\vec{\nabla} p - \rho\vec{\nabla}\Omega\} \tag{6.10}$$

For the case of constant ρ (incompressible fluid), we can write

$$\vec{v} = -\frac{k}{\mu}\rho g\left\{\vec{\nabla}\left(\frac{p}{\rho g}\right) - \frac{1}{g}\vec{\nabla}\Omega\right\} = -\frac{k\rho g}{\mu}\vec{\nabla}\varphi \tag{6.11}$$

where φ is called the head, has dimensions of length, and is given by

$$\varphi = \frac{p}{g\rho} - \frac{\Omega}{g} \tag{6.12}$$

Darcy's law can be written in terms of head φ

$$\vec{v} = -\frac{k\gamma}{\mu}\vec{\nabla}\,.\,\varphi = -K\vec{\nabla}\varphi \tag{6.13}$$

where $\gamma = \rho g$. K is defined as the hydraulic conductivity and is a function of both the intrinsic permeability of the soil and the properties of the fluid.

$$K = -\frac{kg}{\nu} = -\frac{ncd^2}{\mu}\gamma \tag{6.14}$$

($\nu = \mu/\rho$ = dynamic viscosity.)

For the general anisotropic case, K in formula (6.13) is a 3×3 tensor, called the conductivity tensor. If we now assume that there is an orthogonal frame for which the tensor has only three non-zero coefficients on the diagonal, the gradients of φ and the corresponding velocities have the same directions, i.e. parallel to the axis of the frame. This system is called the *principal axis* of permeability. One can now write each velocity component as

$$v_i = -K_{ii}\frac{\partial\varphi}{\partial x_i} \qquad i = 1, 2, 3 \tag{6.15}$$

In general, however, the potential gradient and other seepage velocity vectors are not parallel but related through

$$v_i = -K_{ij}\frac{\partial\varphi}{\partial x_j} \qquad i = 1, 2, 3; j = 1, 2, 3 \tag{6.16}$$

If K is a constant, i.e. the medium is homogeneous, isotropic and the fluid is of constant density and viscosity, one can write (6.13) as

$$\vec{v} = \vec{\nabla}\Phi \tag{6.17}$$

where Φ is a velocity potential given by

$$\Phi = -K\varphi = -K\left(\frac{p}{\gamma} + x_2\right) \tag{6.18}$$

Note that under these conditions we can define also a stream function ψ such that

$$v_1 = -\frac{\partial \psi}{\partial x_2}, \qquad v_2 = \frac{\partial \psi}{\partial x_1} \tag{6.19}$$

The equivalence between groundwater and potential flow is then complete and one may consider the former as if it were a frictionless flow.

The continuity equation for an element of soil filled with a fluid can be written

$$\vec{\nabla} \cdot (\rho \vec{v}) = -\frac{\partial (n\rho)}{\partial t} \tag{6.20}$$

where n is the porosity. If fluid and solid are incompressible this equation gives

$$\vec{\nabla} \cdot (\vec{v}) = 0 \tag{6.21}$$

Substituting Darcy's law [equation (6.13)] we obtain

$$-K\nabla^2 \varphi = \nabla^2 \Phi = 0 \tag{6.22}$$

which is the same formula as the one governing the flow of inviscid and incompressible fluids.

6.2 CONFINED SEEPAGE PROBLEMS

We will consider first the case of two-dimensional incompressible steady flow through orthotropic soil, for which Darcy's law can be written

$$v_1 = -K_{11}\frac{\partial \varphi}{\partial x_1}, \qquad v_2 = -K_{22}\frac{\partial \varphi}{\partial x_2} \tag{6.23}$$

The continuity equation (6.21) becomes

$$\frac{\partial}{\partial x_1}\left(K_{11}\frac{\partial \varphi}{\partial x_1}\right) + \frac{\partial}{\partial x_2}\left(K_{22}\frac{\partial \varphi}{\partial x_2}\right) = 0 \tag{6.24}$$

Boundary conditions are of the type

$$\begin{aligned} &\text{(a)}\ \varphi = \bar{\varphi} \quad \text{on } S_1 \\ &\text{(b)}\ v_n = -\left(K_{11}\frac{\partial \varphi}{\partial x_1}\alpha_{n1} + K_{22}\frac{\partial \varphi}{\partial x_2}\alpha_{n2}\right) = \bar{v}_n \quad \text{on } S_2 \end{aligned} \tag{6.25}$$

where $\bar{v}_n$ is the velocity in the direction normal to the boundary. α_{n1} and α_{n2} are the direction cosines of the normal with respect to x_1 and x_2 (Figure 6.1).

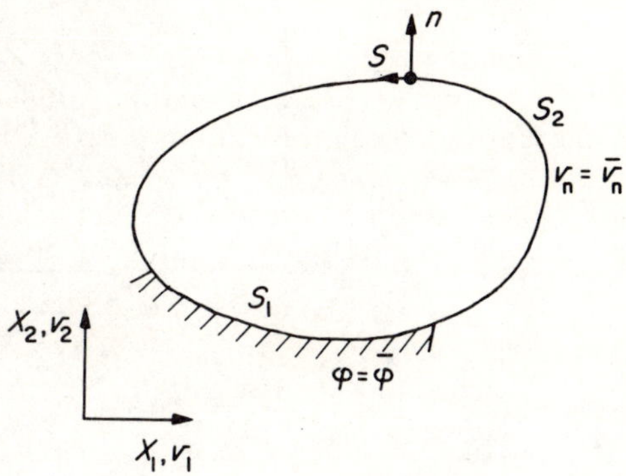

Figure 6.1 Basic definitions

From formulae (6.24) and (6.25), one can write the following statement:

$$\iint \left\{ \frac{\partial}{\partial x_1}\left(K_{11}\frac{\partial \varphi}{\partial x_1}\right) + \frac{\partial}{\partial x_2}\left(K_{22}\frac{\partial \varphi}{\partial x_2}\right)\right\} \delta\varphi \, \mathrm{d}x_1 \, \mathrm{d}x_2$$

$$= \int_{S_2} \{v_n - \bar{v}_n\} \, \delta\varphi \, \mathrm{d}S \tag{6.26}$$

Integrating by parts we obtain

$$\iint \left\{K_{11}\frac{\partial \varphi}{\partial x_1}\frac{\partial \delta\varphi}{\partial x_1} + K_{22}\frac{\partial \varphi}{\partial x_2}\frac{\partial \delta\varphi}{\partial x_2}\right\} \mathrm{d}x_1 \, \mathrm{d}x_2 - \int_{S_2} \bar{v}_n \, \delta\varphi \, \mathrm{d}S = 0 \tag{6.27}$$

The corresponding statement for the axisymmetric case (e.g. flow towards a well as in Figure 6.2) is

$$\iiint \left\{\frac{1}{r}\frac{\partial}{\partial r}\left(rK_{rr}\frac{\partial \varphi}{\partial r}\right) + \frac{\partial}{\partial z}\left(K_{zz}\frac{\partial \varphi}{\partial z}\right)\right\} \delta\varphi \, \mathrm{d}\theta \, r \, \mathrm{d}r \, \mathrm{d}z$$

$$= \int_{S_2} \{-v_n + \bar{v}_n\} \, \delta\varphi \, \mathrm{d}\theta \, \mathrm{d}S \tag{6.28}$$

where

$$v_n = K_{rr}\frac{\partial \varphi}{\partial r}\alpha_{nr} + K_{zz}\frac{\partial \varphi}{\partial z}\alpha_{nz} \tag{6.29}$$

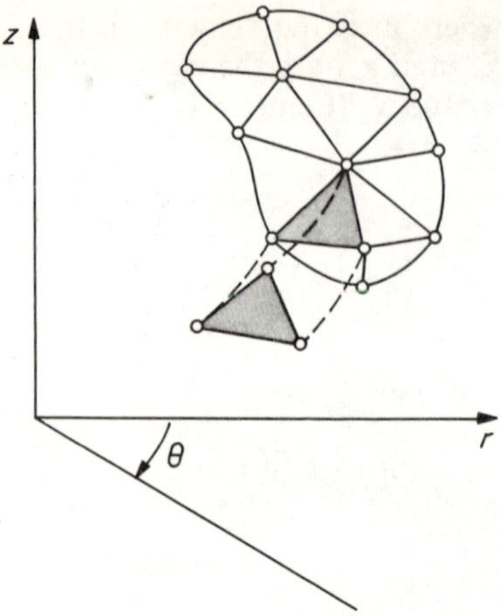

Figure 6.2 Axisymmetric case

Integrating over $\theta\,(0 < \theta < 2\pi)$, we obtain

$$2\pi \iint \left\{ \frac{\partial}{\partial r}\left(r K_{rr} \frac{\partial \varphi}{\partial r} \right) + r \frac{\partial}{\partial z}\left(K_{zz} \frac{\partial \varphi}{\partial z} \right) \right\} \delta\varphi \, \mathrm{d}r \, \mathrm{d}z$$

$$= 2\pi \int_{S_2} (-v_n + \bar{v}_n)\, \delta\varphi \, \mathrm{d}S \tag{6.30}$$

Finally we may integrate (6.30) by parts and the following variational expression results:

$$\iint r\left(K_{rr} \frac{\partial \varphi}{\partial r} \frac{\partial \delta\varphi}{\partial r} + K_{zz} \frac{\partial \varphi}{\partial z} \frac{\partial \delta\varphi}{\partial z} \right) \mathrm{d}r \, \mathrm{d}z - \int \bar{v}_n \, \delta\varphi \, \mathrm{d}S = 0 \tag{6.31}$$

The usual finite element assumption can be made such that

$$\varphi = \boldsymbol{\Phi}^T(r, z)\boldsymbol{\Phi}^n \tag{6.32}$$

The rest of the development is identical as for Laplace's equations shown in Chapter 5. We can also decompose φ in a Fourier series in θ for the more general cylindrical coordinates case.

Example 6.1

Consider a dam resting on isotropic soil as shown in Figure 6.3. The right and left hand side boundaries, as well as the rock under the soil stratum and the dam itself, are assumed to be impervious. Potential boundary conditions were applied at the nodes in contact

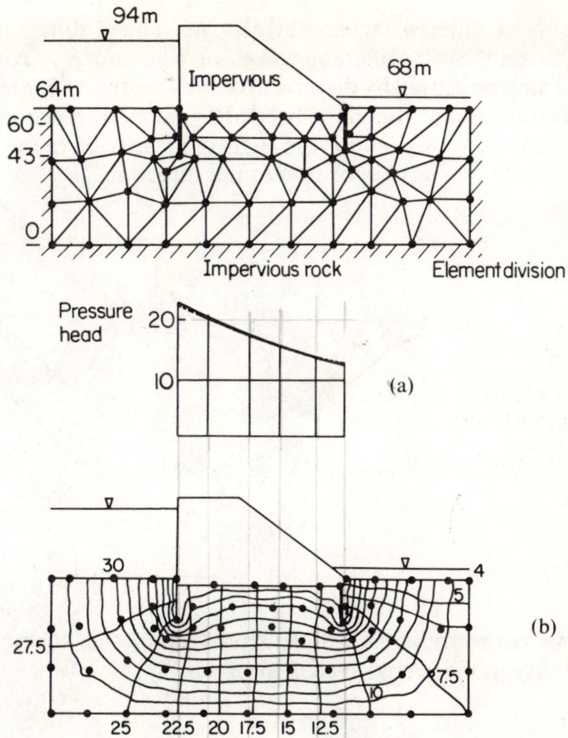

Figure 6.3 Pressure distribution under two-piled dam (three node elements). (a) *Pressures under the dam*: ------ *Lambe and Whitman*[1]; ——— *Finite element*; (b) *Flow lines and equi-potentials* (*computer drawn*)

with the water to the right ($\bar{H} = 30$ m) and to the left ($\bar{H} = 4$ m) of the dam.

The structure is assumed to have two screens at the bottom. Results for the pressure head distribution under the structure are shown in the figure and compared against those obtained by Lambe and Whitman.[1] Note that, although the finite element mesh used is rather coarse, the results are in reasonable agreement.

The problem was solved in terms of equipotentials and later on the flow lines (perpendicular to them) were drawn.

In seepage problems the number of practical cases which involve homogeneous isotropic soil is relatively small. Hence computer programs should have provision for non-homogeneous and at least orthotropic soil properties. This means that the permeabilities $K'_{11}K'_{22}$ in directions X'_1 and X'_2 respectively not only change from soil to soil of different types but also may have different values within the same soil. Different values of K'_{11} and K'_{22} within the same soil may be caused by distortion of the soil strata. The following three parameters are therefore needed to define the properties on each element: K'_{11}, K'_{22} and α, the angle of inclination of the stratum with respect to the vertical or horizontal direction.

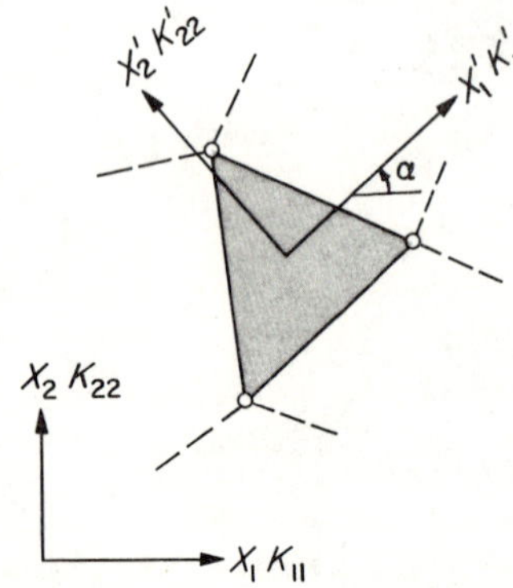

Figure 6.4 Change of coordinates for permeability coefficients

For an orthotropic two-dimensional case, the coefficients of permeability in the global system are given by,

$$\mathbf{K} = \mathbf{R}^T\mathbf{K}'\mathbf{R} \tag{a}$$

or

$$\begin{bmatrix} K_{11} & K_{12} \\ K_{21} & K_{22} \end{bmatrix} = \begin{bmatrix} \cos\alpha & \sin\alpha \\ -\sin\alpha & \cos\alpha \end{bmatrix} \begin{bmatrix} K'_{11} & 0 \\ 0 & K'_{22} \end{bmatrix} \begin{bmatrix} \cos\alpha & -\sin\alpha \\ \sin\alpha & \cos\alpha \end{bmatrix}$$

$$= \begin{bmatrix} K'_{11}\cos^2\alpha + K'_{22}\sin^2\alpha & (K'_{22} - K'_{11})\cos\alpha\sin\alpha \\ (K'_{22} - K'_{11})\cos\alpha\sin\alpha & K'_{11}\sin^2\alpha + K'_{22}\cos^2\alpha \end{bmatrix} \tag{b}$$

Note that now we have a $K_{12} = K_{21}$ permeability coefficient due to the transformation, which implies that in the global system Darcy's law becomes

$$\begin{aligned} v_1 &= -\left(K_{11}\frac{\partial\varphi}{\partial x_1} + K_{12}\frac{\partial\varphi}{\partial x_2}\right) \\ v_2 &= -\left(K_{22}\frac{\partial\varphi}{\partial x_2} + K_{12}\frac{\partial\varphi}{\partial x_1}\right) \end{aligned} \tag{c}$$

The corresponding variational principle (6.27) ought to be modified taking into account cross derivative terms. In order not to work with K_{12}, we can deduce the element matrices in the *local* (K'_{11} and K'_{22} only) system, and once computed they can be assembled with other element matrices without any rotation being needed.

6.3 PROBLEMS INVOLVING FREE SURFACES

Let us now extend the previous analysis to problems with phreatic surfaces. Here the exact position of the surface is not known and its determination becomes part of the analysis of the problem. The condition to find the free surface is simply that at any point on it, the total potential head φ is equal to the free surface *elevation* head H from a reference plane (the atmospheric pressure is taken equal to zero as it is acting everywhere).

We will now consider a general case of seepage flow as seen in Figure 6.5. In it four different types of boundaries are encountered:

1 *Impervious boundary*, such as the surface of soil strata and rocks—AF on the figure—The condition for such boundaries is $v(s) = 0$ on S_2, hence they are streamlines.
2 *Water boundary*. These constitute the upstream and downstream faces (ABC and EF) of the porous domain. Hydrostatic pressure is exerted on these boundaries and the total potential head φ along them can be taken as constant and equal to the elevation of the water face. These boundaries are equipotential lines.
3 *Line of seepage* (*or phreatic surface*). The line of seepage CD is the upper streamline of the flow domain. At every point along this line the pore pressure is atmospheric and therefore the total head equals the elevation head, i.e. $\varphi = H = x_2$ at every point on the free surface. In addition to this condition the flow across the free surface is zero, hence this boundary is a streamline.

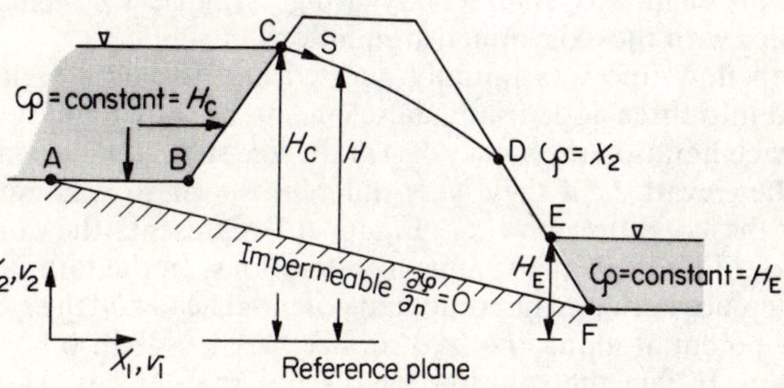

Figure 6.5 Boundaries in seepage

4 *Seepage face.* This represents a boundary where the water seeps out of the soil into air. As the pressure on this surface is both constant and atmospheric the total head equals the elevation ($\varphi = x_2$). This boundary is neither an equipotential nor a streamline.

Example 6.2

Let us consider the flow towards a well. The well has a radius of 1 m and completely penetrates the aquifer (Figure 6.6). The water level in the well is kept constant at 20 m above the impermeable base by pumping at a constant rate.

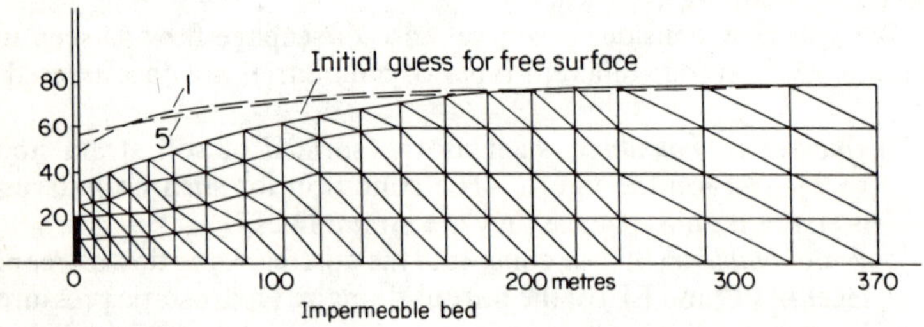

Figure 6.6 Initial element division with various top flow lines. Axisymmetric flow: 1, solution for free surface after one iteration; 5, solution after five iterations

The aquifer consists of homogeneous and isotropic soil ($K_{11} = K_{22}$, $\alpha = 0$) for simplicity. The flow through the aquifer is confined at the bottom by an impermeable bed, while the top water surface remains free. The radius of influence of the well was extended up to 370 m where the ground water level was taken at 80 m above the impervious boundary and the flow was assumed uniform and horizontal.

The problem was analysed by using cylindrical system of co-ordinates with the axisymmetric equations.

A top flow line was initially guessed and the flow region was divided into three node triangular elements, as shown in Figure 6.6. After each iteration the values of φ at the free surface were compared with the elevation; if they were different the mesh was moved to satisfy the condition $\varphi = x_2$. Figure 6.7 represents the computer network of finite elements after five iterations. In the fifth iteration the difference between the computed potential head and the elevation of any potential along the free surface was less than 0.1% of the elevation. In fact the solution for the free surface was reasonably accurate after one or two iterations.

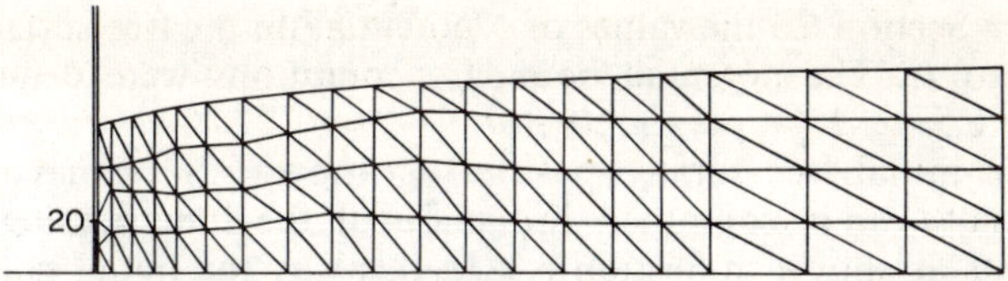

Figure 6.7(a) *Final element division for axisymmetric flow*

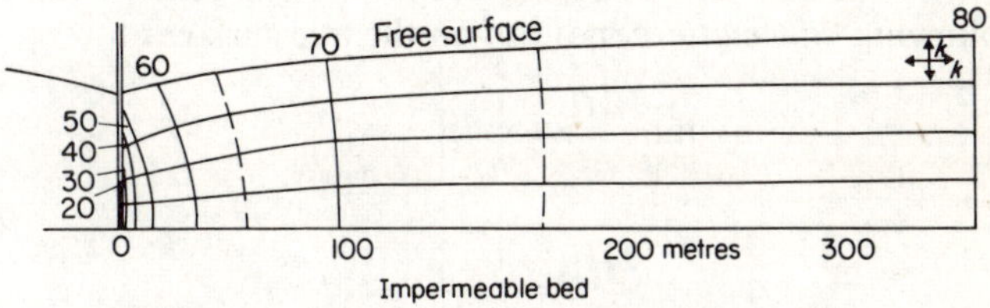

Figure 6.7(b) *Seepage towards an axisymmetric well*

The area through which flow occurs is proportional to the radial distance from the well and the thickness of the flow region. Hence as the flow approaches the well this area becomes much smaller and the potential gradient is considerably increased. Along the wall of the well the hydraulic gradient was estimated to be about three. This is three times higher than the allowable value and therefore protection against 'piping' would be needed in this case.

6.4 TRANSIENT FREE SURFACE FLOW

In some engineering problems we have to determine the transient position of a free surface, such as, for instance, the case of seepage through a dam when the water level in the reservoir suddenly drops.

In this case the free surface will change with time until sufficient time has elapsed for the surface to reach a steady state position, corresponding to the new water elevation behind the dam.

The problem is now defined in terms of Darcy's law plus the continuity equation

$$\vec{\nabla} \cdot \vec{v} = 0 \tag{6.33}$$

$$\bar{v} = -K\vec{\nabla}\varphi \tag{6.34}$$

or

$$v_i = -K_{ii}\frac{\partial \varphi}{\partial x_i} \tag{6.35}$$

As in Section 6.3 the values of φ potential on the free surface are equal to H. The potential boundary conditions were defined in Figure 6.5.

If an initial free surface position is known, the values of the potential φ can be computed throughout all the domain using finite elements as shown in Section 6.3. In this way the initial potential state is defined.

We next consider a drop in the water level; take a Δt increment and follow the free surface during this step. To do this we can write the following '*kinematic*' condition on the free surface:

$$\lim_{\Delta t \to 0} \frac{\Delta H}{\Delta t} = \frac{1}{n} v_2 \Big|_{\text{free surface}}$$

or

$$\frac{\mathrm{D}H}{\mathrm{D}t} = \frac{1}{n} v_2 \Big|_{\text{free surface}} \tag{6.36}$$

The porosity n is introduced in the equation in order to obtain actual average velocities of the fluid as distinct from the v_i, which are Darcy's average velocities.

For the two-dimensional case $\{H = H(x_1, t)\}$, (6.36) expands to

$$\frac{\mathrm{D}H}{\mathrm{D}t} = \frac{\partial H}{\partial t} + \frac{v_1}{n} \frac{\partial H}{\partial x_1} = \frac{v_2}{n} \Big|_{\text{free surface}} \tag{6.37}$$

Thus

$$\begin{aligned} \frac{\partial H}{\partial t} &= -\frac{v_1}{n} \frac{\partial H}{\partial x_1} + \frac{v_2}{n} \\ &= \frac{K_{11}}{n} \frac{\partial \varphi}{\partial x_1} \frac{\partial H}{\partial x_1} - \frac{K_{22}}{n} \frac{\partial \varphi}{\partial x_2} \end{aligned} \tag{6.38}$$

We can now solve the problem using the following two steps:

1 Having some initial H values, determine the φ potential over the domain using a finite element grid as in Section 6.3
2 Calculate the H values after a Δt increment by applying finite differences in (6.38)

Example 6.3

We will now review an earth dam problem solved by Cheng and Li.[2] They modified formula (6.38) by taking into account that the pressure is constant along the phreatic surface S (Figure 6.5), i.e.

$$\rho g[\varphi - x_2] = p_a = \text{atmospheric surface} \tag{a}$$

Hence

$$\frac{1}{\rho g}\,\mathrm{d}p_a = \frac{\partial\varphi}{\partial x_1}\,\mathrm{d}x_1 + \left[\left(\frac{\partial\varphi}{\partial x_2}\right) - 1\right]\mathrm{d}x_2 = 0 \tag{b}$$

for which $x_2 = H$ can be written

$$\frac{\partial\varphi}{\partial x_1} = \left(1 - \frac{\partial\varphi}{\partial x_2}\right)\frac{\partial H}{\partial x_1} \tag{c}$$

This value is now substituted into equation (6.38) giving

$$n\frac{\partial H}{\partial t} = K_{11}\left(1 - \frac{\partial\varphi}{\partial x_2}\right)\left(\frac{\partial H}{\partial x_1}\right)^2 - K_{22}\frac{\partial\varphi}{\partial x_2} \tag{d}$$

Equation (d) can be solved using finite differences in time and space to obtain the value of H. However, Cheng[2] pointed out that the solution of this nonlinear equation is unstable using finite differences. This instability apparently arises because of the shorter wavelengths formed by the nonlinear interaction of longer waves, and it is not always possible to eliminate it by reducing the time step. The problem can still be solved by using some 'smoothing' techniques [e.g. the use of a time–space finite element approximation for equation (d) will smooth the solution], also a simple quasi-linearisation method can be applied.[3] This method consists in expanding the nonlinear term of (d) in a Taylor series:

$$\left(\frac{\partial H}{\partial x_1}\right)^2 = \left(\frac{\partial \hat{H}}{\partial x_1}\right)^2 + 2\left(\frac{\partial \hat{H}}{\partial x_1}\right)\left[\left(\frac{\partial H}{\partial x_1}\right) - \left(\frac{\partial \hat{H}}{\partial x_1}\right)\right] + \ldots \tag{e}$$

where $\partial\hat{H}/\partial x_1$ are known functions, $\hat{H}$ indicates the value of elevation at the $(n-1)$th iteration, and H the same at nth iteration. We can now express (d) using expansion (e), as follows:

$$n\frac{\partial H}{\partial t} = K_{11}\left(1 - \frac{\partial\varphi}{\partial x_2}\right)\left[2\frac{\partial \hat{H}}{\partial x_1}\frac{\partial H}{\partial x_1} - \left(\frac{\partial \hat{H}}{\partial x_1}\right)^2\right] - K_{22}\frac{\partial\varphi}{\partial x_2} \tag{f}$$

Forward difference in time (Euler method) and implicit central difference in space were used to approximate (f). The resulting system of equations was solved and the results iterated until convergence was reached. This was usually possible until only two to four iterations for a required accuracy of

$$\left|\frac{\partial H}{\partial x_1} - \frac{\partial \hat{H}}{\partial x_1}\right| < 10^{-5}$$

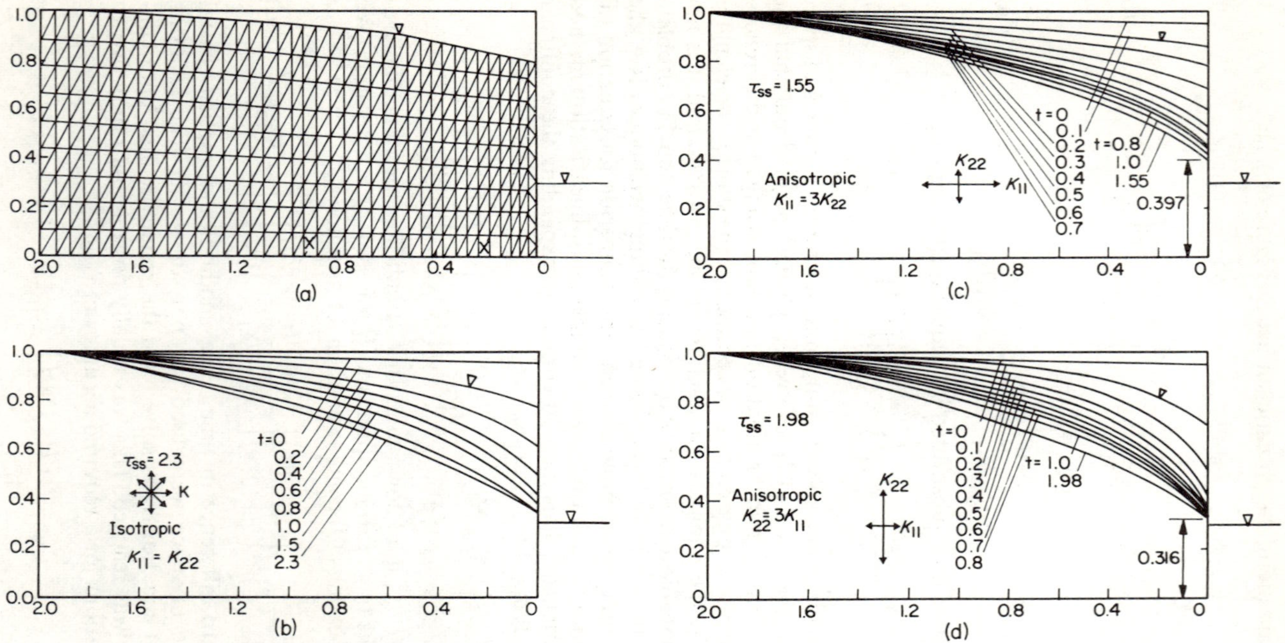

Figure 6.8 (a) *The finite element display of the flow region in the earth dam problem;* (b) *Time development of the free surface in an isotropic earth dam;* (c) *Time development of the free surface in an anisotropic earth dam,* $K_x = 3K_y$*;* (d) *Time development of the free surface in an anisotropic earth dam,* $K_y = 3K_x$

for all x_1 values. Once this state is reached, a new solution for the φ potential can be computed using the finite element mesh which has now to be vertically displaced in order to fit the new phreatic surface.

For the problem shown in Figure 6.8(a), a total of 421 nodes and 702 three node triangular elements were used.[2]

Three different cases were solved:

1 The porous medium was treated as isotropic, $K_{11}/K_{22} = 1$
2 The medium was treated as orthotropic with $K_{11}/K_{22} = \frac{1}{3}$
3 It was taken to be orthotropic with $K_{11}/K_{22} = 3$

Figure 6.8(b) shows the free surface for different times for the above cases. The time was made dimensionless by using a parameter $\tau = (K_{11}/nl)t$, where l is a characteristic length, in this case the head on the left-hand side. The final free surface positions corresponding to the steady state are indicated in the figures.

6.5 CONFINED AQUIFER ANALYSIS

We will now study the case of a saturated soil confined by impervious strata (Figure 6.9). This type of aquifer will be considered

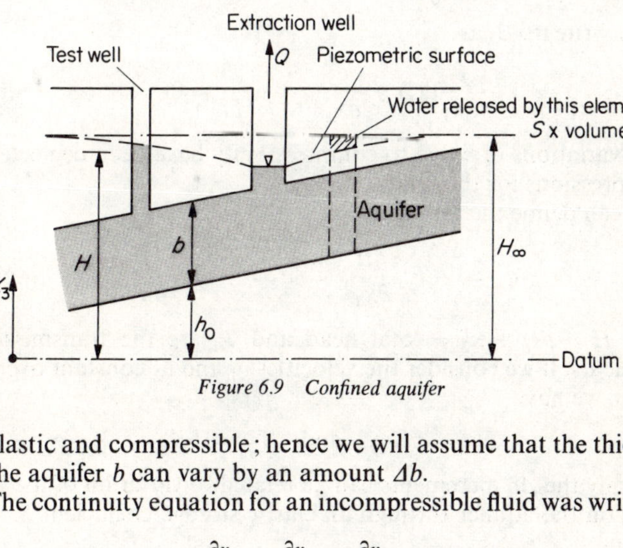

Figure 6.9 Confined aquifer

as elastic and compressible; hence we will assume that the thickness of the aquifer b can vary by an amount Δb.

The continuity equation for an incompressible fluid was written as

$$\frac{\partial v_1}{\partial x_1} + \frac{\partial v_2}{\partial x_2} + \frac{\partial v_3}{\partial x_3} = 0 \tag{6.39}$$

If we now integrate it with respect to x_3 from h_0 to $h_0 + b$, we obtain

$$\int_{h_0}^{h_0+b} \frac{\partial v_1}{\partial x_1} \mathrm{d}x_3 + \int_{h_0}^{h_0+b} \frac{\partial v_2}{\partial x_2} \mathrm{d}x_3 + v_3|_{h_0+b} = 0 \tag{6.40}$$

noting that the vertical velocity at h_0, for smoothly sloping aquifers, can be taken as zero.

The 'kinematic' condition for this case is that at $x_3 = h_0 + b$, the velocity has to be

$$v_3|_{h_0+b} = \frac{\mathrm{D}b}{\mathrm{D}t} = \frac{\partial b}{\partial t} + v_1\frac{\partial b}{\partial x_1} + v_2\frac{\partial b}{\partial x_2} \tag{6.41}$$

Substitution of (6.41) into (6.40) after applying the Leibnitz rule, i.e.

$$\frac{\partial}{\partial x}\int_{h_0}^{\eta} v(x, z)\,\mathrm{d}z = \int_{h_0}^{\eta} \frac{\partial v}{\partial x}\mathrm{d}z + v|_{z=\eta}\frac{\partial \eta}{\partial x} - v|_{z=h_0}\frac{\partial h_0}{\partial x} \tag{6.42}$$

gives the following continuity equation:

$$\frac{\partial}{\partial x_1}\int_{h_0}^{h_0+b} v_1\,\mathrm{d}x_3 + \frac{\partial}{\partial x_2}\int_{h_0}^{h_0+b} v_2\,\mathrm{d}x_3 + \frac{\partial b}{\partial t} = 0 \tag{6.43}$$

Defining q_1 and q_2 as

$$q_1 = \int v_1\,\mathrm{d}x_3 \qquad q_2 = \int v_2\,\mathrm{d}x_3$$

we can write (6.43) as

$$\frac{\partial}{\partial x_1}(q_1) + \frac{\partial}{\partial x_2}(q_2) + \frac{\partial b}{\partial t} = 0 \tag{6.44}$$

Small variations of b due to compressibility have been neglected in the expressions for the limits.

We can define the flux terms as

$$q_1 = -T_{11}\frac{\partial H}{\partial x_1}, \qquad q_2 = -T_{22}\frac{\partial H}{\partial x_2} \tag{6.45}$$

where $H = p/\gamma + x_3$ = total head and T_{ii} are the transmissivity coefficients. If we consider the velocities v_1 and v_2 constant over the section, we have

$$T_{11} = bK_{11} \quad \text{and} \quad T_{22} = bK_{22}$$

Finally the Δb increments can be related to variation of the head acting on the aquifer through an elastic storage coefficient S, such that

$$\Delta b = S\,\Delta H \tag{6.46}$$

From equations (6.45) and (6.46) we can write formula (6.44) as

$$\frac{\partial}{\partial x_1}\left(T_{11}\frac{\partial H}{\partial x_1}\right) + \frac{\partial}{\partial x_2}\left(T_{22}\frac{\partial H}{\partial x_2}\right) = S\frac{\partial H}{\partial t} \tag{6.47}$$

The storage coefficient can be interpreted as the volume of water released by each unit base column of an aquifer when the water table is lowered by a unit length. Any other effects such as water compressibility are generally included in the S coefficient, which is usually found by experiment.

For a confined aquifer, the value of S has been shown to be,[4]

$$S = b\rho g(n\kappa + \beta)$$

where β is the vertical compressibility of the soil (vertical strain of the mass of soil per unit change in pressure) and κ is the compressibility of water (inverse of bulk modulus).

We can also include in equation (6.47) any concentrated or distributed value for a sink, or source, function Q, such as (for a point sink)

$$Q = \sum Q_w(x_1, x_2)\, \Delta_{ij} \tag{6.48}$$

where Δ_{ij} is a step function for a point sink at the given coordinates $\bar{x}_1$ and $\bar{x}_2$

$$\begin{aligned} \Delta_{ij} &= 0 \quad \text{if } x_1 \neq \bar{x}_1 \text{ or } x_2 \neq \bar{x}_2 \\ \Delta_{ij} &= 1 \quad \text{if } x_1 = \bar{x}_1 \text{ and } x_2 = \bar{x}_2 \end{aligned} \tag{6.49}$$

For a distributed sink we have simply

$$Q = \sum Q_w(x_1 x_2)$$

Our governing equation can now be written

$$\mathscr{L}(H) = \frac{\partial}{\partial x_1}\left(T_{11}\frac{\partial H}{\partial x_1}\right) + \frac{\partial}{\partial x_2}\left(T_{22}\frac{\partial H}{\partial x_2}\right) - S\frac{\partial H}{\partial t} - Q = 0 \tag{6.50}$$

with boundary conditions

$$\begin{aligned} &\text{(a)}\ H = \bar{H} \quad \text{on } S_1 \\ &\text{(b)}\ q_n = T_{11}\frac{\partial H}{\partial x_1}\alpha_{n1} + T_{22}\frac{\partial H}{\partial x_2}\alpha_{n2} = \bar{q}_n \quad \text{on } S_2 \end{aligned} \tag{6.51}$$

One can propose an approximate function on each element for H such as

$$H \simeq \sum_{i=1}^{n} H_i(t)\phi_i(x_1 x_2) = \boldsymbol{\phi}\mathbf{H}^n \tag{6.52}$$

and apply Galerkin's method only in space:

$$\iint \mathscr{L}(H)\,\delta H(x_1, x_2)\,\mathrm{d}x_1\,\mathrm{d}x_2 = \int_{S_2} (q_n - \bar{q}_n)\,\delta H(x_1 x_2)\,\mathrm{d}S \tag{6.53}$$

The transmissivity and storage coefficients will be considered to be constant over each element. Equation (6.53) can be written after integration by parts and substitution of (6.52), as

$$\iint T_{11}\left(\frac{\partial \boldsymbol{\phi}}{\partial x_1}\right)^T\left(\frac{\partial \boldsymbol{\phi}}{\partial x_1}\right) + T_{22}\left(\frac{\partial \boldsymbol{\phi}}{\partial x_2}\right)^T\left(\frac{\partial \boldsymbol{\phi}}{\partial x_2}\right)\mathrm{d}x_1\,\mathrm{d}x_2\,\mathbf{H}^n$$

$$+ \iint S\boldsymbol{\phi}^T\boldsymbol{\phi}\,\mathrm{d}x_1\,\mathrm{d}x_2\,\dot{\mathbf{H}}^n$$

$$= -\iint \boldsymbol{\phi}^T Q\,\mathrm{d}x_1\,\mathrm{d}x_2 + \int \boldsymbol{\phi}^T \bar{q}_n\,\mathrm{d}S \tag{6.54}$$

or more briefly

$$\mathbf{K}\mathbf{H}^n + \mathbf{M}\dot{\mathbf{H}}^n = \mathbf{P} \tag{6.55}$$

where **K**, **M** and **P** are element matrices. For the whole continuum one has,

$$\mathscr{K}\mathscr{H} + \mathscr{M}\dot{\mathscr{H}} = \mathscr{P} \tag{6.56}$$

After the boundary conditions $\mathscr{H} = \overline{\mathscr{H}}$ on S_1 are imposed we can integrate equation (6.56) with respect to time. This can be done using explicit or implicit method. Among the explicit methods we have the Runge–Kutta and Euler techniques.

We will solve (6.56) using the trapezoidal rule which is an implicit method that requires the solution of a system of equations at each step. One starts by dividing the time into a series of Δt increments assuming that the *initial* conditions of the problem are known ($\mathscr{H}_0$ = vector at initial time). Hence

$$\dot{\mathscr{H}} = \frac{\mathscr{H}_{\Delta t} - \mathscr{H}_0}{\Delta t}, \qquad \mathscr{H}_{\Delta t/2} = \frac{\mathscr{H}_{\Delta t} + \mathscr{H}_0}{2} \tag{6.57}$$

$\mathscr{H}_{\Delta t}$ and $\mathscr{H}_{\Delta t/2}$ are unknown. Substituting equations (6.57) into the equilibrium equation at time $\Delta t/2$

$$\mathscr{K}\mathscr{H}_{\Delta t/2} + \mathscr{M}\dot{\mathscr{H}}_{\Delta t/2} = \mathscr{P}_{\Delta t/2} \tag{6.58}$$

and assuming

$$\mathscr{P}_{\Delta t/2} = \frac{\mathscr{P}_{\Delta t} + \mathscr{P}_0}{2} \tag{6.59}$$

we obtain

$$\left(\mathscr{K} + \frac{2}{\Delta t}\mathscr{M}\right)\mathscr{H}_{\Delta t} = (\mathscr{P}_{\Delta t} + \mathscr{P}_0) - \left(\mathscr{K} - \frac{2\mathscr{M}}{\Delta t}\right)\mathscr{H}_0 \quad (6.60)$$

For constant Δt, $\mathscr{K}$ and $\mathscr{M}$, we can compute the inverse of the

$$\left(\mathscr{K} + \frac{2}{\Delta t}\mathscr{M}\right)$$

matrix and use it for every time step.

This scheme is also called Crank–Nickolson recurrence formula.

Example 6.4

Pinder[5] has analysed an aquifer at Musquodoboit Harbour, Nova Scotia, by finite elements and has compared results with a finite difference solution previously obtained.

The aquifer is located adjacent to the Musquodoboit river (Figure 6.10) and because of the special geological characteristics of the region—it has a permeable stratum in between two relatively impermeable ones—the aquifer can be considered as confined.

The transmissivity and storage coefficient were obtained from a single 36 hour pumping test. At the pump test site the aquifer is approximately 4800 ft wide and extends along the river 5700 ft. The pumping well and three observation wells were drilled for the test. The discharge was constant and equal to 0·963 cfs.

The finite element discretisation is shown in Figure 6.10. The model consisted of 44 isoparametric elements with 96 nodes and several different types of elements were used. The boundaries of the aquifer were assumed to be impermeable and the river was represented by elongated elements.

The finite element model was then required to simulate the 36 hour pumping test in an analogous manner to the finite difference model. The same aquifer parameters were used.

The computer drawdown plotted in Figure 6.11 indicates a general agreement between the two numerical methods; the greatest discrepancy occurring early in the simulation period. It was impossible to decide which model was more exact as the exact solution of the problem is not known. Field observation values are not comparable as the aquifer parameters were obtained by modifying the pumping test parameter until the finite difference drawdowns were similar to the field values.

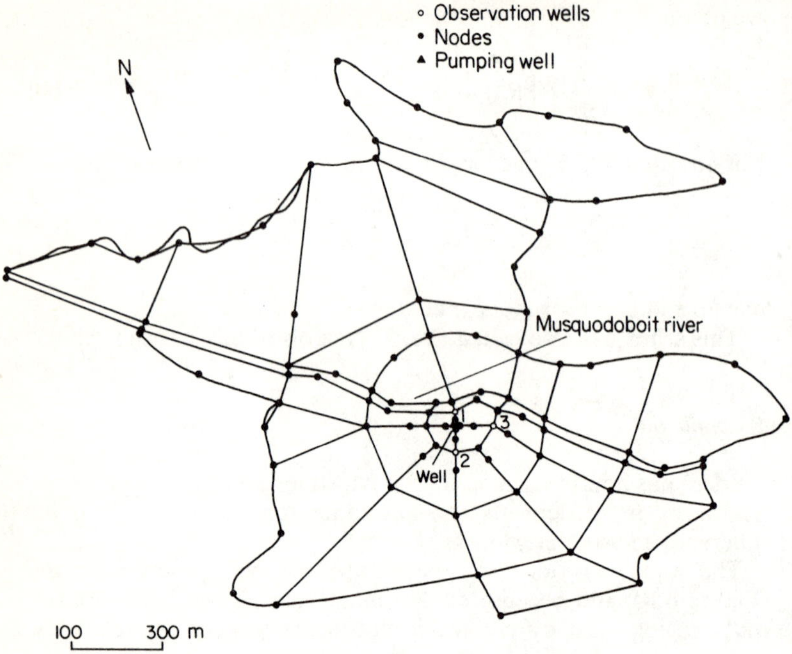

Figure 6.10 Element configuration for Galerkin analysis of Musquodoboit Harbour aquifer

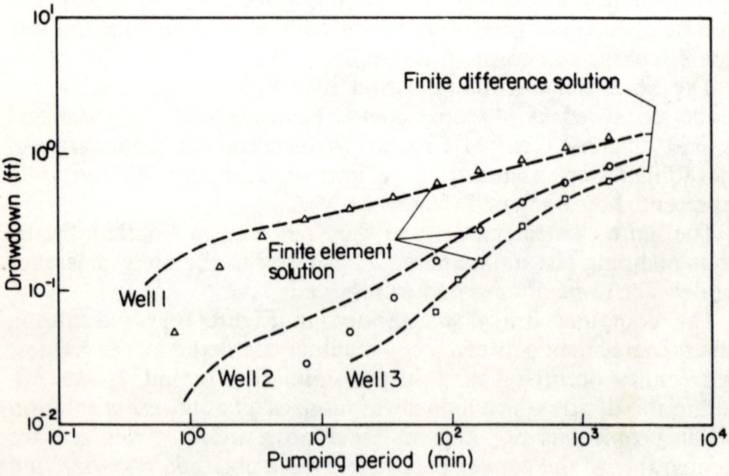

Figure 6.11 Comparison of finite difference and finite element solution at Musquodoboit Harbour, Nova Scotia

6.6 UNCONFINED AQUIFER ANALYSIS

Consider now the case of an aquifer with a free surface (Figure 6.12), the shape of which is not known *a priori*. In order to solve the problem we will assume (Dupuit's assumption):

1 The rate of change of the aquifer depth with respect to x_1 and x_2 and the slope of the aquifer bottom are small. One can then consider H as constant throughout the depth. This gives constant velocities v_1 and v_2 over the depth
2 The compressibility of the water may be neglected.

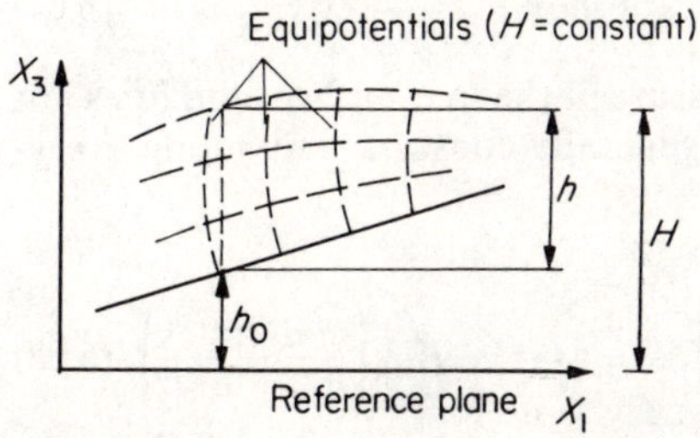

Figure 6.12 Free surface aquifer

The continuity equation is simply

$$\vec{\nabla} \cdot (\vec{v}) = 0 \tag{6.61}$$

We can integrate it with respect to x_3, giving

$$\int_{h_0}^{h_0+h} \frac{\partial v_1}{\partial x_1} \mathrm{d}x_1 + \int_{h_0}^{h_0+h} \frac{\partial v_2}{\partial x_2} \mathrm{d}x_2 + \int_{h_0}^{h_0+h} \frac{\partial v_3}{\partial x_3} \mathrm{d}x_3 = 0 \tag{6.62}$$

or

$$\int_{h_0}^{h_0+h} \frac{\partial v_1}{\partial x_1} \mathrm{d}x_1 + \int_{h_0}^{h_0+h} \frac{\partial v_2}{\partial x_2} \mathrm{d}x_2 + v_3|_{h+h_0} = 0 \tag{6.63}$$

For the free surface we have (kinematic condition)

$$v_3|_{h+h_0} = n\frac{\mathrm{D}h}{\mathrm{D}t} \tag{6.64}$$

Assumption (1) implies

$$q_1 = \int_{h_0}^{h_0+h} v_1 \, \mathrm{d}x_3 = hv_1, \qquad q_2 = \int_{h_0}^{h_0+h} v_2 \, \mathrm{d}x_3 = hv_2 \tag{6.65}$$

Using Leibnitz' rule plus equations (6.65) and (6.64), formula (6.62) can be written

$$\frac{\partial}{\partial x_1}(q_1) + \frac{\partial}{\partial x_2}(q_2) + n\frac{\partial h}{\partial t} = 0 \tag{6.66}$$

Instead of n (porosity) it is usual to use the specific yield of the soil N. This specific yield is defined as the volume of water the saturated soil can yield by gravity. As there is some water that will always remain in the soil, $N < n$.

The elevation head for the free surface, if the bottom slope is small, can be approximated by

$$H = h_0 + h + \frac{p}{\gamma} \tag{6.67}$$

where p is the atmospheric pressure at the free surface, and does not need to be considered as it is generally constant and acting everywhere.

By Darcy's law one has

$$v_1 = -K_{11}\left(\frac{\partial h_0}{\partial x_1} + \frac{\partial h}{\partial x_1}\right), \qquad v_2 = -K_{22}\left(\frac{\partial h_0}{\partial x_2} + \frac{\partial h}{\partial x_2}\right) \tag{6.68}$$

Thus (6.66) becomes

$$\frac{\partial}{\partial x_1}\left\{K_{11}h\left(\frac{\partial h_0}{\partial x_1} + \frac{\partial h}{\partial x_1}\right)\right\} + \frac{\partial}{\partial x_2}\left\{K_{22}h\left(\frac{\partial h_0}{\partial x_2} + \frac{\partial h}{\partial x_2}\right)\right\} = N\frac{\partial h}{\partial t} \tag{6.69}$$

Finally if a sink term Q is considered, one has

$$\frac{\partial}{\partial x_1}\left(K_{11}h\frac{\partial H}{\partial x_1}\right) + \frac{\partial}{\partial x_2}\left(K_{22}h\frac{\partial H}{\partial x_2}\right) - Q = N\frac{\partial H}{\partial t} \tag{6.70}$$

Equation (6.70) is nonlinear and time dependent. In order to find a solution we have to linearise it. One scheme is to assume a value of $h = \hat{h}$ such that

$$\frac{\partial}{\partial x_1}\left(K_{11}\hat{h}\frac{\partial H}{\partial x_1}\right) + \frac{\partial}{\partial x_2}\left(K_{22}\hat{h}\frac{\partial H}{\partial x_2}\right) = Q = N\frac{\partial H}{\partial t} \tag{6.71}$$

Thus we can apply Galerkin's technique in space and obtain a system of finite element equations similar to those obtained before; i.e.

$$\mathscr{K}\mathscr{H} + \mathscr{M}\dot{\mathscr{H}} = \mathscr{P} \tag{6.72}$$

where $\mathscr{K}$ is a function of $\mathscr{H}$.

Equation (6.72) can be integrated in time using an implicit method such as the trapezoidal rule already discussed or any of the explicit techniques. After each Δt we can iterate on the equations until $h = \hat{h}$ to some desired accuracy.

The main disadvantage of Dupuit's approximation is that it fails to take into account a free seepage surface of the type shown in Figure 6.13. In other words we are considering that $h = h_w$ for the well shown in the figure. The error involved in this assumption is generally small and confined to a short distance from the well.

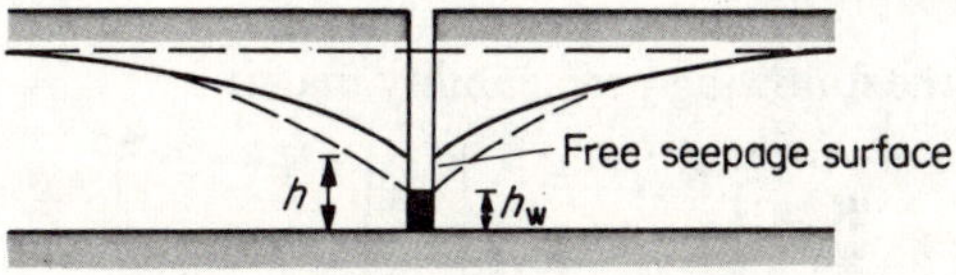

Figure 6.13 Free seepage surface

For the case of horizontal aquifer bottom and steady state flow, equation (6.70) simplifies to

$$\frac{\partial}{\partial x_1}\left(K_{11}h\frac{\partial h}{\partial x_1}\right) + \frac{\partial}{\partial x_2}\left(K_{22}h\frac{\partial h}{\partial x_2}\right) = Q \qquad (6.73)$$

or

$$\frac{\partial}{\partial x_1}\left\{K_{11}\frac{\partial(h^2)}{\partial x_1}\right\} + \frac{\partial}{\partial x_2}\left\{K_{22}\frac{\partial(h^2)}{\partial x_2}\right\} = 2Q \qquad (6.74)$$

Equation (6.74) can be solved directly in terms of h^2.

REFERENCES

1. LAMBE, T. W. and WHITMAN, R. V., *Soil Mechanics*, Wiley (1969)
2. CHENG, R. T. and LI, C., 'On the Solution of Transient Free-Surface Flow Problems in Porous Media by the Finite Element Method', *J. Hydrol.*, **20,** 49–63 (1973)
3. BELLMAN, R. W. and KALABA, R. E., *Quasi-linearization and Non-linear Boundary-value Problems*, Elsevier (1965)
4. DE WIEST, R. (Ed.), *Flow through Porous Media*, Academic Press (1969)
5. PINDER, G. F. and FRIND, E. O., 'Application of Galerkin's Procedure to Aquifer Analysis', *Water Resources Res.*, **8,** No. 1 (Feb. 1972)

BIBLIOGRAPHY

DUPUIT, J., *Etudes Théoriques et Pratiques sur le Mouvement des Eaux dans les Canaux Découverts et à Travers les Terrains Perméables*, 2nd Edition, Dunod, Paris (1863)

HARR, M. E., *Groundwater and Seepage*, McGraw-Hill (1962)

LI, WEN-HSIUNG, *Differential Equations of Hydraulic Transients, Dispersion, and Ground-Water Flow*, Prentice Hall (1972)

NEWMAN, S. P. and WITHERSPOON, P. A., 'Finite Element Method of Analysing Steady Seepage with a Free Surface', *Water Resources Res.*, **6,** No. 3, 889–897 (1970)

JAVANDEL, I. and WITHERSPOON, P. A., 'Application of the Finite Element Method to Transient Flow in Porous Media, *Soc. Petrol. Eng. J.*, **8,** No. 3, 241–252 (1968)

EXERCISES

6-1 Given the following permeability matrix

$$\mathbf{K} = \begin{bmatrix} K_{11} & K_{12} \\ K_{21} & K_{22} \end{bmatrix} = \begin{bmatrix} 6 & -3 \\ -3 & 4 \end{bmatrix}$$

Find the principal directions of orthotropy and the values of $K'_{11} K'_{22}$ corresponding to them.

6-2 Discuss how the Laplace equation computer program of Chapter 3 can be modified to take free surface into account.

Draw a macro-flow chart.

6-3 Isoparametric elements like the ones studied in Chapter 3 are specially well suited to follow the free surface configuration during drawdown. Explain the steps involved in writing such a program for two dimensions using the simple technique of moving the free surface in accordance with its computed normal slope (see later paragraphs of Section 6.4).

6-4 A confined aquifer as shown in Figure 6.14 is located near a constant head shoreline. Water is pumped from a ditch parallel to the shoreline at a rate Q.

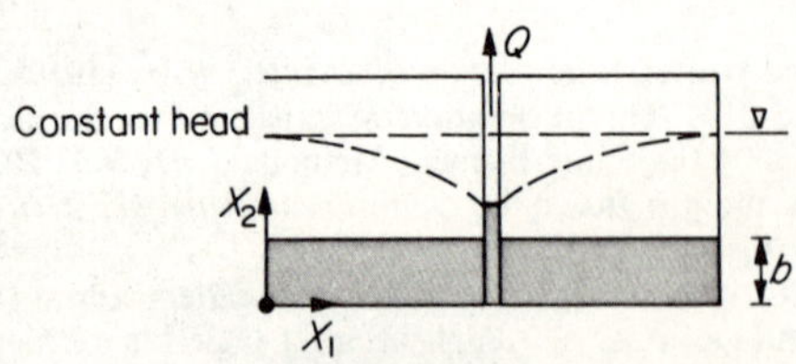

Figure 6.14 Confined aquifer

Determine the equation governing the flow. Find the steady state solution for the constant discharge Q.

6-5 The total head in a ditch as shown in Figure 6.15 varies periodically with

$$H_w = \bar{H} \sin \omega t$$

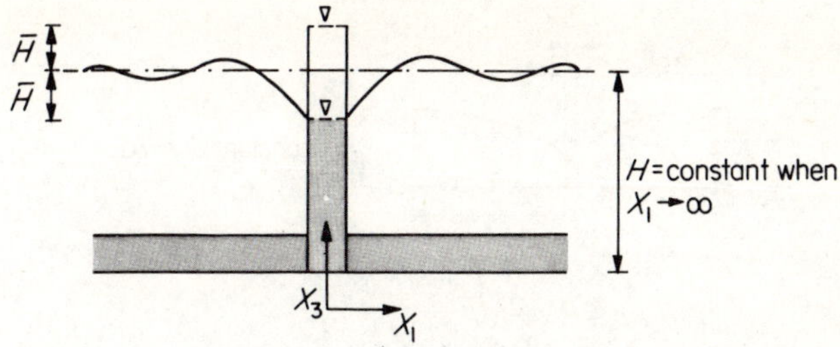

Figure 6.15 Isotropic confined aquifer

Find the expression for H in the isotropic confined aquifer shown in the figure.

6-6 Deduce the continuity equation for an incompressible fluid in a non-deformable orthotropic media when the x_i directions are not principal directions of permeability.

6-7 Write down the three-dimensional orthotropic continuity equation in cylindrical coordinates assuming that the flow varies with respect to x, z and θ.

6-8 Explain how the above three-dimensional equation can be solved by expanding the potentials and boundary conditions using Fourier series in θ.

6-9 Terzaghi's equation of one-dimensional consolidation is

$$\frac{\partial p}{\partial t} = C_v \frac{\partial^2 p}{\partial z^2} \tag{a}$$

where p is the pores pressure, C_v the vertical consolidation coefficient and z the vertical axis. The thickness of the stratum is Z and it is taken to be resting on a rigid and impermeable rock.

1 Write the boundary and initial conditions for equation (a). Consider a load P is applied at $t = 0$ on the surface.
2 Propose a finite element model in space and write the resulting system of equations for a few elements.
3 Discuss how to integrate those equations in time.

6-10 Given the matrix equation

$$\mathbf{K\Phi} + \mathcal{M}\dot{\mathbf{\Phi}} = \mathcal{P}$$

find the recurrence law for integrating it in time using,

1 The trapezoidal rule (6.5)

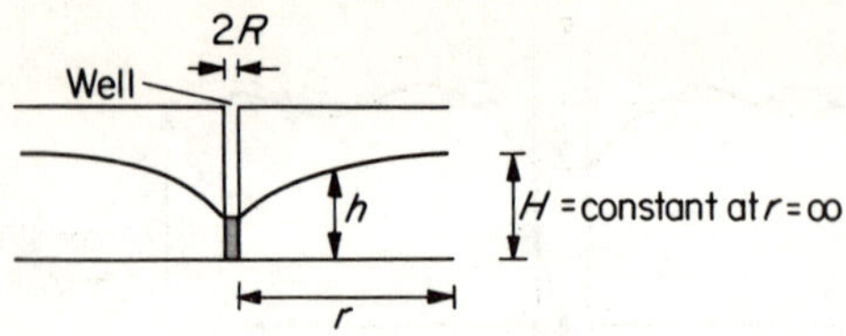

Figure 6.16

2 Galerkin's method (Example 1.15)
3 The Euler scheme

$$\dot{\mathbf{\Phi}}^{(n)} = \frac{\mathbf{\Phi}^{(n)} - \mathbf{\Phi}^{(n-1)}}{2}$$

6-11 Consider a central well as shown in Figure 6.16. Given a discharge Q, show that

$$Q = K\frac{\partial h}{\partial r}2\pi rh$$

if the velocity is assumed constant over the depth h. Compute the drawdown inside the well.

7 Shallow Water Circulation Problems

7.1 SHALLOW WATER EQUATIONS

The present state of the art and the lack of suitable data in many cases do not justify more complex mathematical models for circulation in coastal regions, lakes, etc., than those based on the numerical solution of the shallow water equations. Fully three-dimensional solutions are not warranted at this stage as they would require a large amount of extra data and computer time.

Finite difference solutions for circulation problems have been implemented in the past,[1–6] but they suffer from lack of mesh flexibility and, in some cases, difficulty in the satisfaction of the boundary conditions. Finite elements here allow for great flexibility in the analysis grid, and the advantages of having to satisfy only the essential boundary conditions.

In what follows a consistent derivation of the vertically averaged equations for long wave propagation is presented. The formulation is then applied with finite elements.

The governing equations for the fluid, neglecting temperature effects (Chapter 4), can be written

$$-\frac{\partial p}{\partial x_k} + \frac{\partial \tau_{ik}}{\partial x_i} + \rho b_k = \frac{D(\rho v_k)}{Dt} \qquad (7.1)$$

$$i, k = 1, 2, 3$$

$$\frac{\partial(\rho v_i)}{\partial x_i} + \frac{\partial \rho}{\partial t} = 0 \qquad (7.2)$$

These equations are difficult to apply for the solution of shallow water problems because of

1 the presence of the free surface
2 the variable nature of the boundary when the tide rises and falls
3 the large number of variables in the solution

These difficulties can be solved by simplifying the equations into what are called the shallow water equations. The first simplification we will introduce is to reduce the third momentum relationship to

$$-\frac{\partial p}{\partial x_3} = \rho g \tag{7.3}$$

where the body forces are negative as they act in the opposite direction to the x_3 axis.

Formula (7.3) implies that we have neglected all acceleration terms and corresponding stresses. Integrating (7.3) we have

$$p = \int_{x_3}^{\eta} \rho g \, \mathrm{d}x_3 = \rho g(\eta - x_3) + p_a \tag{7.4}$$

where p_a is the atmospheric pressure acting on the surface of the water and η is the elevation of the free surface (see Figure 7.1).

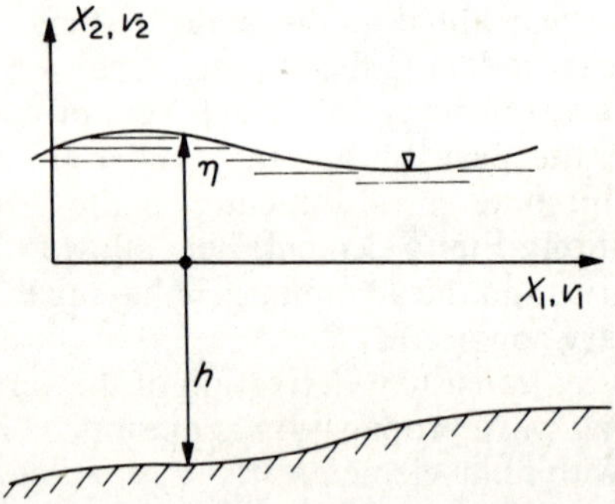

Figure 7.1 Geometrical notation for shallow water equations

The remaining two momentum equations—in the x_1 and x_2 directions—can now be written

$$\frac{\mathrm{D}(\rho v_k)}{\mathrm{D}t} = \frac{\partial(\rho v_k)}{\partial t} + \frac{\partial(\rho v_k v_i)}{\partial x_i} = -\frac{\partial p}{\partial x_k} + \frac{\partial \tau_{ik}}{\partial x_i} + b_k \tag{7.5}$$

The above formula represents two equations, one for $k = 1$ and the other for $k = 2$. The subindex i in this case indicates summation, $i = 1, 2, 3$. v are average velocities, ρ is the variable mass density and τ the sum of viscous plus Reynolds stresses ($\tau_{ik} = \tau_{ki}$).

Our aim is now to integrate (7.2) and (7.5) with respect to x_3. This gives, for the continuity equation,

$$\int_{-h}^{\eta} \left(\frac{\partial(\rho v_i)}{\partial x_i} + \frac{\partial \rho}{\partial t} \right) \mathrm{d}x_3 = 0 \tag{7.6}$$

where h is the depth from a datum surface (not necessarily horizontal) and η the free surface variable (Figure 7.1).

Let us now define a momentum flux variable q_k (mass per unit length and time) such that

$$q_k = \int_{-h}^{\eta} \rho v_k \, \mathrm{d}x_3 = \rho \int_{-h}^{\eta} v_k \, \mathrm{d}x_3 \tag{7.7}$$

Note that $\rho(x_1, x_2)$ is assumed not to be a function of x_3.

In order to integrate (7.6) we need to use the kinematic condition and Leibnitz' rule for partial differentiation of an integral between variable limits. The latter gives, for instance,

$$\frac{\partial}{\partial x_1} \int_{h_1(x_1 x_2)}^{h_2(x_1 x_2)} f(x_1 x_2 x_3) \, \mathrm{d}x_3 = \int_{h_1}^{h_2} \frac{\partial f}{\partial x_1} \mathrm{d}x_3 + f \bigg|_{h_2} \frac{\partial h_2}{\partial x_1} - f \bigg|_{h_1} \frac{\partial h_1}{\partial x_1} \tag{7.8}$$

and similarly for x_2.

The kinematic relationship for the free surface can be written

$$v_3|_{x_3 = \eta} = \frac{\mathrm{D}\eta}{\mathrm{D}t} = \frac{\partial \eta}{\partial t} + v_1 \bigg|_{\eta} \frac{\partial \eta}{\partial x_1} + v_2 \bigg|_{\eta} \frac{\partial \eta}{\partial x_2} \tag{7.9}$$

Applying formulae (7.7)–(7.9) to equation (7.6), we obtain

$$\frac{\partial q_i}{\partial x_i} + \frac{\partial(\rho H)}{\partial t} = 0 \tag{7.10}$$

where $H = h + \eta$.

To integrate the two momentum equations (7.5) with respect to x_3, we define the following instantaneous velocities:

$$\begin{aligned} v_1 &= \bar{v}_1(x_1 x_2 t) + v_1'(x_1 x_2 x_3 t) \\ v_2 &= \bar{v}_2(x_1 x_2 t) + v_2'(x_1 x_2 x_3 t) \end{aligned} \tag{7.11}$$

where $\bar{v}$ denotes the vertically averaged velocities and v' the vertical deviations. Hence

$$\langle v_k \rangle = \int_{-h}^{\eta} v_k \, dx_3 = \frac{1}{\rho} q_k, \qquad \bar{v}_k = \frac{1}{H} \langle v_k \rangle \tag{7.12}$$

as $\langle v_k' \rangle = 0$.

We will assume that the body forces are only those due to Coriolis effects. Thus, for the northern hemisphere,

$$b_1 = \rho f v_2, \qquad b_2 = -\rho f v_1 \tag{7.13}$$

One can also assume that the surface and bottom slopes are small with respect to unity. Hence we can approximate the internal stress components as follows (see Figure 7.2):

$$\begin{aligned} \tau_1|_s &\simeq \left\{ -\tau_{11} \frac{\partial \eta}{\partial x_1} - \tau_{12} \frac{\partial \eta}{\partial x_2} + \tau_{13} \right\}_{\text{surface}} \\ \tau_1|_b &\simeq \left\{ \tau_{11} \frac{\partial h}{\partial x_1} + \tau_{12} \frac{\partial h}{\partial x_2} - \tau_{13} \right\}_{\text{bottom}} \end{aligned} \tag{7.14}$$

and similarly for $\tau_2|_s$ and $\tau_2|_b$. Note that the τs can be interpreted as external force components applied at the top and bottom.

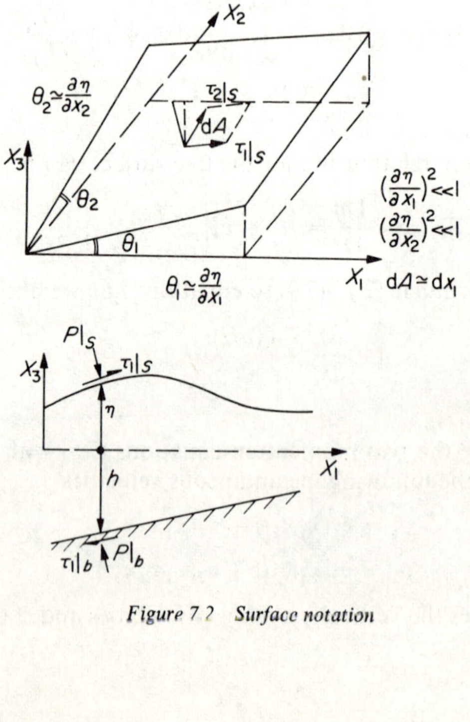

Figure 7.2 Surface notation

We will now substitute (7.11)–(7.13) into the momentum equations integrated with respect to x_3. One also needs to apply Leibnitz' rule plus the kinematic condition to obtain the following result:

$$\frac{\partial q_1}{\partial t} + \frac{\partial}{\partial x_1}\left(\frac{q_1^2}{H}\right) + \frac{\partial}{\partial x_2}\left(\frac{q_1 q_2}{H}\right)$$
$$= -\frac{\partial N_p}{\partial x_1} + \frac{\partial N_{11}}{\partial x_1} + \frac{\partial N_{12}}{\partial x_2} + fq_2 + p\bigg|_s \frac{\partial \eta}{\partial x_1} + \tau_1\bigg|_s$$
$$+ p\bigg|_b \frac{\partial h}{\partial x_1} - \tau_1\bigg|_b \tag{7.15}$$
$$\frac{\partial q_2}{\partial t} + \frac{\partial}{\partial x_1}\left(\frac{q_1 q_2}{H}\right) + \frac{\partial}{\partial x_2}\left(\frac{q_2^2}{H}\right)$$
$$= -\frac{\partial N_p}{\partial x_2} + \frac{\partial N_{22}}{\partial x_2} + \frac{\partial N_{12}}{\partial x_1} - fq_1 + p\bigg|_b \frac{\partial \eta}{\partial x_2} + \tau_2\bigg|_s$$
$$+ p\bigg|_b \frac{\partial h}{\partial x_2} - \tau_2\bigg|_b$$

where

$$N_p = \langle p \rangle = \int_{-h}^{\eta} p \, \mathrm{d}x_3 = \rho g \frac{H^2}{2} + Hp_a$$
$$N_{11} = \langle \tau_{11} \rangle - \langle \rho v_1' v_1' \rangle$$
$$N_{22} = \langle \tau_{22} \rangle - \langle \rho v_2' v_2' \rangle \tag{7.16}$$
$$N_{12} = \langle \tau_{12} \rangle - \langle \rho v_1' v_2' \rangle$$

Furthermore, the N_{ik} terms can be approximated by

$$N_{11} \simeq 2\varepsilon_{11} \frac{\partial q_1}{\partial x_1}$$
$$N_{22} \simeq 2\varepsilon_{22} \frac{\partial q_2}{\partial x_2} \tag{7.17}$$
$$N_{12} \simeq \varepsilon_{12}\left(\frac{\partial q_1}{\partial x_2} + \frac{\partial q_2}{\partial x_1}\right)$$

The ε_{ik} are generalised eddy viscosity coefficients. For isotropic behaviour they become $\varepsilon_{11} = \varepsilon_{22} = \varepsilon_{12} = \varepsilon$.

The bottom shear stresses are usually given by the following relationships:

$$\begin{aligned} \tau_1|_b &= \left(\frac{g}{c^2}\right)\frac{1}{\rho}\frac{q_1(q_1^2 + q_2^2)^{\frac{1}{2}}}{H^2} \\ \tau_2|_b &= \left(\frac{g}{c^2}\right)\frac{1}{\rho}\frac{q_1(q_1^2 + q_2^2)^{\frac{1}{2}}}{H^2} \end{aligned} \tag{7.18}$$

where ρ is the density of the water, g the acceleration due to gravity and c the friction factor, or Chezy's coefficient. Note that g/c^2 is a dimensionless quantity.

The shear components on the surface of the water are generally due to wind and can be expressed as

$$\begin{aligned} \tau_1|_s &= \gamma^2 \rho_a W^2 \cos\theta \\ \tau_2|_s &= \gamma^2 \rho_a W^2 \sin\theta \end{aligned} \tag{7.19}$$

where W is the velocity of the wind, ρ_a the density of air and θ the angle between the x_1 axis and the direction of the wind. γ^2 is a dimensionless coefficient called the wind stress coefficient; its value is given as approximately 0·0026 in Reference 4.

Equation (7.15) can now be written

$$\begin{aligned} \frac{\partial q_1}{\partial t} + \frac{\partial}{\partial x_1}\left(\frac{q_1^2}{H}\right) + \frac{\partial}{\partial x_2}\left(\frac{q_1 q_2}{H}\right) &= \frac{\partial}{\partial x_1}(N_{11} - N_p) + \frac{\partial N_{12}}{\partial x_2} + B_1 \\ \frac{\partial q_2}{\partial t} + \frac{\partial}{\partial x_1}\left(\frac{q_1 q_2}{H}\right) + \frac{\partial}{\partial x_2}\left(\frac{q_2^2}{H}\right) &= \frac{\partial}{\partial x_2}(N_{22} - N_p) + \frac{\partial N_{12}}{\partial x_1} + B_2 \end{aligned} \tag{7.20}$$

where B_1 and B_2 terms are

$$\begin{aligned} B_1 &= fq_2 + \gamma^2 \rho_a W^2 \cos\theta - \left(\frac{g}{c^2}\right)\frac{1}{\rho}\frac{q_1(q_1^2 + q_2^2)^{\frac{1}{2}}}{H^2} + p_a\frac{\partial H}{\partial x_1} \\ &\quad + \rho g H \frac{\partial h}{\partial x_1} \\ B_2 &= -fq_1 + \gamma^2 \rho_a W^2 \sin\theta - \left(\frac{g}{c^2}\right)\frac{1}{\rho}\frac{q_2(q_1^2 + q_2^2)^{\frac{1}{2}}}{H^2} + p_a\frac{\partial H}{\partial x_2} \\ &\quad + \rho g H \frac{\partial h}{\partial x_2} \end{aligned} \tag{7.21}$$

BOUNDARY CONDITIONS

To solve the resulting system of equations (7.20) under condition (7.10) we need to establish the necessary boundary conditions. Consider that the S boundary (Figure 7.3) consists of two parts, land type boundaries S_1 and ocean type S_2.

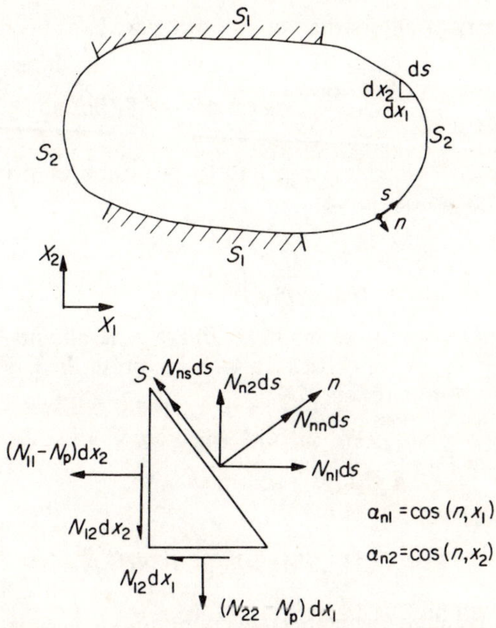

Figure 7.3 Boundary definitions

On the boundary reference system s–n, the momentum flux can be written as

$$q_n = \int_{-h}^{n} \rho v_n \, \mathrm{d}x_3 = \alpha_{n1} q_1 + \alpha_{n2} q_2$$

$$q_s = \int_{-h}^{n} \rho v_s \, \mathrm{d}x_3 = -\alpha_{n2} q_1 + \alpha_{n1} q_2 \tag{7.22}$$

where $\alpha_{n1} = \cos(n, x_1)$, $\alpha_{n2} = \cos(n, x_2)$.

For the force resultants we have

$$N_{n1} = \alpha_{n1}(N_{11} - N_p) + \alpha_{n2} N_{12}$$

$$N_{n2} = \alpha_{n1} N_{12} + \alpha_{n2}(N_{22} - N_p) \tag{7.23}$$

We can transform again to calculate N_{nn} and N_{ns}

$$\begin{aligned} N_{nn} &= \alpha_{n1} N_{n1} + \alpha_{n2} N_{n2} \\ N_{ns} &= -\alpha_{n2} N_{n1} + \alpha_{n1} N_{n2} \end{aligned} \tag{7.24}$$

On land type boundaries we usually have

$$q_n = 0 \quad \text{on } S_1 \tag{7.25}$$

However, if a river enters the body of water we can specify its mass input as

$$\left.\begin{aligned} q_n &= \bar{q}_n = \{q\}_{\text{river inflow}} \\ q_s &= 0 \end{aligned}\right\} \text{on part of } S_1 \text{ boundary} \tag{7.26}$$

On an ocean boundary one has to prescribe in principle the normal and tangential forces

$$\left.\begin{aligned} N_{nn} &= \bar{N}_{nn} \\ N_{ns} &= \bar{N}_{ns} \end{aligned}\right\} \quad \text{on } S_2 \tag{7.27}$$

but as the eddy viscosity terms in (7.20) are generally neglected, the tangential forces or velocities cannot be prescribed. Hence our boundary conditions reduce to

$$\begin{aligned} & q_n = 0 \quad \text{or} \quad q_n = \bar{q}_n \quad \text{on } S_1 \\ \text{plus} \quad & N_{nn} = \bar{N}_{nn} = -N_p \quad \text{on } S_2 \end{aligned} \tag{7.28}$$

7.2 FINITE ELEMENT FORMULATION

In order to formulate our finite element model, let us write equations (7.20) plus boundary conditions (7.28) and (7.10) plus condition (7.26) in the following weighted residual way:

$$\iint \left\{ \frac{\partial q_1}{\partial t} + \frac{\partial}{\partial x_1}\left(\frac{q_1^2}{H}\right) + \frac{\partial}{\partial x_2}\left(\frac{q_2 q_1}{H}\right) + \frac{\partial N_p}{\partial x_1} - B_1 \right\} \delta q_1 \, dA$$

$$= \int_{S_2} \alpha_{n1}(N_p - \bar{N}_p)\, \delta q_1 \, dS \tag{7.29}$$

$$\iint \left\{ \frac{\partial q_2}{\partial t} + \frac{\partial}{\partial x_1}\left(\frac{q_2 q_1}{H}\right) + \frac{\partial}{\partial x_2}\left(\frac{q_2^2}{H}\right) + \frac{\partial N_p}{\partial x_2} - B_2 \right\} \delta q_2 \, dA$$

$$= \int_{S_2} \alpha_{n2}(N_p - \bar{N}_p)\, \delta q_2 \, dS \tag{7.30}$$

$$\iint \left\{ \frac{\partial q_1}{\partial x_1} + \frac{\partial q_2}{\partial x_2} + \frac{\partial(\rho H)}{\partial t} \right\} \delta H \, dA = \int_{S_1} (q_n - \bar{q}_n)\, \delta H \, dS \tag{7.31}$$

where $\delta q_1\ \delta q_2\ \delta H$ are variations which satisfy the boundary conditions in mass flux and elevation. Note that if the elevation is known on S_2 the boundary conditions $N_p = \bar{N}_p$ will be identically satisfied [see first of equations (7.16)], hence the boundary terms on S_2 do not need to be included in (7.30) and (7.31).

The B_i and $\partial N_p/\partial x_i$ terms can be combined to give

$$
\begin{aligned}
B_1^* &= B_1 - \frac{\partial N_p}{\partial x_1} \\
&= fq_2 + \gamma^2 \rho_a W^2 \cos\theta - \left(\frac{g}{c^2}\right)\frac{1}{\rho}\frac{q_1(q_1^2+q_2^2)^{\frac{1}{2}}}{H^2} - \rho g H \frac{\partial (H-h)}{\partial x_1} \\
&\quad - H\frac{\partial p_a}{\partial x_1} \\
B_2^* &= B_2 - \frac{\partial N_p}{\partial x_2} \\
&= -fq_1 + \gamma^2 \rho_a W^2 \sin\theta - \left(\frac{g}{c^2}\right)\frac{1}{\rho}\frac{q_1(q_1^2+q_2^2)^{\frac{1}{2}}}{H^2} - \rho g H \frac{\partial (H-h)}{\partial x_2} \\
&\quad - H\frac{\partial p_a}{\partial x_2}
\end{aligned}
\tag{7.32}
$$

The continuity equation (7.31) can be integrated by parts to render a simpler expression, i.e.

$$
\iint \left\{ q_1 \frac{\partial \delta H}{\partial x_1} + q_2 \frac{\partial \delta H}{\partial x_2} - \frac{\partial(\rho H)}{\partial t}\delta H \right\} \mathrm{d}A = \int_{S_1} \bar{q}_n\, \delta H\, \mathrm{d}S \tag{7.33}
$$

which has then to be solved together with the following final momentum equations:

$$
\iint \left\{ \frac{\partial q_1}{\partial t} + \frac{\partial}{\partial x_1}\left(\frac{q_1^2}{H}\right) + \frac{\partial}{\partial x_2}\left(\frac{q_1 q_2}{H}\right) - B_1^*\, \delta q_1 \right\} \mathrm{d}A = 0 \tag{7.34}
$$

$$
\iint \left\{ \frac{\partial q_2}{\partial t} + \frac{\partial}{\partial x_2}\left(\frac{q_2 q_1}{H}\right) + \frac{\partial}{\partial x_2}\left(\frac{q_2^2}{H}\right) - B_2^*\, \delta q_2 \right\} \mathrm{d}A = 0 \tag{7.35}
$$

The above variational statements are the starting point for the finite element models which we are now going to develop. Let us assume that the same interpolation functions apply for q_i and H, i.e.

$$
q_1 = \boldsymbol{\phi}^T \mathbf{q}_1^n, \qquad q_2 = \boldsymbol{\phi}^T \mathbf{q}_2^n, \qquad H = \boldsymbol{\phi}^T \mathbf{H}^n \tag{7.36}
$$

Substituting these values into the momentum and continuity

equations (7.33–7.35) we can obtain,

$$\left.\begin{aligned}\delta \mathbf{q}_1^{n,T}\left(\mathbf{M}\frac{\partial \mathbf{q}_1^n}{\partial t} - \mathbf{F}_1\right) &= 0\\ \delta \mathbf{q}_2^{n,T}\left(\mathbf{M}\frac{\partial \mathbf{q}_2^n}{\partial t} - \mathbf{F}_2\right) &= 0\\ \delta \mathbf{H}^{n,T}\left(\mathbf{M}^*\frac{\partial \mathbf{H}^n}{\partial t} - \mathbf{F}_H\right) &= 0\end{aligned}\right\} \tag{7.37}$$

where

$$\left.\begin{aligned}\mathbf{M} &= \iint \boldsymbol{\phi}\boldsymbol{\phi}^T\,\mathrm{d}A, \qquad \mathbf{M}^* = \iint \rho\boldsymbol{\phi}\boldsymbol{\phi}^T\,\mathrm{d}A\\ \mathbf{F}_1 &= \iint (\boldsymbol{\phi}B_1^* - \boldsymbol{\phi}A_1)\,\mathrm{d}A\\ \mathbf{F}_2 &= \iint (\boldsymbol{\phi}B_2^* - \boldsymbol{\phi}A_2)\,\mathrm{d}A\\ \mathbf{F}_H &= \iint (\boldsymbol{\phi}_{,1}q_1 + \boldsymbol{\phi}_{,2}q_2)\,\mathrm{d}A - \int_{S_1} \boldsymbol{\phi}\bar{q}_n\,\mathrm{d}S\end{aligned}\right\} \tag{7.38}$$

A_1 and A_2 stand for the advective part of the particle derivatives, i.e.

$$\left.\begin{aligned}A_1 &= \frac{\partial}{\partial x_1}\left(\frac{q_1^2}{H}\right) + \frac{\partial}{\partial x_2}\left(\frac{q_1 q_2}{H}\right)\\ A_2 &= \frac{\partial}{\partial x_1}\left(\frac{q_1 q_2}{H}\right) + \frac{\partial}{\partial x_2}\left(\frac{q_2^2}{H}\right)\end{aligned}\right\} \tag{7.39}$$

Equations (7.37) are a highly nonlinear system of equations valid over one element. We can then assemble (7.37) for the whole continuum and write the results

$$\mathscr{M}\dot{\mathscr{Q}} = \mathscr{F}(\mathscr{Q}_0) \tag{7.40}$$

an expression for which we assume the mass flux and elevation boundary conditions has been taken into account. $\mathscr{M}$ is the global mass matrix and $\dot{\mathscr{Q}}$ represents the derivatives with respect to time of the mass flux and elevation of all the nodes. All the other terms are included in $\mathscr{F}$ and computed at $t = t_0$, or, when an iterative method is applied at the end of the time step, they are the values obtained at the previous iteration. Integration of (7.40) may be attempted using an explicit method of solution, such as the Runge–Kutta or

Euler technique, or an implicit method, e.g. the trapezoidal rule, Galerkin's method, etc.

7.3 NUMERICAL INTEGRATION SCHEMES

Efficient stable numerical integration schemes are essential in shallow water studies since a typical problem will involve several hundred node points and integrations over at least one tidal cycle. Complex multi-step methods, although more accurate, require considerably more computation time and storage. Therefore, in addition to the schemes already seen, we will now study the stability and accuracy of relatively simple implicit schemes.

Explicit stability criteria for general finite element formulations have not yet been developed. The difficulty is due to the arbitrariness of the coefficient matrices (i.e. the elements are confined to a zone adjacent to the diagonal but their magnitudes may be irregular) and also for the present case the skew symmetry of the Coriolis and surface elevation terms. One generally has to resort to approximate stability measures based on norms but we make no attempt here to resolve this problem.

The simplest scheme is the trapezoidal rule. Its one-dimensional form is

$$\frac{\mathrm{d}y}{\mathrm{d}t} = f(y, t)$$

$$y_{n+1} - y_n = \frac{\Delta t}{2}(f_{n+1} + f_n) + E\,\Delta t \tag{7.41}$$

$$E = \frac{1}{12}(\Delta t)^2 \left|\frac{\mathrm{d}^2 f}{\mathrm{d}t^2}\right|_{\xi} \qquad t_n < \xi < t_{n+1}$$

Iteration is then required since the forcing terms in the shallow water equations are nonlinear. During the iteration we may include a relaxation factor to accelerate convergence. Convergence is defined by the percentage change in the Euclidian norms for the surface elevation and velocity vectors.

A second method examined is the third-order predictor–corrector iterative scheme,

$$\frac{\mathrm{d}y}{\mathrm{d}t} = f(y, t)$$

Predictor:

$$f_{n+1} = 3f_n - 3f_{n-1} + f_{n-2}$$

Corrector:

$$
\begin{aligned}
y^* - y_n &= \frac{\Delta t}{12}(5f_{n+1}^{j-1} + 8f_n - f_{n-1}) + E\,\Delta t \\
y_{n+1}^j &= \theta y^* + (1-\theta)y_{n+1}^{j-1} \qquad (7.42) \\
E &= \tfrac{1}{12}(\Delta t)^3 \left|\frac{d^3 f}{dt^3}\right|_\xi \qquad t_n < \xi < t_{n+1}
\end{aligned}
$$

where θ is a relaxation factor.

This scheme is not self-starting and requires more storage than the trapezoidal rule. However, it is more accurate and usually converges faster. The Euclidian norms can again be taken as the convergence criteria.

The predictor–corrector scheme may be coupled with the following version of the fourth-order Runge–Kutta method:

$$
\begin{aligned}
\frac{dy}{dt} &= f(y, t) \\
k_1 &= \Delta t\,.\,f(y_n, t_n) \\
k_2 &= \Delta t\,.\,f(y_n + 0{\cdot}4k_1, t_n + 0{\cdot}4\,\Delta t) \qquad (7.43) \\
k_3 &= \Delta t\,.\,f(y_n + 0{\cdot}296978k_1 + 0{\cdot}158760k_2, t_n + 0{\cdot}455737\,\Delta t) \\
k_4 &= \Delta t\,.\,f(y_n + 0{\cdot}218100k_1 - 3{\cdot}050965k_2 + 3{\cdot}832864k_3, t_n + \Delta t) \\
y_{n+1} &= y_n + 0{\cdot}174750k_1 - 0{\cdot}551481k_2 + 1{\cdot}205535k_3 + 0{\cdot}171185k_4 \\
E &= 0(\Delta t^4)
\end{aligned}
$$

This scheme has the lowest bound on the error for this family of Runge–Kutta methods.[7]

The solution of a given problem can start with an optional number of integration steps using the Runge–Kutta method (minimum three time steps), and then shift to the predictor–corrector method. At any time step it is possible to change back to the Runge–Kutta method to take advantage of its better accuracy. This flexible formulation also makes it very easy to increase or decrease the time increment Δt, if so desired.

The trapezoidal rule, which has been discussed in Chapter 6, and Galerkin in time are two important implicit methods. Both techniques imply the solution of a whole system of equations but have the advantage of allowing for the use of larger time increments than those of the explicit methods.

Example 7.1

The numerical integration schemes discussed above were used in a circulation model for Massachusetts Bay.[8] After some preliminary tests it was decided to use the Runge–Kutta integration scheme.

The shallow water equations used were equivalent to those discussed above, but convective terms were neglected in the momentum equations, which is an accepted simplification in many shallow water problems.[5]

The geographic boundaries and the finite element grid are shown in Figure 7.4. Since very few actual data were available, a model yielding only the gross circulation was appropriate at the time. A fairly coarse grid of 74 elements and 53 nodes was laid out, reflecting

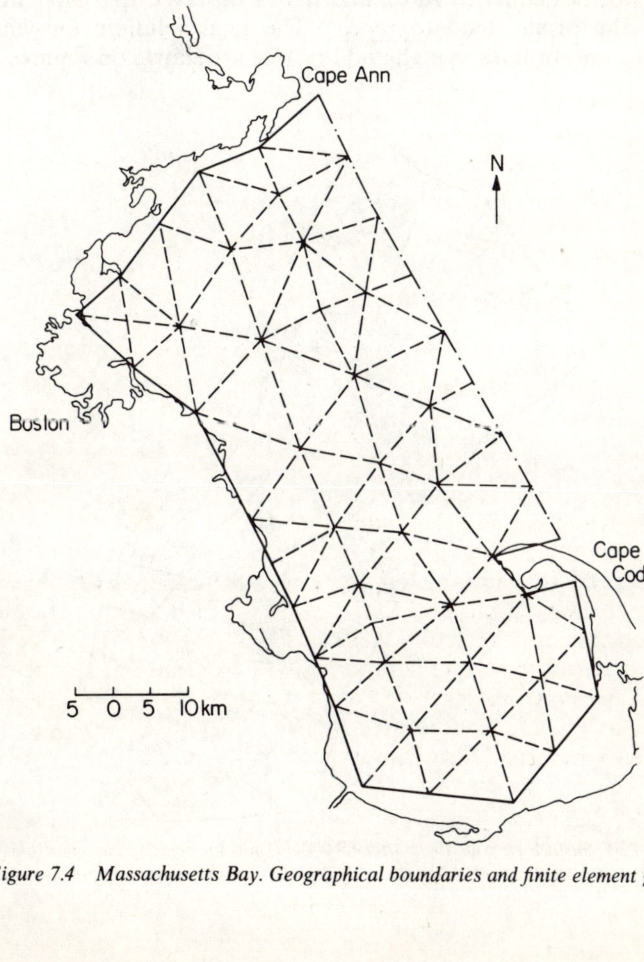

Figure 7.4 Massachusetts Bay. Geographical boundaries and finite element grid

somewhat the varying bottom topography. The tidal ranges for the two shore nodes at the extremities of the ocean boundary were obtained from tide tables,[9] and the tide level was assumed to vary linearly in between. The Coriolis parameter was determined for a latitude of 42°N as $f = 0{\cdot}973 \times 10^{-4}\ s^{-1}$. No attempts were made to model lateral inflows, and both velocities on the land boundaries were taken to be zero.

An initial solution with a small constant value of the friction coefficient was carefully examined in order to estimate new improved coefficients for each element, so that the tidal ranges and lag times at the shore points more closely match available tide table data. In estimating these coefficients a strong correlation with local depth was assumed. The model was started with a 'flat' condition (elevation and velocities equal to zero), and it was observed that after three cycles the results tend to repeat. The final solution, for which surface contour lines at high and low tide are shown on Figures 7.5

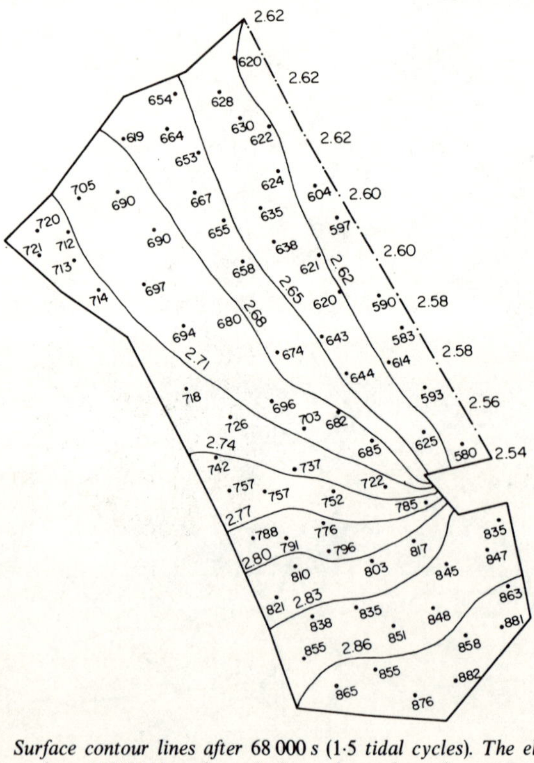

Figure 7.5 Surface contour lines after 68 000 s (1·5 *tidal cycles). The elevations are given in metres above MLW. Note that only decimals are shown for centroidal elevations*

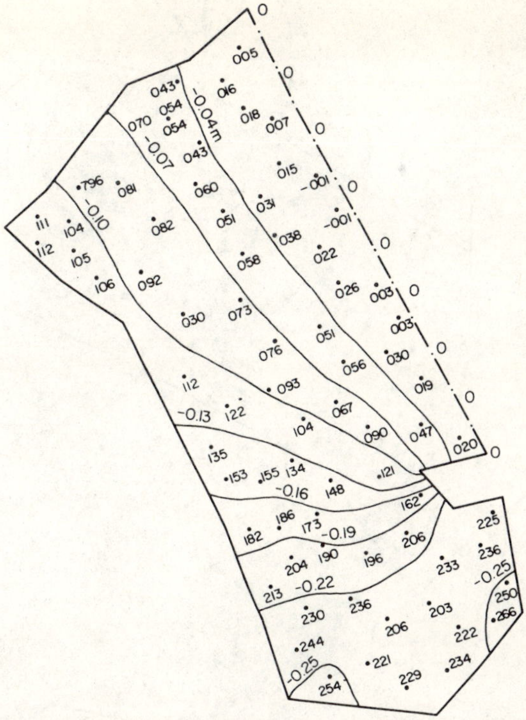

Figure 7.6 Surface contour lines after 90 000 s (two tidal cycles). Centroidal elevations are given in metres below MLW

and 7.6, had g/c^2 varying between 0·0025 and 0·0011. However, to really tune the model, current records at several points are desirable.

The calculated slack water velocities are shown in Figures 7.7 and 7.8, and typical time histories of surface elevations and velocities are plotted in Figures 7.9 and 7.10.

The criterion used to determine the time step was

$$\Delta t < \frac{\Delta x}{\sqrt{2c}} = \frac{6000}{\sqrt{219}} = 223\ \mathrm{s}$$

where $c = \sqrt{(gh)}$ = celerity. Hence a Δt of 200 seconds was selected.

The predictor–corrector scheme was applied to the same problem but exhibited gradual instability after one tidal cycle (44 600 s). When Δt was reduced to 150 s, comparable results were obtained for more than two tidal cycles. However, 5 per cent more computing time was required.

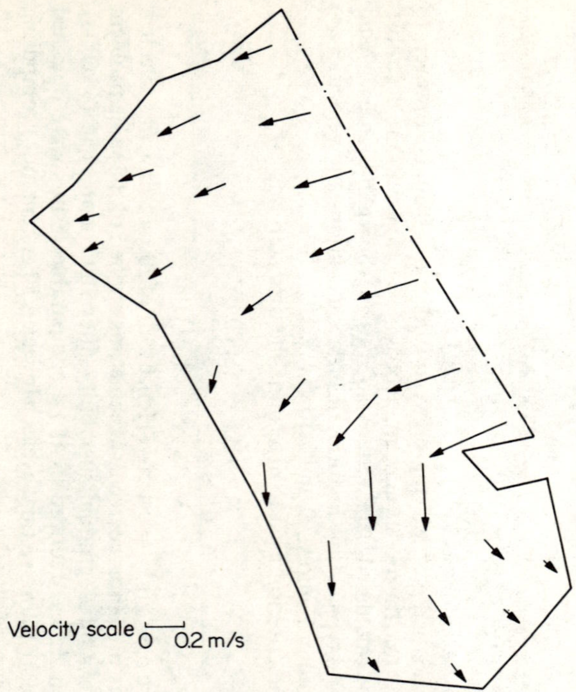

Figure 7.7 Computed currents after 56 000 *s* (1·25 *tidal cycles*)

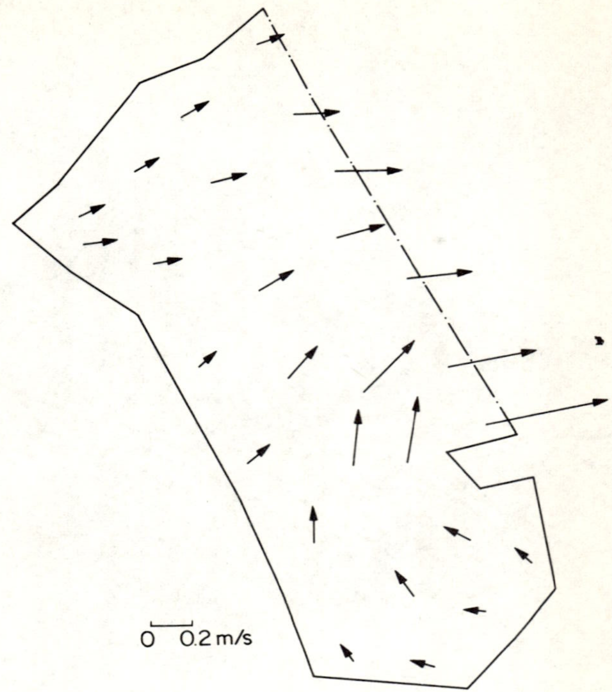

Figure 7.8 Computed currents after 78 000 *s* (1·75 *tidal cycles*)

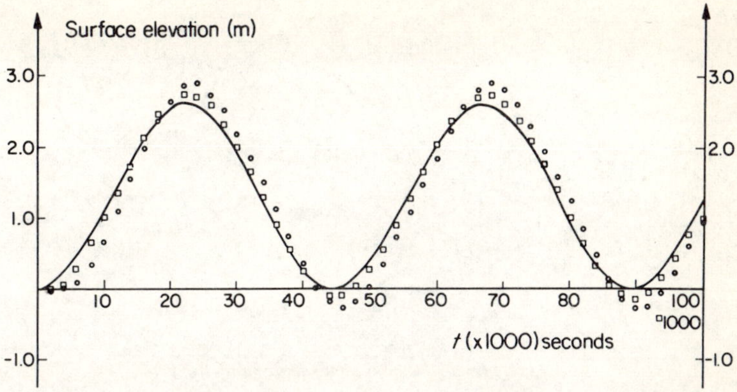

Figure 7.9 Time history of computed elevations at Boston and in Cape Cod Bay. □ *Boston*; ○ *Cape Cod Bay (Barnstable)*; — 1·30 *m* (1 − cos ωt)

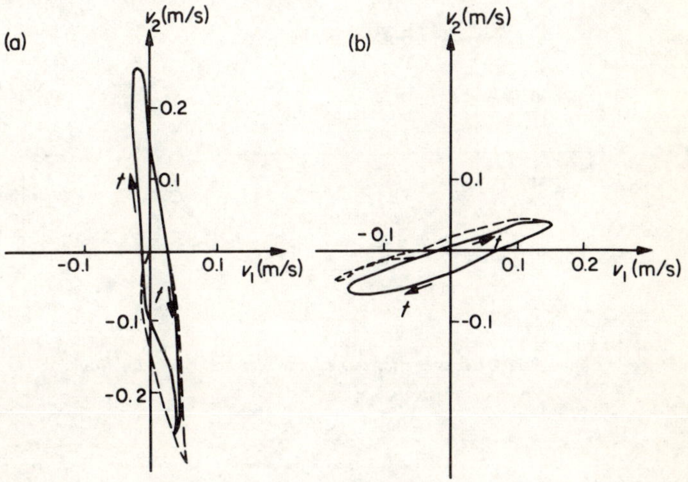

Figure 7.10 Time history of computed currents (a) *Centre of Cape Cod Bay*; (b) 15 *km east of Boston*

Example 7.2

A simple mathematical tidal model for the Solent was developed using finite elements (Figure 7.11).[10] The model used 86 three node triangular elements with 58 nodes. The elevation H and two flux parameters, q_1 and q_2, were taken as nodal variables. This gives a total of 174 unknowns for the model (Figure 7.13).

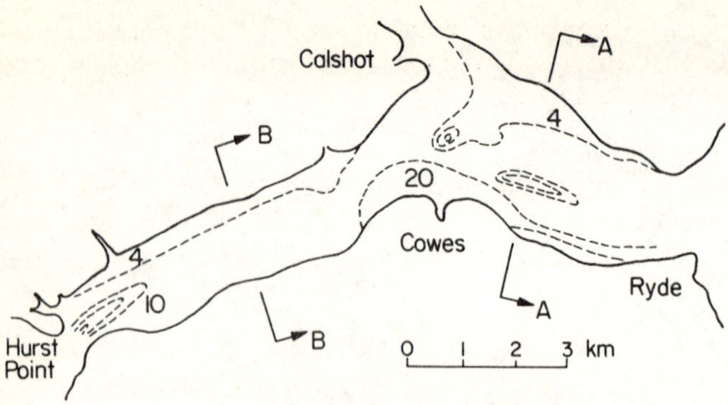

Figure 7.11 The Solent (depth in metres)

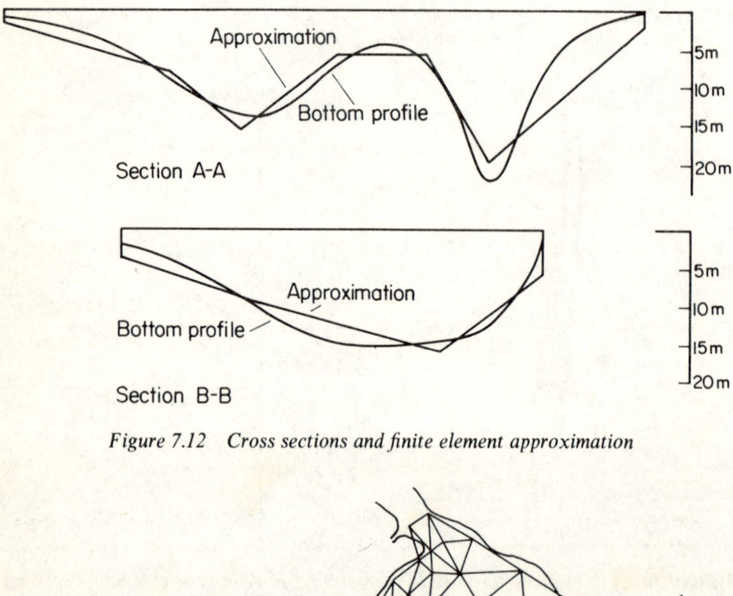

Figure 7.12 Cross sections and finite element approximation

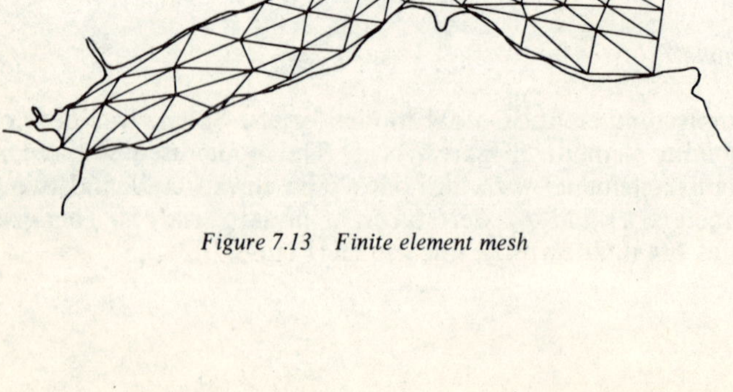

Figure 7.13 Finite element mesh

The ocean type boundary conditions (elevation only) (Figure 7.14) are obtained from the Admiralty Tide Tables for the Solent area. The model can take into consideration wind effects, bottom friction and Coriolis, although the latter was ignored for the Solent.

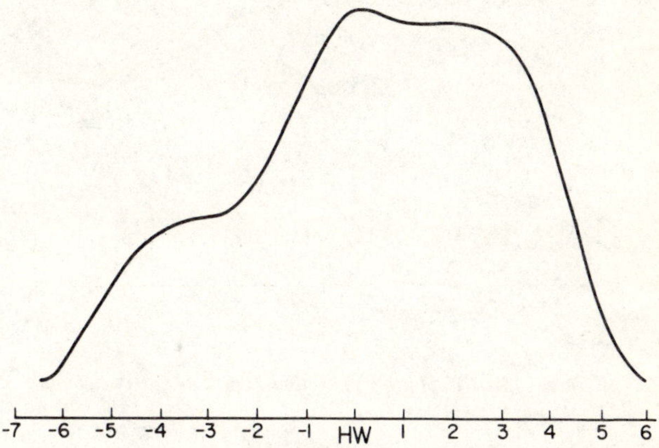

Figure 7.14 Typical spring tidal curve for Southampton

The time integration was carried out using fourth-order Runge–Kutta and the model was started from initial conditions of zero elevation and flux. The time step was selected using the Frederick–Courant–Levy finite difference criterion

$$\Delta t \leqslant \frac{\Delta x}{\sqrt{2}c} \tag{a}$$

where c is the celerity equal to $\sqrt{(gh)}$. h was taken to be the depth in the deepest part of the Solent.

The solution was started with $\Delta t = 30\,\text{s}$ for the first tidal cycle; it was increased to 120 s for the second cycle. This value agrees with Δt obtained from formula (a), which gives 20 s for the worst element. Note that the formula is based on a Euler-type integration and that for fourth-order Runge–Kutta the values should be multiplied by approximately four.

The motion of shallow waters is strongly dependent on the bottom topography. Hence the finite element mesh was chosen in such a way as to represent the complex channels and shallow flats.

For this example the derivatives of velocities in certain regions may be important, and hence convective terms were not neglected.

The model was started from the low water conditions with all qs and Hs equal to zero (cold starting). The computation was carried out for two complete cycles under neither wind nor Coriolis forces. For the bottom friction a coefficient of $c = 10\,\mathrm{m}^{1/2}\,\mathrm{s}^{-1}$ was chosen.

Typical results are shown in Figures 7.15 and 7.16. The velocities in Figure 7.15 agree reasonably well with those shown in the

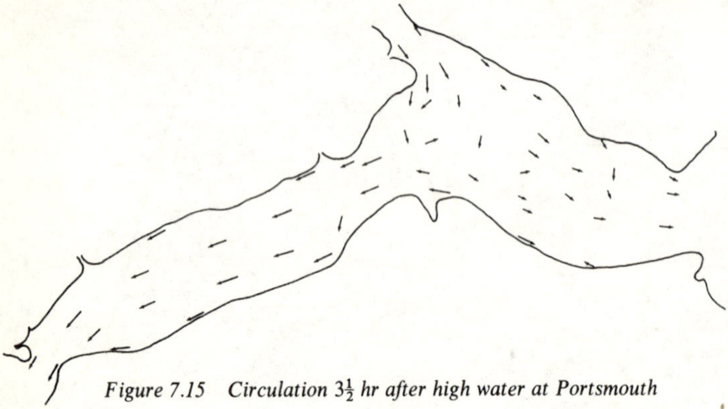

Figure 7.15 Circulation $3\frac{1}{2}$ hr after high water at Portsmouth

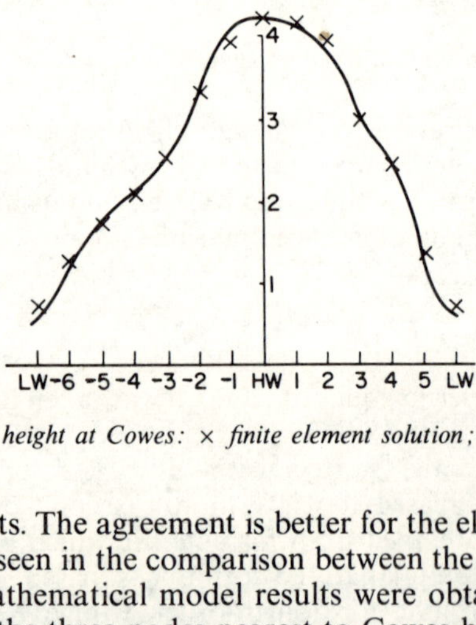

Figure 7.16 Tidal height at Cowes: × finite element solution; — tidal height from Admiralty charts

Admiralty charts. The agreement is better for the elevations (Figure 7.16) as can be seen in the comparison between the water heights at Cowes. The mathematical model results were obtained by finding the average of the three nodes nearest to Cowes harbour entrance (Figure 7.13).

7.4 LAKE CIRCULATION

Flow in lakes, cooling ponds and other water bodies can be approximated to provide an initial estimate of the circulation, which can then be checked against the full shallow water equations. This flow is then governed by linearised equations which are obtained by neglecting the inertia terms in the momentum equations; i.e.

$$\begin{aligned} -fq_2 + \rho g H \frac{\partial \eta}{\partial x_1} + (\tau_1|_s - \tau_1|_b) &= 0 \\ fq_1 + \rho g H \frac{\partial \eta}{\partial x_2} + (\tau_2|_s - \tau_2|_b) &= 0 \end{aligned} \tag{7.44}$$

and the time dependent terms in the continuity formula

$$\frac{\partial q_1}{\partial x_1} + \frac{\partial q_2}{\partial x_2} = 0 \tag{7.45}$$

If the η values are much smaller than the h we can write $H \simeq \eta$ in (7.44), hence

$$\begin{aligned} -fq_2 + \rho g h \frac{\partial \eta}{\partial x_1} + (\tau_1|_s - \tau_1|_b) &= 0 \\ fq_1 + \rho g h \frac{\partial \eta}{\partial x_2} + (\tau_2|_s - \tau_2|_b) &= 0 \end{aligned} \tag{7.46}$$

The momentum components can now be defined as

$$\frac{q_i}{H} \simeq \frac{q_i}{h} = \frac{\rho}{H} \int_{-h}^{\eta} v_i \, \mathrm{d}x_3 \tag{7.47}$$

The $\tau|_s$ terms are due to wind stresses and the $\tau|_b$ terms are bottom friction stress components. The latter are assumed here to be linearly proportional to the mean momentum components:

$$\tau_1|_b = \gamma q_1, \qquad \tau_2|_b = \gamma q_2 \tag{7.48}$$

We can now cross-differentiate equations (7.46) assuming that the derivatives of h are negligible (i.e. the bottom slope is small), and afterwards subtract both equations. This gives, taking continuity into consideration, the following equation:

$$\left(\frac{\partial \tau_1|_s}{\partial x_2} - \frac{\partial \tau_2|_s}{\partial x_1}\right) = \gamma \left\{\frac{\partial q_1}{\partial x_2} - \frac{\partial q_2}{\partial x_1}\right\} \tag{7.49}$$

One can propose a stream function such as

$$q_1 = \frac{\partial \psi}{\partial x_2}, \qquad q_2 = -\frac{\partial \psi}{\partial x_1} \tag{7.50}$$

Whence formula (7.49) becomes

$$W = \gamma \nabla^2 \psi \tag{7.51}$$

where

$$W = \left(\frac{\partial \tau_1|_s}{\partial x_2} - \frac{\partial \tau_2|_s}{\partial x_1} \right)$$

The boundary conditions associated with this equation are

$$\frac{\partial \psi}{\partial n} = 0 \quad \text{on 'land' boundaries}$$
$$\psi = \bar{\psi} \quad \text{on 'inlet' boundaries} \tag{7.52}$$

Equation (7.51) plus conditions (7.52) can be written in variational form and the finite element technique can then be applied. The programs already developed for inviscid flow can be used.

Note that in the above formulation we have included the Coriolis parameter but assumed that it is constant for all the lake, i.e. the lake is small enough to allow the neglect of local variations in the Coriolis forces.

Example 7.3

Cheng[11] has produced numerical results for wind circulation in Lake Erie using the above formulation. He employs the mesh shown in Figure 7.17, consisting of 516 three node triangular elements and

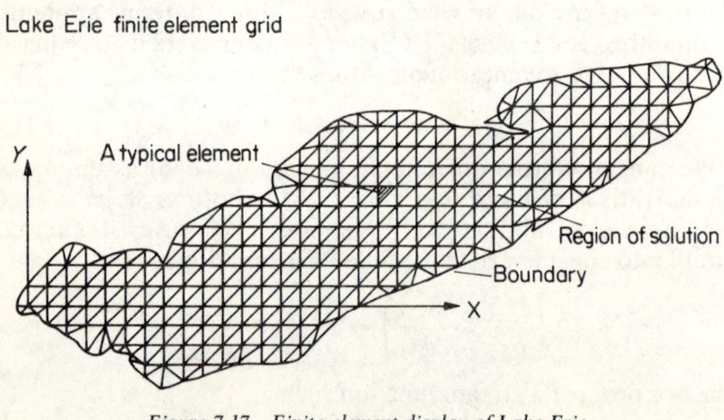

Figure 7.17 Finite element display of Lake Erie

308 nodal points. As a first numerical example he calculates the stream lines for the flow in and out of the lake without wind effects, taking $\psi = 0$ for the south shore of the lake and $\psi = 1$ for the north shore (Figure 7.18). For this case the governing equation becomes a

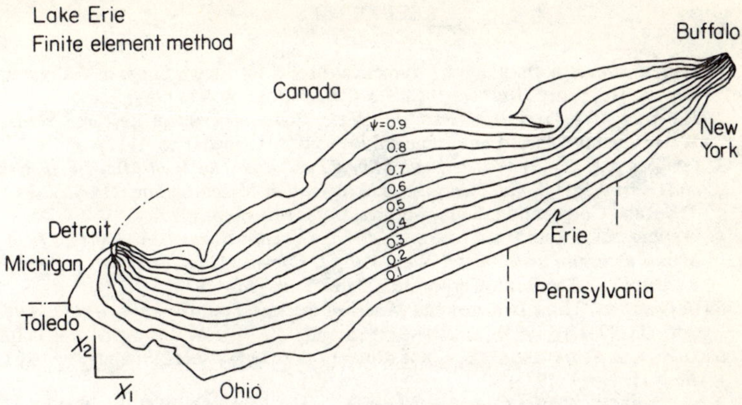

Figure 7.18 Steady state Detroit River to Niagara River flow pattern in Lake Erie based on potential flow

Laplace equation. He then considers the right-hand side term in formula (7.51) equal to 1, X_1 and X_2, which allows him to superimpose three different sets of results in order to obtain any solution of the type

$$\nabla^2\psi = A + BX_1 + CX_2 \tag{a}$$

Note that the right-hand side of equation (a) represents a quadratic wind stress distribution.

In this way Cheng is able to present different mean circulation streamlines driven by quadratic wind stress distribution, as for instance the one shown in Figure 7.19.

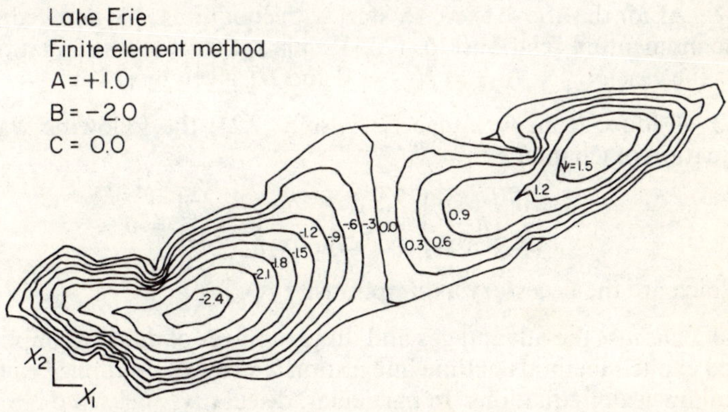

Figure 7.19 Two-celled mean circulation driven by a quadratic wind stress distribution $W = A + BX_1 + CX_2$

REFERENCES

1. REID, R. O. and BODINE, B. R., 'Numerical Model for Storm Surge in Galveston Bay', *J. Waterways Harbors Div., A.S.C.E.*, **94**, No. WW1 (1968)
2. *Estuarine Modeling: An Assessment*, Water Pollution Control Research Series, Water Quality Office, Environmental Protection Agency (Feb. 1971)
3. LEENDERSTSE, J. J., *A Water-Quality Simulation Model for Well-Mixed Estuaries and Coastal Seas, Vol. I, Principles of Computation*, Memorandum RM-6230-RC, The Rand Corporation, Santa Monica, California (Feb. 1970)
4. LEENDERSTSE, J. J. and GRITTON, E. C., *A Water-Quality Simulation Model for Well-Mixed Estuaries and Coastal Seas, Vol. II, Computational Procedures*, Report R-708-NYC, The Rand Corporation, New York (July 1971)
5. HEAPS, N. S., 'Three Dimensional Model of the Irish Sea', *Geophys. J. Roy. Astr. Soc.*, **35**, 99–120 (1973)
6. ABBOTT, M. B., DANESGAARD, A. and RODENHUIS, G. S., 'System 21, Jupiter', *J. Hyd. Res.*, **11**, No. 1 (1973)
7. RALSTON, A., *A First Course in Numerical Analysis*, McGraw-Hill (1965)
8. CONNOR, J. J. and WANG, J., 'Finite Element Modeling of Hydrodynamic Circulation', in *Numerical Methods in Fluid Dynamics*, C. Brebbia and J. J. Connor (Eds.), Pentech Press (1974)
9. U.S. Department of Commerce, NOAA, Tide Tables, East Coast of North and South America (1973)
10. BREBBIA, C. A. and ADEY, R., 'Circulation Problems', Proc. Finite Element Seminar, Chilton, UK, Atlas Computer Laboratory (1975)
11. CHENG, R. T. and TUNG, C., 'Wind Driven Lake Circulation by the Finite Element Method', *Proc. 13th Conference Great Lakes* (1970)

EXERCISES

7-1 Using Leibnitz' rule and the kinematic condition, check that equation (7.10) can be derived from (7.6).

7-2 As for the above exercise, start with equations (7.5) and deduce the momentum relationships (7.21); then specialise these equations for the case $N_{11} = N_{22} = N_{12} = 0$ and N_p given by (7.16).

7-3 Deduce from formulae (7.10) and (7.21) the following wave equation of Chapter 5:

$$\frac{\partial}{\partial x_1}\left(h\frac{\partial \eta}{\partial x_1}\right) + \frac{\partial}{\partial x_2}\left(h\frac{\partial \eta}{\partial x_2}\right) - \frac{\partial^2 \eta}{\partial t^2} = 0 \tag{a}$$

Which are the necessary assumptions?

7-4 Discuss the advantages and disadvantages of different implicit and explicit methods of time integration which can be applied to the shallow water equations. In particular, discuss

(a) Euler's method

(b) The Runge–Kutta method
(c) The trapezoidal rule
(d) Galerkin's method

7-5 Specialise the shallow water equations for a one-dimensional river of variable depth.

7-6 Modify the Laplace equation program developed in Chapter 2 in order to solve wind driven circulation problems.

7-7 A lake with an island (Figure 7.20) is subjected to a wind such that W in equation (7.51) is given by

$$W = A + BX_2 \tag{a}$$

Discuss a finite element mesh (three node triangles) to solve the problem assuming that the depth of the lake is constant.

For which wind distribution would you expect two circulation regions, one around the island, the other to the right-hand side.

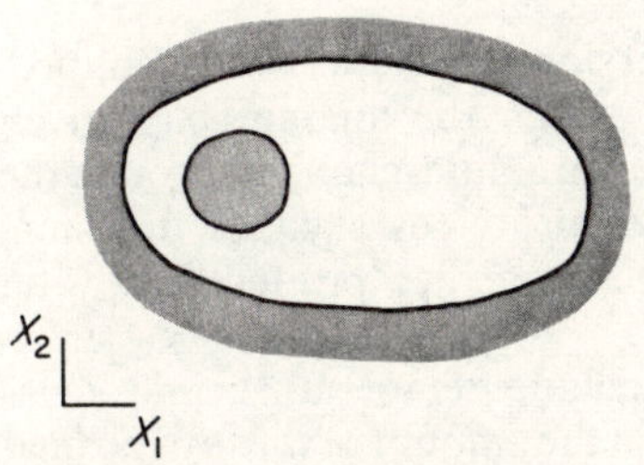

Figure 7.20 Lake with an island

7-8 How would you account for river inflow in Exercise 7.7?

7-9 Consider a rectangular 'lake' type body of water 300 km long from north to south and 200 km wide. The mean latitude of the lake is 30° and its depth 10 m. The average velocities due to inflows and outlets are in the order of 0·10 m/s. Are Coriolis induced currents going to be significant?

8 Dispersion Problems

8.1 INTRODUCTION

In this chapter the problem of dispersion in general will be studied. Dispersion is the phenomenon by which the amount of a given variable, such as a substance dissolved in a fluid, changes through the domain. This phenomenon involves processes such as diffusion, convection, decay, etc., which added together produce the total dispersion.

Classical dispersion processes include temperature and mass transfer. In recent years the increase in the amount of waste deposited in estuaries from industry and urban developments has led to concern regarding the quality of water in these areas. It has therefore become important to predict the effect of any proposed development on water quality. The same argument may be applied to dispersion of atmospheric pollutants in the air. These processes are governed by the mass transfer equation, which is similar to the energy equation previously deduced (see Chapter 4).

Let us first consider the process of molecular diffusion in a fluid. Assume that the concentration of substance 1 in fluid 2 is the variable under study, and the fluid is at rest. The random motion of the molecules is such that the net motion will occur from points of high to points of low concentration. Fick's law for diffusion of 1 can be expressed in terms of mass flux $\vec{q}$ as

$$\vec{q} = -\rho K^m \vec{\nabla}\left(\frac{\rho_1}{\rho}\right) \tag{8.1}$$

where ρ_1 is the mass of substance 1 per unit volume of the mixture diffusing through a liquid of total density ρ (density of component 1 plus 2, i.e. $\rho = \rho_1 + \rho_2$); the value of K^m depends on many parameters. In formula (8.1) the diffusion is assumed to be isotropic, i.e. equal in all directions. A more general expression for the components of $\vec{q}$ is

$$q_j = -\rho K_{ij}^m \left(\frac{\partial(\rho_1/\rho)}{\partial x_j} \right) \tag{8.2}$$

The quantity ρ_1/ρ will be called θ, the concentration of component 1 in the 'host' fluid. Thus (8.2) becomes

$$q_j = -\rho K_{ij}^m \left(\frac{\partial \theta}{\partial x_j} \right) \tag{8.3}$$

The mixing processes in fluids are very much affected by turbulence; i.e. the velocities at a point are random in such a fashion that the fluid can be visualised as having an average velocity plus some standard deviation component which creates eddies of different sizes. This hypothesis will be used in the mass transfer equation in order to obtain a diffusion term similar to the one in (8.3). Hence q_j for turbulent flow will be expressed as (see 8.2)

$$q_j = -\rho(K_{ij}^m + K_{ij}^t) \left(\frac{\partial \theta}{\partial x_j} \right) \tag{8.4}$$

where K_{ij}^t is the coefficient of turbulent diffusion. Its determination in practice is very difficult, especially in those problems where the introduction of the diffusing fluid in the host fluid alters the turbulence (for instance 'plumes' in seepage outlets). Those problems will not be considered here.

8.2 THE MASS TRANSFER EQUATION

If we consider (Figure 8.1) a volume at times t and $t + \Delta t$, the amount of dissolved substance flowing into and out of the volume is governed by the following relationship:

$$\lim_{\Delta t \to 0} \left\{ \frac{\Delta \iiint \rho \theta \, \mathrm{d}V}{\Delta t} \right\} = \frac{\mathrm{D}}{\mathrm{D}t} \iiint \theta \rho \, \mathrm{d}V \tag{8.5}$$

Equation (8.5) represents the increase in the amount of substance contained in V.

If the mass enclosed by the control volume can be considered constant (i.e. the change in concentration does not alter the total mass

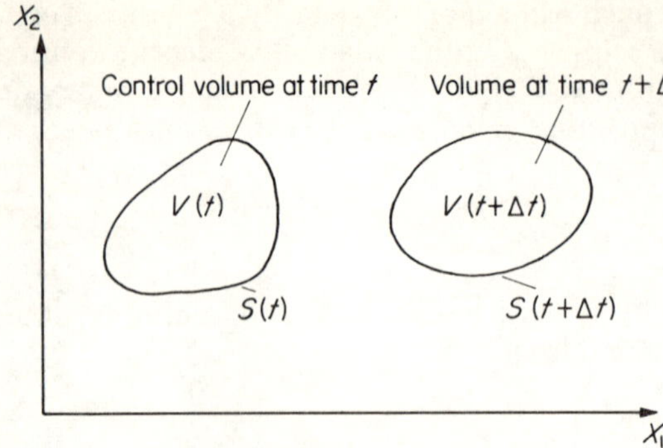

Figure 8.1 Volume at t and t + Δt

appreciably), we have [see (4.20)]

$$\frac{\mathrm{D}}{\mathrm{D}t}\iiint \theta\rho\,\mathrm{d}V = \iiint \frac{\mathrm{D}\theta}{\mathrm{D}t}\rho\,\mathrm{d}V \tag{8.6}$$

$\mathrm{D}/\mathrm{D}t$ is the total derivative which involves the v_i velocity components in the x_i directions.

The rate of input of the substance across the boundary in terms of flux plus any external input can be written

$$\iiint \rho p\,\mathrm{d}V - \iint q_n\,\mathrm{d}S \tag{8.7}$$

where p is the external input term (distributed input rate per unit mass) and q_n is the normal flux across the boundary.

We can write the equilibrium relationship in terms of (8.6) and (8.7) as

$$\iiint \rho\frac{\mathrm{D}\theta}{\mathrm{D}t}\mathrm{d}V = \iiint \rho p\,\mathrm{d}V - \iint_S q_n\,\mathrm{d}S \tag{8.8}$$

Applying Gauss's theorem one transforms the S term in (8.8) into a term in V such that

$$\iiint \left\{\rho\frac{\mathrm{D}\theta}{\mathrm{D}t} - \rho p + \frac{\partial q_i}{\partial x_i}\right\}\mathrm{d}V = 0 \tag{8.9}$$

As (8.9) is valid for any control volume, we have

$$\rho\frac{\mathrm{D}\theta}{\mathrm{D}t} - \rho p + \frac{\partial q_i}{\partial x_i} = 0 \tag{8.10}$$

which is the equilibrium equation governing the problem.

For molecular diffusion, the relationship between the qs and θ is given by formula (8.2):

$$q_i = -\rho K_{ij}^m \left(\frac{\partial \theta}{\partial x_j} \right) \tag{8.11}$$

Generally the principal directions of the K_{ij}^m components are assumed to coincide with the directions of the axis. This gives the orthotropic coefficients of molecular diffusion

$$q_i = -\rho K_{ii}^m \frac{\partial \theta}{\partial x_i} \tag{8.12}$$

The equilibrium equation (8.10) can now be written in terms of θ:

$$\rho \frac{D\theta}{Dt} - \rho p - \frac{\partial}{\partial x_i}\left(\rho K_i^m \frac{\partial \theta}{\partial x_i} \right) = 0 \tag{8.13}$$

or

$$\rho \left\{ \frac{\partial \theta}{\partial t} + \frac{\partial}{\partial x_i}(v_i \theta) \right\} = \rho p + \rho \frac{\partial}{\partial x_i}\left(K_i^m \frac{\partial \theta}{\partial x_i} \right) \tag{8.14}$$

Turbulence is usually accounted for by expressing the variables in (8.13) as a sum of an *average* value plus a *deviation*:

$$\begin{aligned} v_i &= v_i' + v_i'' \\ \theta_i &= \theta_i' + \theta_i'' \end{aligned} \tag{8.15}$$

where v_i' and θ_i' are averaged values of velocities and concentration and v_i'', θ_i'' the corresponding deviations.

The average values are defined as

$$\theta' = \langle \theta \rangle = \frac{1}{T} \int_{\xi}^{\xi + T} \theta \, d\xi \tag{8.16}$$

As a consequence

$$\langle \theta'' \rangle = 0, \qquad \langle \theta' \rangle = \theta' \tag{8.17}$$

Thus for instance,

$$\langle \theta' v'' \rangle = \theta' \langle v'' \rangle = 0$$

We can now substitute (8.15) into the instantaneous mass transfer equation (8.13) and obtain

$$\rho \left\{ \frac{\partial(\theta' + \theta'')}{\partial t} + \frac{\partial}{\partial x_i}[(v_i' + v_i'')(\theta' + \theta'')] \right\} = \rho p + \rho \frac{\partial}{\partial x_i}\left(K_{ii}^m \frac{\partial(\theta' + \theta'')}{\partial x_i} \right) \tag{8.18}$$

Integrating over the total ensemble, (8.18) becomes

$$\rho\left\{\frac{\partial\theta'}{\partial t} + v_i'\frac{\partial\theta'}{\partial x_i} + \frac{\partial}{\partial x_i}\langle v_i''\theta''\rangle\right\} = \rho p + \rho\frac{\partial}{\partial x_i}\left(K_{ii}^m\frac{\partial\theta'}{\partial x_i}\right) \quad (8.19)$$

The deviation term in (8.19) is usually treated in a similar way as equation (8.12), i.e.

$$\langle v_i''\theta''\rangle = -K_{ii}^t\frac{\partial\theta'}{\partial x_i} \quad (8.20)$$

where the K_{ii}^t terms are a set of empirical coefficients called eddy diffusion coefficients. The validity of these coefficients in a turbulence field is questionable but unavoidable due to the present state of the art.

The molecular diffusivity coefficient K_{ii}^m can be added to K_{ii}^t and this gives a K_{ii} total diffusivity coefficient. Thus (8.19) can be written

$$\rho\left\{\frac{\partial\theta}{\partial t} + v_i\frac{\partial\theta}{\partial x_i}\right\} = \rho p + \rho\frac{\partial}{\partial x_i}\left(K_{ii}\frac{\partial\theta}{\partial x_i}\right) \quad (8.21)$$

where θ' and v_i' averages have been written as θ and v_i for simplicity. Note that equation (8.21) is similar to the energy equation deduced in Chapter 4.

BOUNDARY CONDITIONS

The boundary conditions required to solve equation (8.21) can be of two types (Figure 8.2):

(a) Applied values of concentration ($\bar{\theta} = \theta$, where the $\bar{\theta}$ terms are the known values), on the S_1 part of the boundary.

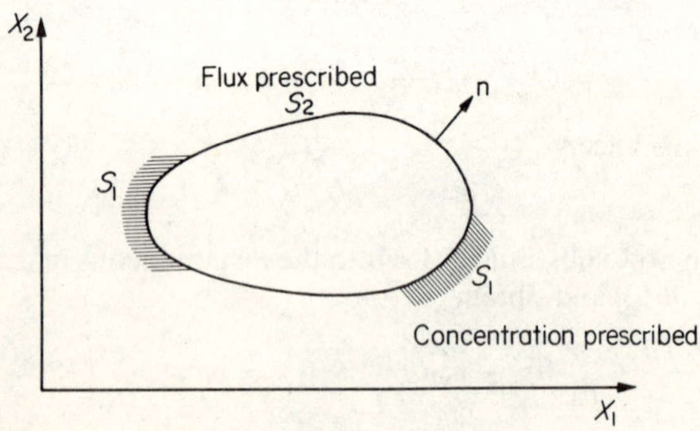

Figure 8.2 Boundary conditions

(b) Applied flux values ($\bar{q}_n = \alpha_{ni} q_i$, where $\bar{q}_n$ is the normal flux across the boundary and α_{ni} the direction cosines of the normal with respect to the x_i axis), on the S_2 part of the boundary.

Note that the total boundary S is $S = S_1 + S_2$.

SOURCES AND SINKS

Sources and sinks terms acting on the volume are represented in equation (8.21) by the p term. In general we also want to introduce a decay term by dividing p into a distributed source rate and a decay term:

$$p = f - \beta\theta \tag{8.22}$$

where f is a distributed source and β a decay constant.

Equation (8.22) represents the linear decay case, which is very convenient in mathematical modelling; the other usual type of decay is the time exponential law

$$\theta = \theta_0 \, e^{-t/\gamma}$$

where θ_0 is the initial concentration and γ is a constant. Only linear decay was used in the following.

We may represent absorption at the boundary by rewriting the prescribed flux terms in terms of the θ_e value outside the boundary and θ the value at the boundary. This gives

$$\bar{q}_n = -\gamma(\theta - \theta_e) + \alpha_{ni} q_i \tag{8.23}$$

where γ is a boundary absorption coefficient, and

$$q_i = -\rho K_{ii} \frac{\partial \theta}{\partial x_i}$$

A point sink can be defined as

$$Q = \sum Q_w(x_i) \, \Delta_i \tag{8.24}$$

where $Q_w(x_i)$ is the sink at position x_i, and Δ is the delta function (equal to 1 for x_i coordinates, zero otherwise).

VARIATIONAL STATEMENT

By weighting the equilibrium equation (8.21) and the boundary conditions by a variation of θ, we can write the following Galerkin type expression:

$$\iiint\left\{\frac{\partial q_i}{\partial x_i} - \rho f + \rho\beta\theta + \rho\frac{D\theta}{Dt}\right\}\delta\theta\,dV$$
$$= \iint_{S_2}\{\alpha_{ni}q_i - \gamma(\theta - \theta_e) - \bar{q}_n\}\,\delta\theta\,dS + \sum_i Q_w\,\Delta_i\,\delta\theta \qquad (8.25)$$

where the θ terms are assumed to satisfy the conditions $\theta = \bar{\theta}$ on S_1.

Integrating (8.25) by parts, we obtain

$$\iiint\left[-q_i\frac{\partial\delta\theta}{\partial x_i} + \rho\left\{\frac{D\theta}{Dt} - f + \beta\theta\right\}\delta\theta\right]dV$$
$$= \iint_{S_2}\{-\gamma(\theta - \theta_e) - \bar{q}_n\}\,\delta\theta\,dS + \sum_i Q_w\,\Delta_i\,\delta\theta$$

In terms of θ, one has

$$\iiint\left[\rho K_{ii}\left(\frac{\partial\theta}{\partial x_i}\right)\left(\frac{\partial\delta\theta}{\partial x_i}\right) + \rho\left\{-f + \beta\theta + \frac{D\theta}{Dt}\right\}\delta\theta\right]dV$$
$$= -\iint_{S_2}\{\bar{q}_n - \gamma(\theta_e - \theta)\}\,\delta\theta\,dS + \sum_i Q_w\,\Delta_i\,\delta\theta \qquad (8.26)$$

FINITE ELEMENT FORMULATION

By assuming that the concentration remains constant in the x_3 direction we can reduce the problem to a two-dimensional domain. Thus (8.26) becomes

$$\iint\left[\rho h K_{ii}\left(\frac{\partial\theta}{\partial x_i}\right)\left(\frac{\partial\delta\theta}{\partial x_i}\right) + \rho h\left\{-f + \beta\theta + \frac{D\theta}{Dt}\right\}\delta\theta\right]dA$$
$$= -\int_{S_2} h\{\bar{q}_n - \gamma(\theta_e - \theta)\}\,\delta\theta\,dS + \sum_i Q_w\,\Delta_i\,\delta\theta \qquad (8.27)$$

where h is the depth. We introduce an expansion for θ that is valid on each element

$$\theta = \boldsymbol{\phi}^T\boldsymbol{\theta}^n \qquad (8.28)$$

where $\boldsymbol{\phi}(x_1x_2)$ is an interpolation function and $\boldsymbol{\theta}^n$ is the vector of nodal unknowns.

After substitution into (8.27) we obtain for each element

$$\delta\boldsymbol{\theta}^{n,T}\{\mathbf{K}\boldsymbol{\theta}^n + \mathbf{A}\boldsymbol{\theta}^n + \mathbf{M}\dot{\boldsymbol{\theta}}^n - \mathbf{Q}\} = \mathbf{0} \tag{8.29}$$

where

$$\mathbf{K} = \iint h\rho\left[K_{11}\left(\frac{\partial\boldsymbol{\phi}}{\partial x_1}\right)\left(\frac{\partial\boldsymbol{\phi}}{\partial x_1}\right)^T + K_{22}\left(\frac{\partial\boldsymbol{\phi}}{\partial x_2}\right)\left(\frac{\partial\boldsymbol{\phi}}{\partial x_2}\right)^T + \beta\boldsymbol{\phi}\boldsymbol{\phi}^T\right]\mathrm{d}A + \int h\gamma\boldsymbol{\phi}\boldsymbol{\phi}^T\,\mathrm{d}S$$

$$\mathbf{A} = \iint h\rho\left\{v_1\boldsymbol{\phi}\left(\frac{\partial\boldsymbol{\phi}}{\partial x_1}\right)^T + v_2\boldsymbol{\phi}\left(\frac{\partial\boldsymbol{\phi}}{\partial x_2}\right)^T\right\}\mathrm{d}A \tag{8.30}$$

$$\mathbf{M} = \iint h\rho\boldsymbol{\phi}\boldsymbol{\phi}^T\,\mathrm{d}A$$

$$\mathbf{Q} = \iint h\rho f\boldsymbol{\phi}\,\mathrm{d}A - \int h(\bar{q}_n\boldsymbol{\phi} - \gamma\theta_e\boldsymbol{\phi})\,\mathrm{d}S + \sum_i Q_w\Delta_i\boldsymbol{\phi}$$

For all the elements we obtain

$$\mathscr{K}\boldsymbol{\theta} + \mathscr{A}\boldsymbol{\theta} + \mathscr{M}\dot{\boldsymbol{\theta}} = \mathscr{Q} \tag{8.31}$$

where the θ terms satisfy the $\boldsymbol{\theta} = \bar{\boldsymbol{\theta}}$ conditions.

8.3 DIFFUSION PROBLEMS

Let us consider here the case of having a diffusion-only problem for which the convective terms in (8.31) will be neglected, i.e.

$$\mathscr{K}\boldsymbol{\theta} + \mathscr{M}\dot{\boldsymbol{\theta}} = \mathscr{Q} \tag{8.32}$$

This is the case when there is no flow of the material (e.g. temperature diffusion in solid, concentration changes in still waters, etc.).

A system of equations like that can be integrated on time by using a simple implicit integration scheme such as the trapezoidal rule of Chapter 7, which can be written

$$\boldsymbol{\theta}_t - \boldsymbol{\theta}_0 = \frac{\Delta t}{2}(\dot{\boldsymbol{\theta}}_t + \dot{\boldsymbol{\theta}}_0) \tag{8.33}$$

where the $\boldsymbol{\theta}_t$ terms are the (unknown) values at the end of a Δt step and $\boldsymbol{\theta}_0$ the initial (known) values.

Substituting (8.33) into (8.32) we obtain the following formula:

$$\left(\mathscr{K} + \frac{2}{\Delta t}\mathscr{M}\right)\boldsymbol{\theta}_t = \mathscr{Q}_t + \mathscr{M}\dot{\boldsymbol{\theta}}_0 - \frac{2}{\Delta t}\mathscr{M}\boldsymbol{\theta}_0 \tag{8.34}$$

Applying (8.32) at $t = t_0$ we can write the final recurrence formula

$$\left(\mathscr{K} + \frac{2}{\Delta t}\mathscr{M}\right)\boldsymbol{\theta}_t = \mathscr{Q}_t + \mathscr{Q}_0 - \left(\mathscr{K} - \frac{2}{\Delta t}\mathscr{M}\right)\boldsymbol{\theta}_0 \tag{8.35}$$

Some other time integration schemes, such as explicit ones, could also have been used, but the advantage of the trapezoidal rule is its stability, which permits using large Δt steps for problems with constant $\mathscr{K}$, $\mathscr{M}$ matrices. If these matrices are variable it may be necessary to reduce the time step in order to take their variations into account and an explicit method may then be more appropriate.

Next we could be interested in the steady state solution only, for which case the time integration is no longer necessary as equation (8.32) simply reduces to

$$\mathscr{K}\boldsymbol{\theta} = \mathscr{Q}$$

Example 8.1

As heat conduction is a diffusion process, the above formulae and the heat conduction equations are similar. Now for instance, the flux formula

$$q = -K_{ii}\frac{\partial \theta}{\partial x_i} \tag{a}$$

still applies, and is such that

- q represents the amount of heat flowing per unit area, in calories, joules etc.
- K_{ii} is the thermal conductivity in the x_i direction, and
- θ is the temperature.

The steady state heat flow for the pipe described in Figure 8.3 was analysed using three node elements. The temperature distribution was obtained assuming temperatures of 100° inside the pipe and 0° outside. The results for lines of equal temperature and flow lines are plotted in the figure.

The same problem was then solved for transient temperature conditions, using equation (8.35). Results for nodes 24 and 14 are plotted in Figure 8.4 where we can see that the results converge to the steady state solution.

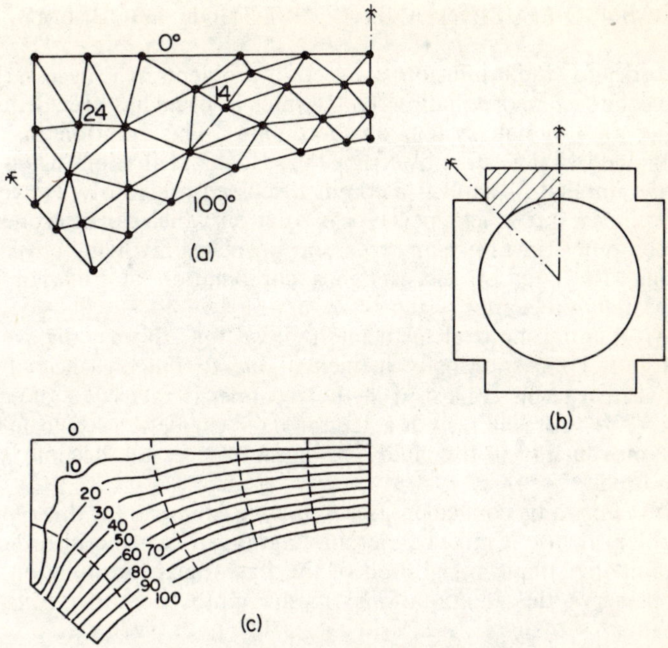

Figure 8.3 Temperature distribution problems for a pipe. (a) *Finite element discretisation*; (b) *The complete pipe*; (c) *Isothermal and flow lines*

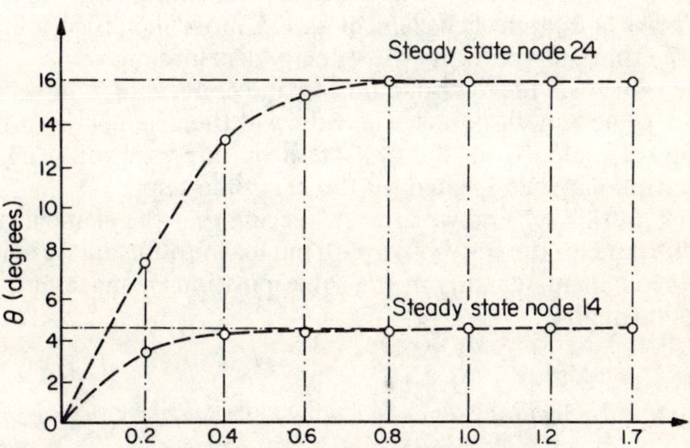

Figure 8.4 Transient temperature solution

8.4 DIFFUSION AND CONVECTION PROBLEMS

In order to study diffusion–convection problems, it is necessary to solve the system of equations (8.31) which as usual has been deduced using an Eulerian system of coordinates. This solution may be obtained by integrating over time the system of equations using one of the implicit or explicit methods discussed previously. However, we will now introduce a new type of solution, which can be economic in computer time for many transient problems involving advective terms. The solution consists in a combination of Eulerian and Lagrangian reference systems.

An Eulerian finite element mesh allows for arbitrary distortions but it is computationally inefficient due to the presence of the convective terms contained in the asymmetric matrix $\mathscr{A}$ of equation (8.31). Alternatively if a Lagrangian scheme is used to follow the movement of the fluid the mesh can easily become very distorted.

It seems to be convenient (see following example) for this type of problem to use a mixed Eulerian–Lagrangian scheme in order to exploit the simple integration of the Lagrangian formulation but to preserve the computational mesh employed in the Eulerian scheme.

For the Lagrangian scheme (no advective terms) we may have a recurrence relation such as (8.35). In order to introduce convection into this transient state equation it may be assumed in many diffusion problems that, for a small Δt, the diffusion is independent of the convection and the matrices in equation (8.35) are constant. The convection is calculated afterwards by assuming that there is a fixed mass of fluid in each element and its movements are computed over the time step knowing the velocity distribution.

The value of θ due to both diffusion and convection is now known at a set of new mesh points. The values at the original mesh points can be calculated from the new mesh by interpolation. The same integration may be repeated for the next time step.

Thus, in this scheme we are considering that the element moves with the fluid for the convective part, but for the diffusion we consider that the element remains in the same position in space and the θ passes through it.

Example 8.2: Accuracy and efficiency of the scheme

To check the efficiency and accuracy of the above scheme, a conventional Eulerian iterative scheme was also developed using the trapezoidal integration formula directly in equation (8.31). This gives

$$\left(\mathscr{K} + \frac{2}{\Delta t}\mathscr{M}\right)\mathbf{\theta}_t = (\mathbf{Q}_t + \mathbf{Q}_0) - \left(\mathscr{K} - \frac{2}{\Delta t}\mathscr{M}\right)\mathbf{\theta}_0 - (\mathscr{A}_0\mathbf{\theta}_0 + \mathscr{A}_t\mathbf{\theta}_t) \tag{a}$$

At each time step an estimate is made for the $\mathbf{\theta}_t\mathscr{A}_t$ term and then the equation is iterated until the required accuracy is achieved.

The above Eulerian and the mixed schemes were applied to a one-dimensional dispersion problem for which an analytical solution exists.

The results presented in Figure 8.5 and Table 8.1 demonstrate the accuracy and convergence of the Lagrangian–Eulerian scheme and show comparison studies conducted with the iterative scheme [equation (a)]. Six node triangular elements were used.

From these and other similar tests it was concluded that the following relationship—originally derived for the stability analysis of one-dimensional finite difference meshes—could be used to estimate the time step required:

$$\frac{v\,\Delta t}{\Delta x} \leqslant K \qquad (K < 1) \tag{b}$$

where v is the velocity, Δt the time step and Δx the characteristic length of the element (in our case the distance between nodes). At

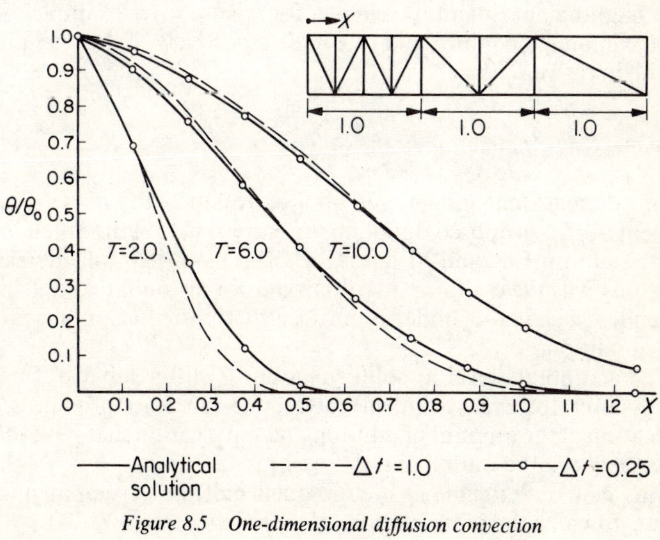

Figure 8.5 One-dimensional diffusion convection

Table 8.1 COMPARISON OF RESULTS FOR ONE-DIMENSIONAL DIFFUSION–CONVECTION PROBLEM

Distance from load X	*Analytic solution*	*Lagrangian–Eulerian scheme*			*Eulerian scheme with iteration*		
		$\Delta t = 0.25$	$\Delta t = 0.5$	$\Delta t = 1.0$	$\Delta t = 0.25$	$\Delta t = 0.5$	$\Delta t = 1.0$
0·000	1·0000	1·0000	1·0000	1·0000	1·0000	1·0000	1·0000
0·125	0·8986	0·9036	0·9127	0·9331	0·8979	0·8969	0·8949
0·250	0·7534	0·7600	0·7677	0·7846	0·7525	0·7503	0·7478
0·375	0·5817	0·5844	0·5876	0·5975	0·5792	0·5756	0·5679
0·500	0·4092	0·4094	0·4075	0·4045	0·4056	0·4012	0·3931
0·625	0·2603	0·2553	0·2482	0·2330	0·2559	0·2515	0·2413
0·750	0·1487	0·1411	0·1321	0·1132	0·1438	0·1401	0·1326
0·875	0·759	0·0685	0·0608	0·0448	0·0719	0·0693	0·0641
1·000	0·0346	0·0279	0·0225	0·0139	0·0308	0·0295	0·0276
1·250		0·0037	0·0028	0·0024	0·0038	0·00376	0·0040
1·500		0·0003	0·0003	0·0002	0·0010	0·0010	0·0010
1·750		0·0003	0·0003	0·0002	0·0003	0·0003	0·0003
2·000		0·0002	0·0001	0·0001	0·0001	0·0001	0·0001
2·500		0·0000	0·0000	0·0000	0·0000	0·0000	0·0000
3·000		0·0000	0·0000	0·0000	0·0000	0·0000	0·0000

Applied concentration $\theta_0 = 1{\cdot}0$ at $X = 0$, $K_x = 0{\cdot}01$ units^2/s, $V = 0{\cdot}05$ units/s.

the same time the accuracy of the scheme is governed by the ratio between the diffusion and the convected parts which implies that in a convection-only problem the size of the elements has to tend to zero in order to avoid errors due to numerical diffusion.

It is interesting to remark that the time and execution stores of the Lagrangian–Eulerian scheme for the above example were approximately one third and one half respectively of those required by the iterative scheme.

Example 8.3: Application in the Solent

A two-dimensional model was applied to study the dispersion of effluent from a proposed outfall in the eastern part of the Solent near Southampton (England) (Figure 8.6). The main item of interest in the study was the coliform distribution from the outfall and its time dependent behaviour under typical (in the present example spring) tide conditions.

The acceptable level of coliform count is still a subject of much discussion. However, coliform bacteria can at least provide some indication of the amount of dilution and purification that have taken place between the outfall and the point in question.

This new outlet came as a consequence of the development of a future town with a population of approximately 600 000 by the

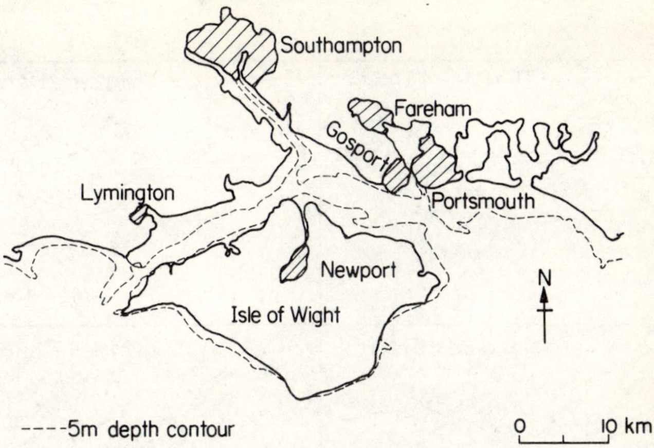

Figure 8.6 The Solent system

year 2001. The development area is to be situated between Southampton and Portsmouth, west of Fareham and a new treatment works plan has been proposed to deal with the sewage produced by this area. The new works were assumed to discharge the sewage effluent from an outfall approximately one kilometre off the coast.

The eastern part of the Solent was then subdivided into 209 six node elements (with 466 nodes) with the smallest element in the area surrounding the outfall (Figure 8.7).

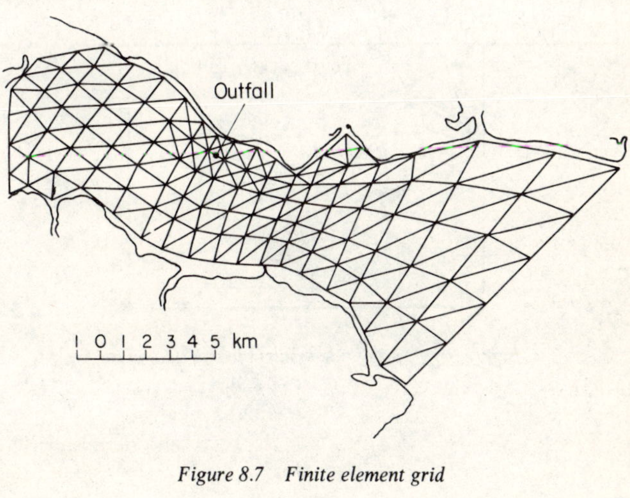

Figure 8.7 Finite element grid

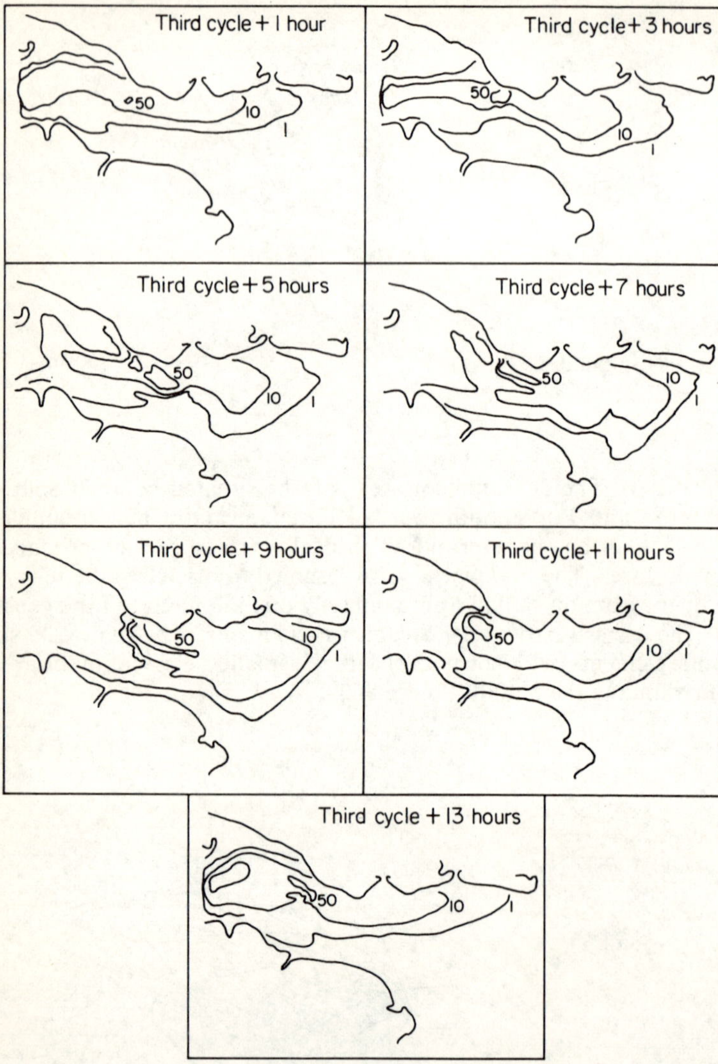

Figure 8.8 Spring tide results (fourth tidal cycle)

The surface velocities in the x_1, x_2 directions during the 13 hours of the spring tide were obtained from an extensive study of the Solent and given as input data. The results are interpolated for the appropriate Δt (in this case 600 seconds).

The turbulent diffusivity coefficients (K_1 and K_2) depend in general on position and time. In the computer program two coefficients were considered: the longitudinal one, in the direction of the nodal velocity vector, and the transverse one in the direction normal to this vector. From experimental observations with dyes, the following values were taken:

$$K_{\text{longitudinal}} = 5\ \text{m}^2/\text{s}$$
$$K_{\text{transverse}} = 0{\cdot}05\ \text{m}^2/\text{s}$$

valid for calm conditions.

In order to represent the decay of the coliform population in the mathematical model it was necessary to define a first-order rate of decay coefficient. However, as no experimental data were available for the conditions in the Solent, a value of $0{\cdot}4 \times 10^{-5}$ per second (0·33 per day) was used, which has been applied for similar conditions by other researchers.

The discharge from the outfall was defined by $Q \approx 6000$ litres/s with a coliform density of $1{\cdot}67 \times 10^6$ per 100 ml, remaining constant over the tidal cycles.

The results shown in Figure 8.8 are for the fourth spring tidal cycle of the model. They are essentially similar to those obtained up to the seventh cycle and indicate that, for the type of decay used in this model, four cycles are sufficient. The contours plotted by the computer are concentrations of 50, 10 and 1×10^3 coliforms per 100 ml for a depth of 1 m. They still have to be divided by the representative depth.

Example 8.4

A finite element dispersion model using 3-node triangular elements has been applied[1] to predict the sediment dispersion in Massachusetts Bay due to an offshore mining operation. Velocities were found using the computer program described in Example 7.1, and correspond to a tidal cycle plus a 10-knot wind from the west. This single tidal cycle of current information is repeated for every tidal cycle that the dispersion model is run. (It was fortunate that many tidal cycles of dispersion information could be obtained from only one cycle of currents, because the dispersion model is considerably less expensive than the circulation model in real-time execution. This is chiefly because the time-step used in the dispersion scheme can be about fifteen times greater than that in the circulation model.)

The time integration was performed using the trapezoidal rule in the following form. Write equations (8.31) as

$$\mathcal{M}\dot{\boldsymbol{\theta}} = \mathcal{P} \tag{a}$$

and apply the trapezoidal rule, i.e.

$$\mathcal{M}(\boldsymbol{\theta}_t - \boldsymbol{\theta}_0) = \frac{\Delta t}{2}(\mathcal{P}_t + \mathcal{P}_0) \tag{b}$$

Since the value of $\mathcal{P}$ depends on $\boldsymbol{\theta}$ this expression cannot be solved explicitly for $\boldsymbol{\theta}_t$ and it is necessary to solve by iteration. The tolerance for convergence is compared to the normalised root-mean-square error of all the nodal values. Thus the solution proceeds to the next step, when

$$\text{tolerance} > \frac{\sqrt{\left(\sum_{j=1}^{n} \{\theta_t^{(iH)} - \theta_t^{(i)}\}^2_{\text{node } j}\right)}}{\sqrt{\left(\sum_{j=1}^{n} \{\theta_t^{(i)}\}^2_{\text{node } j}\right)}} \tag{c}$$

where i indicates the iteration step.

The source used for this problem was a continuous load of 10^7 g/s distributed over the four nodes adjacent to the 's' in Figure 8.9. A value of 1500 seconds was used for the time step. The results of two runs are shown in Figure 8.10 for time equal to seven tidal cycles after the start of the source. In Figure 8.10(a) the dispersion process is isotropic with a coefficient value of $D_x = D_y = 100\ \text{m}^2/\text{s}$. In Figure 8.10(b) longitudinal and transverse coefficients of magnitude of 300 and 100 m^2/s respectively are used to demonstrate another way of applying the dispersion coefficients. Additional studies were made in Massachusetts Bay, including some verification with field experiments.

The negative values of concentration that occasionally arise (see Figure 8.10) are characteristic of a problem that occurs in cases of low dispersion. When the dispersion coefficients are low compared to the velocity field, high concentration gradients result in numerical error. This can be remedied by refining the grid or by distributing the source over a wider area. The exact lower limit of the dispersion coefficients that can be used without resulting in this error thus depends upon the loading strategy, the magnitude of the velocity and the fineness of the grid.

The maximum time step for solution stability is also a function of the grid size and is proportional to the smallest grid element in the

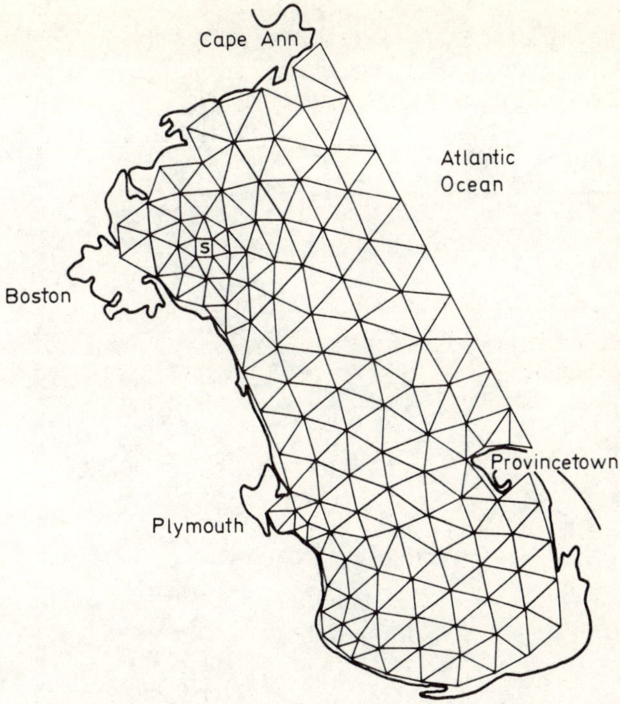

Figure 8.9 Finite element grid for Massachusetts Bay

region. The relationships for this program were empirically found to be

$$D\Delta t/(\Delta x)^2 < 1/10, \qquad v\Delta t/\Delta x < 1/10 \tag{d}$$

where Δt is the time step, Δx is the length of the smallest element, D is the dispersion coefficient and v is the velocity. Δt must meet both requirements of (d); however the coefficient, 10, is not exact. The criterion varies somewhat from problem to problem. Once a stable time step is found, further reduction will not appreciably increase solution accuracy.

Given a certain velocity, v, the grid length, Δx, will limit the range of allowable dispersion coefficients, D. If the grid size is not small enough for the dispersion coefficients, the solution may converge, but non-physical results will occur (such as oscillatory behaviour or negative concentrations). Through an empirical study of the solutions, it has been found that stability is achieved only when

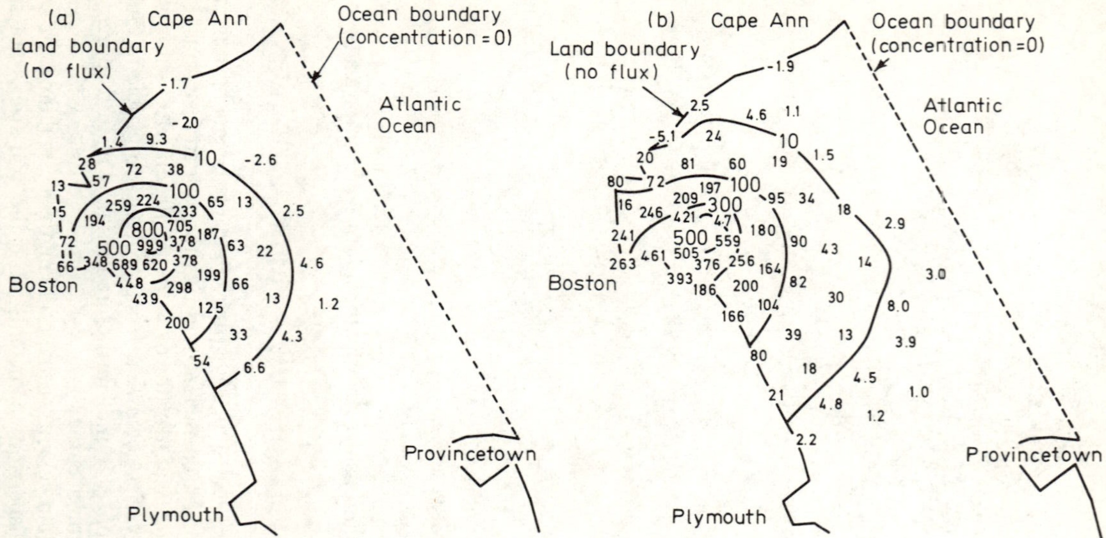

Figure 8.10 Dispersion results for Massachusetts Bay with isoquants of concentration (*isoquant values are in* g/m^3): *(a)* $K_x = K_y = 100\ m^2/s$; *(b)* $K_{long} = 300\ m^2/s$, $K_{transverse} = 100\ m^2/s$

the following condition holds:

$$v\Delta x/D < 2 \tag{e}$$

The value 2 in this equation is not meant to be exact. The requirements vary slightly from problem to problem. Of course the time-step criterion of equations (d) must also hold.

8.5 NONLINEAR DIFFUSION

In many diffusion problems the diffusion constants depend on the values of the concentration functions, hence the problem is non-linear.

Consider for instance the steady state non-convective diffusion equation

$$p + \frac{\partial}{\partial x_i}\left(K_{ii}(\theta)\frac{\partial \theta}{\partial x_i}\right) = 0 \tag{8.36}$$

where the K_{ii} coefficients are functions of θ. We have, after applying finite elements, a nonlinear system of equations; i.e.

$$\mathscr{K}(\boldsymbol{\theta})\boldsymbol{\theta} = \mathscr{Q} \tag{8.37}$$

In order to solve (8.37) we can start with an initial value of θ^i and find an improved $\boldsymbol{\theta}^{i+1}$ approximation. The recurrence equation is

$$\mathscr{K}(\boldsymbol{\theta}^i)\boldsymbol{\theta}^{i+1} = \mathscr{Q} \tag{8.38}$$

Alternatively we can accelerate the convergence of the solution by using Newton–Raphson's method. Here we define a new function from the results of i approximation

$$\boldsymbol{\psi}^i = \mathscr{K}(\boldsymbol{\theta}^i)\boldsymbol{\theta}^i - \mathscr{Q} = \mathscr{K}^i\boldsymbol{\theta}^i - \mathscr{Q} \neq 0 \tag{8.39}$$

Now we impose the condition that

$$\boldsymbol{\psi}^{i+1} = \boldsymbol{\psi}^i + \delta\boldsymbol{\psi}^i = 0 \tag{8.40}$$

which gives the recurrence relation

$$\delta\boldsymbol{\psi}^i = -\boldsymbol{\psi}^i \tag{8.41}$$

where

$$\delta\boldsymbol{\psi}^i = \mathscr{K}(\boldsymbol{\theta}^i)\,\delta\boldsymbol{\theta}^i = -\boldsymbol{\psi}^i \tag{8.42}$$

From equation (8.42) the value of $\delta\boldsymbol{\theta}^i$ can be obtained. Thus

$$\boldsymbol{\theta}^{i+1} = \boldsymbol{\theta}^i + \delta\boldsymbol{\theta}^i \tag{8.43}$$

The convergence of the solution is now second-order.

REFERENCE

1. LEIMKUHLER, W. F., CONNOR, J. J., WANG, J. D., CHRISTODOULOU, G. and SUNDGREN, S., 'A Two-Dimensional Finite Element Dispersion Model', in *Modeling 1975*, Am. Soc. Civ. Engrs (1975)

BIBLIOGRAPHY

ADEY, R. and BREBBIA, C., 'Finite Element Solution for Effluent Dispersion', *Numerical Methods in Fluid Dynamics*, C. Brebbia and J. J. Connor (Eds.), Pentech Press, 1974

BREBBIA, C. A., 'Some Applications of Finite Elements for Flow Problems', in *Variational Methods in Engineering, Vol.* 1, C. Brebbia and H. Tottenham (Eds.), Southampton University Press (1973).

SMITH, I. M., FARRADAY, R. V. and O'CONNOR, B. A., 'Rayleigh–Ritz and Galerkin Finite Elements for Diffusion–Convection Problems', *Water Resource. Res.*, **9**, No. 3 (June, 1973)

ZIENKIEWICZ, O. C. and CHEUNG, Y. K., 'Finite Elements in the Solution of Field Problems', *The Engineer*, September 24 (1965)

EXERCISES

8-1 Consider the two-dimensional heat transfer problem of Figure 8.9. Taking symmetry into account, propose a three node triangular mesh of less than 100 nodes to solve the problem.

8-2 Deduce equation (8.10) using Reynolds transport theorem as given in Chapter 4 [equation (4.15)].

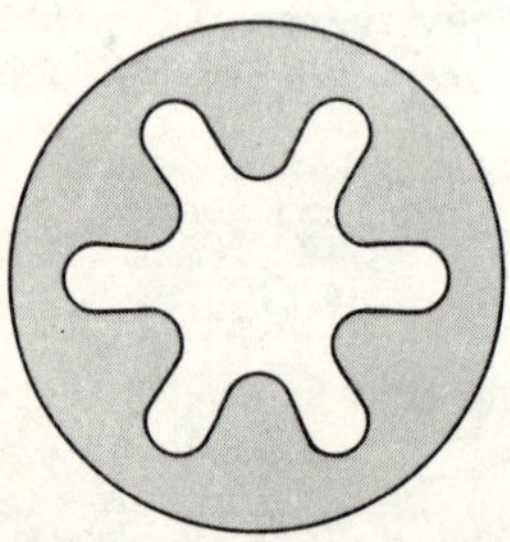

Figure 8.9

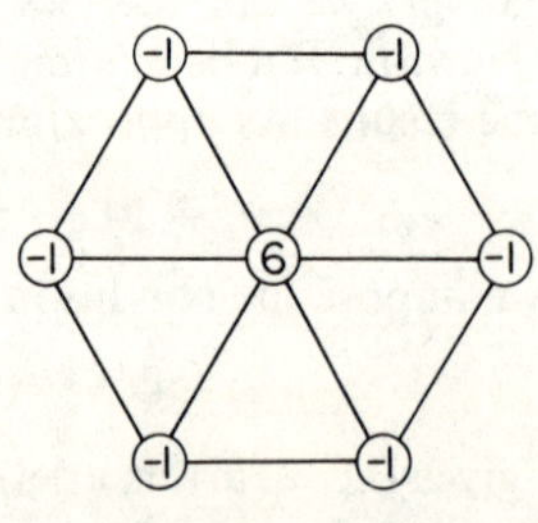

Figure 8.10

8-3 A finite difference approximation for the Laplace equation on a regular hexagonal mesh gives the grid in Figure 8.10. Prove that the same relationship may be obtained using three node equilateral triangular finite elements.

8-4 Deduce the temperature dispersion equation from the energy equation (as in Chapter 4). Which are the necessary assumptions?

8-5 A pollutant is distributed on a one-dimensional channel according to the following (steady state) formula

$$K\frac{\partial^2\theta}{\partial x_1^2} = v\frac{\partial\theta}{\partial x_1}$$

where v is a constant velocity in the x_1 direction, θ is the concentration and K is the diffusion coefficient.

Find the distribution of θ for a mass p (per unit length and per unit time) of pollutant at $x_1 = 0$.

8-6 Deduce the governing equations for mass transfer when the density of the mixture varies with concentration. *Hint*: see the deduction of Reynolds theorem [(4.15) in Chapter 4].

8-7 Write the mass transfer equation (8.21) in cylindrical coordinates.

8-8 Consider the time dependent heat conduction equation

$$\frac{\partial\theta}{\partial t} = \frac{K}{\rho C_p}\left(\frac{\partial^2\theta}{\partial x^2}\right) \tag{a}$$

where K is the thermal conductivity, ρ the mass density and C_p the specific heat. Propose a simple one-dimensional finite element model in space and solve in time using Galerkin's method. Write the corresponding matrices and the final recurrence formula.

8-9 Show that the one-dimensional dispersion equation for a river can be written as

$$\frac{\partial\theta}{\partial t} + v\frac{\partial\theta}{\partial x} = \frac{1}{A}\frac{\partial}{\partial x}\left(AK_1\frac{\partial\theta}{\partial x}\right) - \beta\theta$$

where v is the mean velocity along the river, x the variable along the length of the river, A the (variable) cross sectional area and β the first-order decay constant.

8-10 Compute the loss of heat through two 10 cm wide brick walls separated by 8 cm of insulating material. Assume first that the insulating material is vermiculite and second that it is cellular plastic. The heat flux is given by

$$q = -K\frac{\partial\theta}{\partial x}$$

where

$$K_{\text{vermiculite}} = 0{\cdot}065 \text{ watts/m°C}$$
$$K_{\text{cellular plastic}} = 0{\cdot}037 \text{ watts/m°C}$$
$$K_{\text{brickwork}} = 0{\cdot}806 \text{ watts/m°C}$$

8-11 Write the time dependent diffusion equation with nonlinear $K_i = K_i(\theta)$ coefficients. Discuss the advantages of several explicit and implicit formulations to solve the equations.

8-12 Based on the Laplace program of Chapter 2, develop an algorithm to solve the nonlinear diffusion problem—equation (8.37)—using Newton–Raphson's method.

9 Viscous Incompressible Flow Problems

9.1 INTRODUCTION

In this chapter the solution of the full set of equations for viscous incompressible flow is attempted using finite elements. The solution is valid only for low values of Reynolds number, but it may be possible in the future to extend it to higher values if an accurate way of considering the boundary layer effects can be found.[1]

The governing equations for viscous incompressible flow can be expressed as functions of velocities and pressure, with the velocities required to satisfy the incompressibility condition. Alternatively the two-dimensional equations can be written using a stream function ψ which identically satisfies continuity. Now the momentum equations are combined to produce a higher order governing equation, which requires continuity of the function and its derivatives. For three-dimensional applications the situation is more complex as three stream functions are required.

In order to avoid working with a higher-order ψ function, one can use, in addition to the stream function, the vorticity ω as an unknown. In this way continuity of ψ and ω is necessary between elements, but their derivatives do not need to be continuous.

In what follows we will review the above formulations with special reference to the transient case and the development of the Von Karman vortex street.

9.2 BASIC PRINCIPLES

The governing equations for the flow of an incompressible Newtonian fluid are summarised below for reference (see Chapter 4):

Momentum equations

$$-\frac{1}{\rho}\frac{\partial p}{\partial x_i} + b_i + \frac{1}{\rho}\frac{\partial \tau_{ij}}{\partial x_j} = \frac{\mathrm{D}v_i}{\mathrm{D}t} \qquad i,j = 1,2,3 \tag{9.1}$$

Strain rate—velocity relations

$$e_{ij} = \frac{1}{2}\left(\frac{\partial v_i}{\partial x_j} + \frac{\partial v_j}{\partial x_i}\right) \tag{9.2}$$

Stress–strain rate relations (with Stokes condition)

$$\tau_{ij} = 2\mu e_{ij} - \tfrac{2}{3}\mu e_v \delta_{ij} \tag{9.3}$$

Continuity equation

$$e_v = \frac{\partial v_i}{\partial x_i} = 0 \tag{9.4}$$

Boundary conditions on S_v, or part of the boundary where velocities are prescribed

$$\begin{aligned} v_n &= \bar{v}_n \\ v_s &= \bar{v}_s \end{aligned} \tag{9.5}$$

Conditions on S_p, or part of the boundary where the forces are prescribed

$$\begin{aligned} -p + 2\mu\frac{\partial v_n}{\partial n} &= \bar{p}_n \\ \mu\left(\frac{\partial v_s}{\partial n} + \frac{\partial v_n}{\partial s}\right) &= \bar{p}_s \end{aligned} \tag{9.6}$$

Equations (9.1) and (9.4) are the governing differential equations of the problem, and (9.5) and (9.6) are the boundary conditions.

A variational statement which corresponds to this set of equations is obtained from the Principle of Virtual Power (Chapter 4):

$$\iiint \left[\{-p\, \delta e_v - e_v\, \delta p + \tau_{ij}\, \delta e_{ij}\} + \rho \left\{ -b_i + \frac{\mathrm{D}v_i}{\mathrm{D}t} \right\} \delta v_i \right] \mathrm{d}V$$

$$= \iint_{S_p} \{\bar{p}_n\, \delta v_n + \bar{p}_s\, \delta v_s\}\, \mathrm{d}S \tag{9.7}$$

valid for arbitrary δv_i, δp in V, arbitrary δv_n, δv_s on S_p and $\delta v_n = \delta v_s = 0$ on S_v.

We can verify that applying Gauss's theorem in formula (9.7) in order to eliminate the derivatives of δv_i leads to the continuity, momentum equations and force boundary conditions.

For the two-dimensional case, (9.7) can be written as

$$\iint \left[\{-p\, \delta e_v - e_v\, \delta p + \tau_{11}\, \delta e_{11} + \tau_{22}\, \delta e_{22} + \tau_{12}\, \delta \gamma_{12}\} \right.$$

$$\left. + \rho \left\{ -b_1 + \frac{\mathrm{D}v_1}{\mathrm{D}t} \right\} \delta v_1 + \rho \left\{ -b_2 + \frac{\mathrm{D}v_2}{\mathrm{D}t} \right\} \delta v_2 \right] \mathrm{d}A$$

$$= \int (\bar{p}_n\, \delta v_n + \bar{p}_s\, \delta v_s)\, \mathrm{d}S \tag{9.8}$$

where

$$\gamma_{12} = \frac{\partial v_1}{\partial x_2} + \frac{\partial v_2}{\partial x_1} = 2e_{12}$$

and the continuity equation can now be written

$$e_v = \frac{\partial v_1}{\partial x_1} + \frac{\partial v_2}{\partial x_2} = 0 \tag{9.9}$$

We assume that the velocities satisfy the boundary conditions on S_v, which can be written

$$\begin{aligned} v_n &= \alpha_{n1} v_1 + \alpha_{n2} v_2 = \bar{v}_n \\ v_s &= -\alpha_{n2} v_1 + \alpha_{n1} v_2 = \bar{v}_s \end{aligned} \tag{9.10}$$

9.3 STREAM FUNCTION—VORTICITY APPROACH

Following the strategy employed earlier for inviscid flow, we express the two-dimensional velocity components in terms of a stream function ψ such that

$$v_1 = \frac{\partial \psi}{\partial x_2}, \qquad v_2 = -\frac{\partial \psi}{\partial x_1} \tag{9.11}$$

On the boundary we have $v_n = \partial\psi/\partial s$ and $v_s = -\partial\psi/\partial n$. The continuity equation is now identically satisfied.

One can cross differentiate the two momentum equations in order to eliminate the term in p and this gives, after subtracting one from the other,

$$\frac{\partial b_1}{\partial x_2} - \frac{\partial b_2}{\partial x_1} + \frac{\mu}{\rho}\nabla^4\psi = \frac{\mathrm{D}(\nabla^2\psi)}{\mathrm{D}t} \tag{9.12}$$

The essential boundary conditions for this case are the function and its normal derivative:

$$\left.\begin{aligned} v_n = \bar{v}_n &\Rightarrow \psi_B = \psi_A + \int_A^B \bar{v}_n \,\mathrm{d}S \\ v_s = \bar{v}_s &\Rightarrow \frac{\partial \psi}{\partial n} = -v_s \end{aligned}\right\} \text{ on } S_v \tag{9.13}$$

An alternative form of (9.12) is obtained using the vorticity component:

$$\omega = \frac{1}{2}\left(\frac{\partial v_2}{\partial x_1} - \frac{\partial v_1}{\partial x_2}\right) = -\tfrac{1}{2}\nabla^2\psi \tag{9.14}$$

Substituting (9.14) into (9.12) we obtain the vorticity equation

$$\frac{\mu}{\rho}\nabla^2\omega + \frac{1}{2}\left(\frac{\partial b_2}{\partial x_1} - \frac{\partial b_1}{\partial x_2}\right) = \frac{\mathrm{D}\omega}{\mathrm{D}t} \tag{9.15}$$

Let us consider the stream function formulation by substituting the velocities (9.11) and the variations

$$\delta v_1 = \frac{\partial(\delta\psi)}{\partial x_2}, \qquad \delta v_2 = -\frac{\partial(\delta\psi)}{\partial x_1} \tag{9.16}$$

into the virtual power expression (9.8). This gives

$$\begin{aligned} &\iint \left[\frac{\mu}{\rho}\left\{4\frac{\partial^2\psi}{\partial x_1\,\partial x_2}\frac{\partial^2\delta\psi}{\partial x_1\,\partial x_2} + \left(\frac{\partial^2\psi}{\partial x_2^2} - \frac{\partial^2\psi}{\partial x_1^2}\right)\left(\frac{\partial^2\delta\psi}{\partial x_2^2} - \frac{\partial^2\delta\psi}{\partial x_1^2}\right)\right\}\right. \\ &\qquad \left. + \left\{-b_1 + \frac{\mathrm{D}}{\mathrm{D}t}\left(\frac{\partial\psi}{\partial x_2}\right)\right\}\frac{\partial\delta\psi}{\partial x_2} + \left\{b_2 + \frac{\mathrm{D}}{\mathrm{D}t}\left(\frac{\partial\psi}{\partial x_1}\right)\right\}\frac{\partial\delta\psi}{\partial x_1}\right]\mathrm{d}A \\ &\qquad = \int_{S_p}\frac{1}{\rho}\left(\bar{p}_n\frac{\partial\delta\psi}{\partial s} - \bar{p}_s\frac{\partial\delta\psi}{\partial n}\right)\mathrm{d}S \end{aligned} \tag{9.17}$$

In addition we prescribe ψ and its normal derivative on S_v, according to equation (9.13). Integrating (9.17) by parts, in order to have all the terms multiplied by $\delta\psi$, results in equation (9.12) and the following force boundary conditions:

$$\begin{aligned} \bar{p}_s &= \mu\left(\frac{\partial^2\psi}{\partial s^2} - \frac{\partial^2\psi}{\partial n^2}\right) \\ \frac{1}{\rho}\frac{\partial \bar{p}_n}{\partial s} &= b_s - \frac{\mu}{\rho}\nabla^2\left(\frac{\partial\psi}{\partial n}\right) + \left(\frac{\mathrm{D}}{\mathrm{D}t}\right)\left(\frac{\partial\psi}{\partial n}\right) \end{aligned} \tag{9.18}$$

If the stream function and its normal derivative $\partial\psi/\partial n$ are prescribed on the entire boundary, we have the following variational statement:

$$\int \left\{-\nabla^2\left(\frac{\partial\psi}{\partial t}\right) + \frac{\partial\psi}{\partial x_2}\nabla^2\left(\frac{\partial\psi}{\partial x_1}\right) - \frac{\partial\psi}{\partial x_1}\nabla^2\left(\frac{\partial\psi}{\partial x_2}\right) - \frac{\mu}{\rho}\nabla^4\psi\right\}\delta\psi\,\mathrm{d}A = 0 \tag{9.19}$$

where body forces have been neglected.

Formula (9.19) is the basis for finite element models involving the stream function ψ. Once the element matrices are determined, we sum (9.19) over all the elements and generate the system equations in terms of nodal values of ψ. The final set of equations can be written as

$$\mathscr{K}\boldsymbol{\psi} + \mathscr{M}\dot{\boldsymbol{\psi}} = \mathscr{P} + \mathscr{P}_c \tag{9.20}$$

where $\mathscr{P}_c$ contains the nonlinear convective terms, $\mathscr{K}$ the viscosity terms and $\mathscr{M}$ is a mass matrix.

In the following two examples, equation (9.20) will be applied to Stokesian flow ($\mathscr{P}_c$ and $\dot{\boldsymbol{\psi}}$ are zero) and steady flow (only $\dot{\boldsymbol{\psi}} = 0$).

Example 9.1

Tong[2] has analysed the case of two-dimensional Stokesian flow, for which D/Dt terms are negligible due to the high viscosity of the fluid. For this case, when body forces are neglected, equation (9.17) becomes

$$\begin{aligned} \iint \frac{\mu}{\rho}\left\{4\frac{\partial^2\psi}{\partial x_1\,\partial x_2}\frac{\partial^2\delta\psi}{\partial x_1\,\partial x_2} + \frac{\partial^2\psi}{\partial x_2^2}\frac{\partial^2\delta\psi}{\partial x_2^2} + \frac{\partial^2\psi}{\partial x_1^2}\frac{\partial^2\delta\psi}{\partial x_1^2} \right. \\ \left. - \frac{\partial^2\psi}{\partial x_1^2}\frac{\partial^2\delta\psi}{\partial x_2^2} - \frac{\partial^2\psi}{\partial x_2^2}\frac{\partial^2\delta\psi}{\partial x_1^2}\right\}\mathrm{d}x_1\,\mathrm{d}x_2 \\ = \int_{S_p}\left(\bar{p}_n\frac{\partial\delta\psi}{\partial s} - \bar{p}_s\frac{\partial\delta\psi}{\partial n}\right)\mathrm{d}S \end{aligned} \tag{a}$$

If the velocities are known on all the S boundaries, half the first term in (a) is integrated by parts (half of it with respect to x_1 and the other half with respect to x_2) and also the negative (fourth and fifth terms in (a)) are integrated by parts, we obtain

$$\iint \frac{\mu}{\rho}\left\{2\frac{\partial^2\psi}{\partial x_1\,\partial x_2}\frac{\partial^2\delta\psi}{\partial x_1\,\partial x_2}+\frac{\partial^2\psi}{\partial x_2^2}\frac{\partial^2\delta\psi}{\partial x_2^2}+\frac{\partial^2\psi}{\partial x_1^2}\frac{\partial^2\delta\psi}{\partial x_1^2}\right\}\mathrm{d}x_1\,\mathrm{d}x_2=0 \tag{b}$$

This gives a variational functional of the form

$$F=\frac{1}{2}\iint\frac{\mu}{\rho}\left\{\left(\frac{\partial^2\psi}{\partial x_1^2}\right)^2+2\left(\frac{\partial^2\psi}{\partial x_1\,\partial x_2}\right)^2+\left(\frac{\partial^2\psi}{\partial x_2^2}\right)^2\right\}\mathrm{d}x_1\,\mathrm{d}x_2 \tag{c}$$

The boundary conditions in terms of stream functions are

$$\frac{\partial\psi}{\partial n}=-v_s,\qquad \psi=\psi_A+\int_A^B\bar{v}_n\,\mathrm{d}S \tag{d}$$

Equation (c) is the same variational functional as for plate bending, for which case also the function ψ needs to be continuous up to its first derivatives. Tong[2] analysed viscous flow in a channel using a 'hybrid' formulation instead of a ψ-only approach. This formulation, which was developed for rectangular plate bending elements, gives very accurate results. The description of the hybrid model is beyond the scope of this text but it is sufficient to say that similar results can be obtained using a second-order continuity function for ψ (see Sections 3.5 and 3.6).

Tong uses the model to analyse the effect of a rectangular obstruction in a channel. In Figure 9.1 the geometry of the problem is shown,

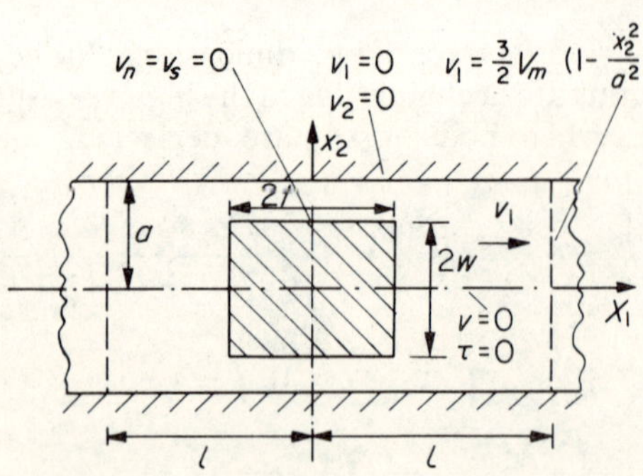

Figure 9.1(a) Channel with obstruction. V_m *= average velocity for Couette flow*

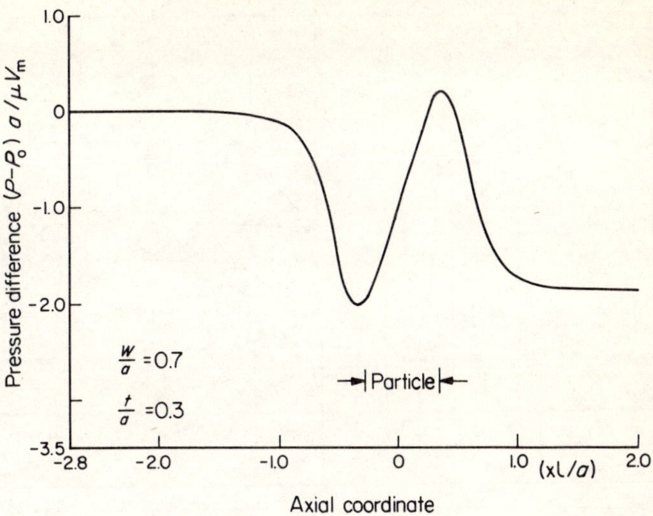

Figure 9.1(b) Pressure difference on the wall for one particle. V_m = average velocity

together with the difference in pressure at the wall for the cases when the particle is present and when it is not.

Example 9.2

Olson[3] has presented results for two-dimensional steady flow using a quintic triangular element for ψ which has proved to be very accurate (see Section 3.6). This element ensures continuity of ψ and its derivatives between adjacent elements and has 18 degrees of freedom ($\psi, \psi_{,1}, \psi_{,2}, \psi_{,11}, \psi_{,12}, \psi_{,22}$ at each of the three vertices).

The variational statement for steady flow in the absence of body forces can be written [see equation (9.19)]

$$\int \left\{ -\frac{\mu}{\rho}\nabla^4\psi + \frac{\partial\psi}{\partial x_2}\nabla^2\left(\frac{\partial\psi}{\partial x_1}\right) - \frac{\partial\psi}{\partial x_1}\nabla^2\left(\frac{\partial\psi}{\partial x_2}\right)\right\} \delta\psi \, \mathrm{d}A = 0 \qquad \text{(a)}$$

Once the finite element model was implemented in statement (a), the Newton–Raphson method was used to solve the resulting non-linear system of equations until convergence was achieved.

The model was applied for the solution of the flow over a circular cylinder, as shown in Figure 9.2. Zero velocities on the cylinder were satisfied at the node points. The natural boundary conditions, which are approximated by the finite element solution, were zero shear stress along the symmetry line and zero p_n derivatives along the top

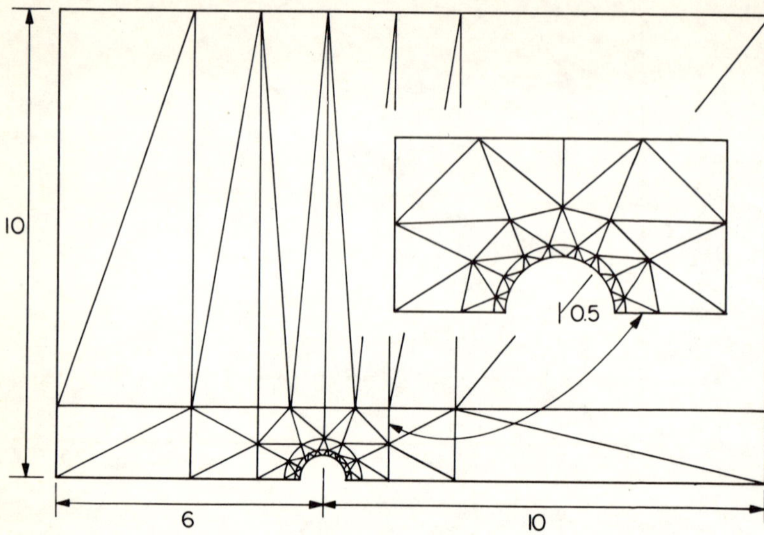

Figure 9.2(a) *Finite element grid for flow over a circular cylinder*

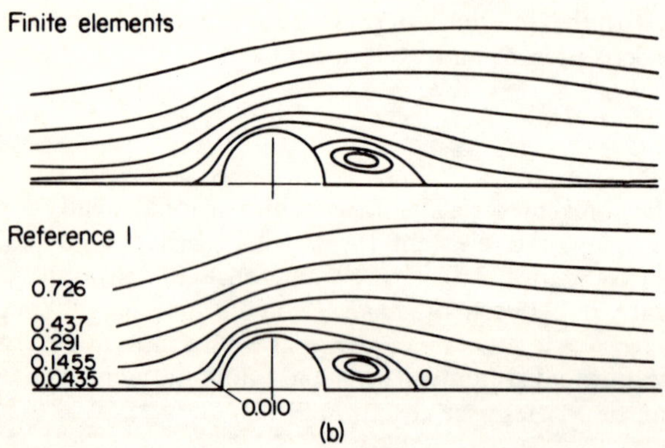

Figure 9.2(b) *Streamlines over cylinder, $Re = 20$. ($-0{\cdot}004$, $-0{\cdot}0029$)*

and downstream edges. The streamlines obtained for $Re - 20$ are in agreement with a previous finite difference solution.[4]

Lieber *et al.*[5] have used equation (9.19) as the basis of another finite element model involving only the stream function ψ. They employ a non fully compatible expansion previously used for plate bending. Results for the steady flow in a constricted channel are presented with Reynolds numbers up to 130. They solve the resulting nonlinear system of equations using an iterative Gauss–Seidel type scheme.

Example 9.3

A transient two-dimensional flow solution can be developed using as variables the stream function ψ and vorticity ω. This approach has the advantage over the stream function only formulation in that it allows for the use of simple first-order continuity interpolation functions. One starts by considering equations (9.14) and (9.15) without body forces; i.e.

$$\nu\nabla^2\omega = \frac{D\omega}{Dt} \tag{a}$$

$$\omega = -\tfrac{1}{2}\nabla^2\psi \tag{b}$$

The natural boundary conditions corresponding to (a) and (b) are

$$\left.\begin{aligned} \nu\frac{\partial\omega}{\partial n} &= g_\omega \quad \text{on } S_\omega \\ \frac{1}{2}\frac{\partial\psi}{\partial n} &= g_\psi \quad \text{on } S_\psi \end{aligned}\right. \tag{c}$$

We can write equations (a) to (c) in the following Galerkin type statements:

$$\left.\begin{aligned} &\iint\left\{\frac{\partial\omega}{\partial t} + \frac{\partial\psi}{\partial x_2}\frac{\partial\omega}{\partial x_1} - \frac{\partial\psi}{\partial x_1}\frac{\partial\omega}{\partial x_2} - \nu\nabla^2\omega\right\}\delta\omega\,dx_1\,dx_2 \\ &\qquad = \int_S\left(g_\omega - \nu\frac{\partial\omega}{\partial n}\right)\delta\omega\,dS \\ &\iint\{\tfrac{1}{2}\nabla^2\psi + \omega\}\,\delta\psi\,dx_1\,dx_2 = \int_S\left(\frac{1}{2}\frac{\partial\psi}{\partial n} - g_\psi\right)\delta\psi\,dS \end{aligned}\right. \tag{d}$$

Integrating the ∇^2 terms by parts gives

$$\left.\begin{aligned} &\iint\left\{\left(\frac{\partial\omega}{\partial t} + \frac{\partial\psi}{\partial x_2}\frac{\partial\omega}{\partial x_1} - \frac{\partial\psi}{\partial x_1}\frac{\partial\omega}{\partial x_2}\right)\delta\omega + \nu\left(\frac{\partial\delta\omega}{\partial x_1}\frac{\partial\omega}{\partial x_1}\right.\right. \\ &\qquad \left.\left. + \frac{\partial\delta\omega}{\partial x_2}\frac{\partial\omega}{\partial x_2}\right)\right\}dx_1\,dx_2 = \int_S g_\omega\,\delta\psi\,dS \\ &\iint\left\{-\frac{1}{2}\frac{\partial\delta\psi}{\partial x_1}\frac{\partial\psi}{\partial x_1} - \frac{1}{2}\frac{\partial\delta\psi}{\partial x_2}\frac{\partial\psi}{\partial x_2} + \omega\,\delta\psi\right\}dx_1\,dx_2 = \int_S g_\psi\,\delta\psi\,dS \end{aligned}\right. \tag{e}$$

We next divide the continuum into elements and assume on each of them that the ω and ψ functions can be approximated by

$$\omega = \boldsymbol{\phi}^T\boldsymbol{\omega}^n, \qquad \psi = \boldsymbol{\phi}^T\boldsymbol{\psi}^n \tag{f}$$

$\boldsymbol{\phi}$ is the interpolation function (we assume that the same function applies for ω and ψ for simplicity) and $\boldsymbol{\omega}^n$, $\boldsymbol{\psi}^n$ are the nodal values of vorticity and stream function.

Substitution of (f) into (e) gives

$$\begin{aligned} \delta\boldsymbol{\psi}^{n,T}\{\mathbf{M}\dot{\boldsymbol{\omega}}^n + \mathbf{A}\boldsymbol{\omega}^n + \nu\mathbf{K}\boldsymbol{\omega}^n - \mathbf{B}_\omega\} &= 0 \\ \delta\boldsymbol{\omega}^{n,T}\{\tfrac{1}{2}\mathbf{K}\boldsymbol{\psi}^n - \mathbf{B}_\psi - \mathbf{M}\boldsymbol{\omega}^n\} &= 0 \end{aligned} \tag{g}$$

where

$$\mathbf{M} = \iint \boldsymbol{\phi}\boldsymbol{\phi}^T \,\mathrm{d}x_1\,\mathrm{d}x_2$$

$$\mathbf{A} = \iint \boldsymbol{\phi}\{\boldsymbol{\phi}_{,2}^T\boldsymbol{\psi}^n\boldsymbol{\phi}_{,1}^T - \boldsymbol{\phi}_{,1}^T\boldsymbol{\psi}^n\boldsymbol{\phi}_{,2}^T\}\,\mathrm{d}x_1\,\mathrm{d}x_2$$

$$\mathbf{K} = \iint \{\boldsymbol{\phi}_{,1}\boldsymbol{\phi}_{,1}^T + \boldsymbol{\phi}_{,2}\boldsymbol{\phi}_{,2}^T\}\,\mathrm{d}x_1\,\mathrm{d}x_2$$

$$\mathbf{B}_\omega = \int_{S_\omega} \boldsymbol{\phi} g_\omega \,\mathrm{d}S, \qquad \mathbf{B}_\psi = \int_{S_\psi} \boldsymbol{\phi} g_\psi \,\mathrm{d}S$$

Equation (g) holds for any arbitrary variation of $\delta\omega$ and $\delta\psi$. Hence we have for one element

$$\begin{aligned} \mathbf{M}\dot{\boldsymbol{\omega}}^n + \mathbf{A}\boldsymbol{\omega}^n + \nu\mathbf{K}\boldsymbol{\omega}^n &= \mathbf{B}_\omega \\ \tfrac{1}{2}\mathbf{K}\boldsymbol{\psi}^n &= \mathbf{B}_\psi + \mathbf{M}\boldsymbol{\omega}^n \end{aligned} \tag{h}$$

We can now assemble all the elements together and obtain for the whole continuum

$$\mathscr{M}\dot{\boldsymbol{\Omega}} + \mathscr{A}\boldsymbol{\Omega} + \nu\mathscr{K}\boldsymbol{\Omega} = \mathscr{B}_\omega \tag{i}$$

$$\mathscr{K}\boldsymbol{\Phi} = \mathscr{B}_\psi + \mathscr{M}\boldsymbol{\Omega} \tag{j}$$

Next one has to integrate equation (i) in time, preferably using an explicit method as the terms are highly nonlinear.

The first step is the solution of (j) using the set of initial vorticities Ω_0. Then the stream functions are evaluated at $t_0 + \Delta t$ integrating equation (i) in time. In order to solve again equation (j) one needs to know the vorticity along a no-slip boundary, for which the stream function is constant. To find this value we can integrate equation (b) at some distance normal to the wall (note that it becomes an ordinary differential equation). Using a linear variation element, such as the three node triangle, the integral may be evaluated directly and gives (Figure 9.3)

$$\Omega_w = -\left[\frac{3(\Phi_w - \Phi)}{2l^2} + \frac{\Omega}{2}\right] \tag{k}$$

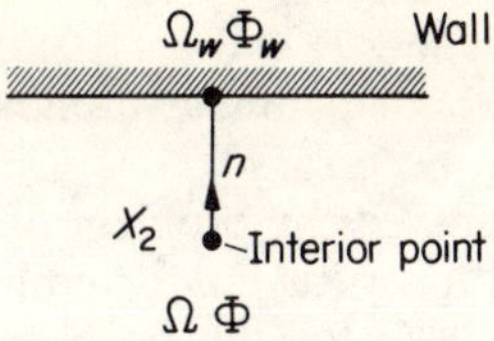

Figure 9.3 Boundary integration

where l is the distance normal to the wall, Φ_w and Ω_w are values at the wall, and Φ and w are values at an interior point situated at a distance l perpendicular to the wall.

The steps of the method can be summarised with the following diagram:

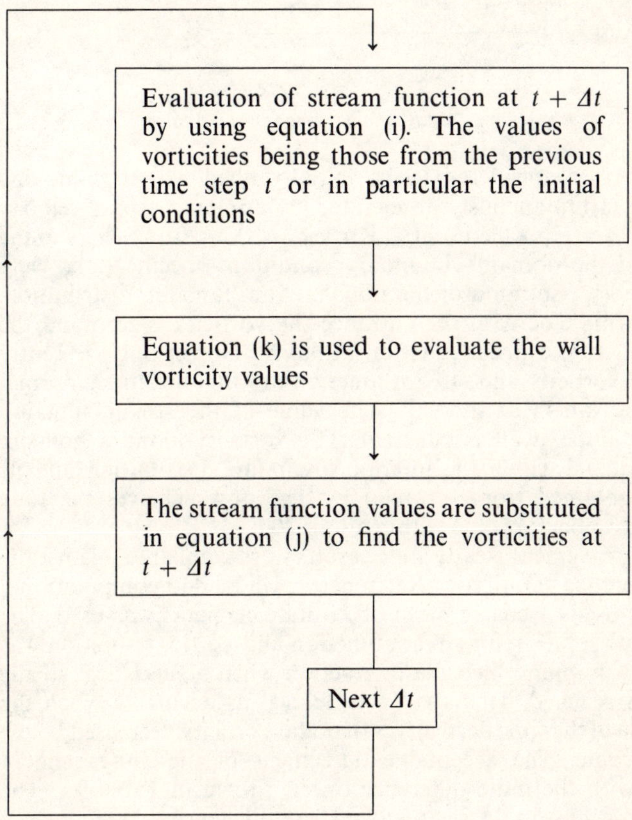

The above procedure has to be repeated for as many Δt as required. Note that with this procedure we are uncoupling equations (i) and

(j) by assuming they apply for different times (t and $t + \Delta t$). If we wish to correct the error introduced in this way one can iterate at the end of each Δt, but this is not usually necessary.

Using the above formulation Baker[6] solved the problem of flow between parallel plates separated by 2 ft for a Reynolds number of 200. The geometry and finite element mesh for the problem are shown in Figure 9.4. Because of symmetry, only one half of the problem needs to be studied. Six node finite elements were used.

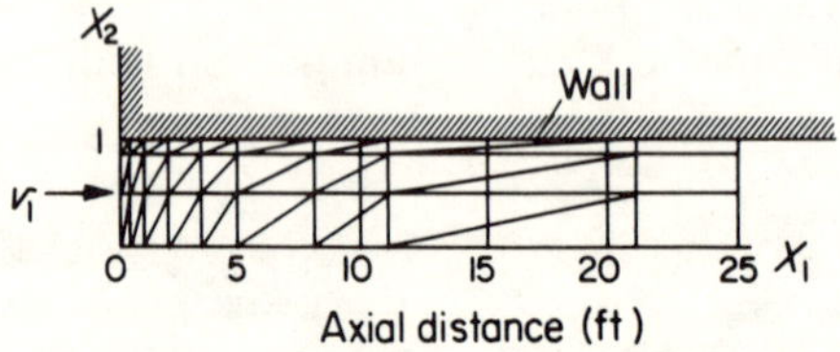

Figure 9.4 Duct flow finite element discretisation

The duct is considered to be initially filled with non-moving fluid and instantaneously accelerated by applying a unit velocity upstream. The equivalent initial vorticity is $\Omega_0 \equiv 0$ anywhere in the interior of the domain—boundary conditions specifications were determined for solution of the initial stream function distribution. For uniform velocity at the entrance the vorticity is zero and the stream function a linear function of the X_2 axis. Along the centre plane the vorticity and stream function are null and their normal derivatives vanish at the exit. The value of the stream function along the upper wall is constant. The vorticity along a non-slip surface depends upon the interior streamline distribution and its value is obtained from formula (k). The flow achieves the fully developed velocity profile within about 20 ft.

The finite element steady state results were compared against the finite difference solution of Reference 7, where 440 computational cells were used. A coarse mesh of 39 finite elements was used. The steady state solution for stream function and vorticity solution was formed by stopping the time interaction when a 'statistical steady state' was reached. This state was defined as occurring when the magnitude of the time derivatives of nodal vorticity decreased below a small value. The acceptable agreement of the finite element solution with the finite difference one is shown in Table 9.1. The agreement is best in the region away from the corner.

More recently Baker[8] extended this formulation to cover the case of solving the momentum plus energy equation using a highly stable explicit integration scheme.

Table 9.1 STEADY-STATE STREAM FUNCTION AND VORTICITY DISTRIBUTIONS, $Re = 200$[6]

Node coordinates		*Stream function*		*Vorticity*	
y	*x*	*Finite element*	*Finite difference*	*Finite element*	*Finite difference*
0·5	0·325	0·595	0·539	0·343	0·039
	1·0	0·641	0·589	0·606	0·262
	2·0	0·656	0·623	0·731	0·623
	3·3	0·667	0·644	1·05	0·881
	4·8	0·675	0·659	1·09	1·10
	8·0	0·669	0·674	1·16	1·27
	11·0	0·684	0·681	1·31	1·44
	21·1	0·688	0·685	0·92	1·50
0·9	0·325	0·967	0·957	3·60	5·12
	1·0	0·980	0·976	3·77	4·26
	2·0	0·982	0·980	3·61	3·57
	3·3	0·983	0·982	3·11	3·20
	4·8	0·983	0·983	3·09	3·04
	8·0	0·983	0·984	3·15	2·88
	11·0	0·984	0·985	2·96	2·76
	21·1	0·983	0·985	3·06	2·70
1·0	0·325	1·0	1·0	8·14	9·73
	1·0	1·0	1·0	4·04	5·05
	2·0	1·0	1·0	3·55	4·44
	3·3	1·0	1·0	3·57	4·14
	4·8	1·0	1·0	3·58	3·71
	8·0	1·0	1·0	3·60	3·28
	11·0	1·0	1·0	3·30	3·12
	21·1	1·0	1·0	3·51	3·06

Example 9.4

The same formulation as described in Example 9.3 has been applied to study the flow in a channel of finite width with a rectangular obstruction.[9] The channel is 4·5 m long and 1 m wide. The obstruction was situated 1·05 m from the channel entrance and was 0·4 m long by 0·166 m [Figure 9.5(a)]. The initial viscosity was taken as 0·01666 m^2/s.

The fluid was instantaneously accelerated by applying a uniform velocity of 2·0 m/s upstream of the channel. The vorticity initial condition was $\Omega_0 = 0$, i.e. vanishing everywhere in the interior of the domain. The boundary conditions are as follows: For uniform flow at the channel entrance the vorticity vanishes and the stream function is a linear function of the X_2 coordinate. The normal derivatives of the vorticity and stream function both vanish at the

Figure 9.5(a) Finite element mesh

exit enforcing parallel (but not necessarily fully developed) flow. The values of the stream function along the upper and lower wall and along the boundary of the obstruction are constant and equal to 2·0, 0·0 and 1·0 m^2/s respectively.

Figure 9.5(b) shows the streamlines and vorticities in the region close to the obstruction at a time $t = 2{\cdot}85$ s. This was the time required to reach the 'statistically steady state' as defined by Baker[6]—i.e. when the largest value of $\partial\Omega/\partial t$ at any node decreases to 0·04. The Reynolds number for this example taken with respect to the width of the obstruction was equal to 20 and the initial vorticity distributed equal to zero. At $t = 2{\cdot}85$ s the viscosity was instantaneously decreased by 60 per cent giving a Reynolds number of 50. The time step was reduced in an inversely proportional ratio to the Reynolds number. Figure 9.5(c) shows the stream function and vorticity at time $t = 11{\cdot}5$ s when the statistically steady state has been reached. At this time the viscosity was again reduced by a further 50 per cent within 0·33 s. Figure 9.5(d) shows the streamlines and vorticity at $t = 17{\cdot}5$ s. At this stage the value of $\partial\Omega/\partial t$ showed no

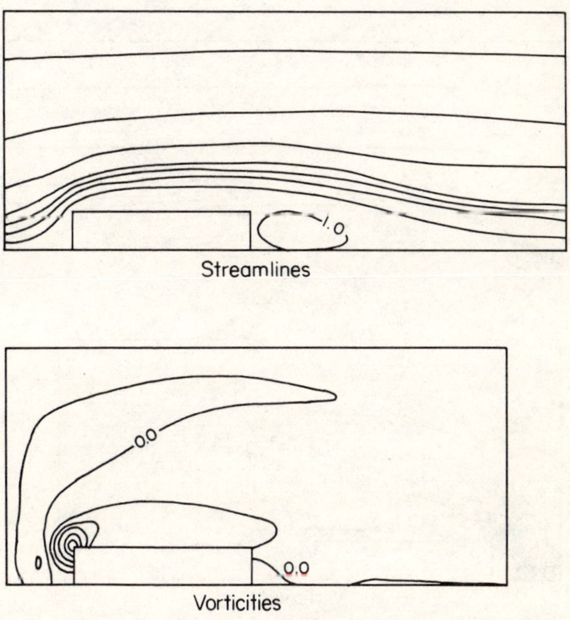

Figure 9.5(b) *Re*: 20

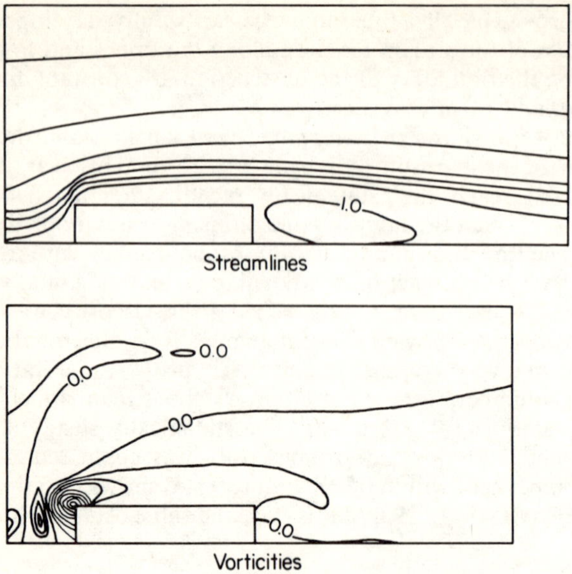

Figure 9.5(c) *Re*: 50

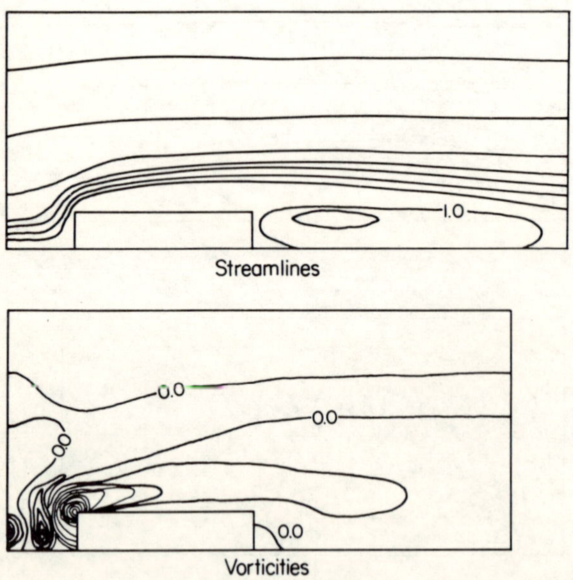

Figure 9.5(d) *Re*: 100

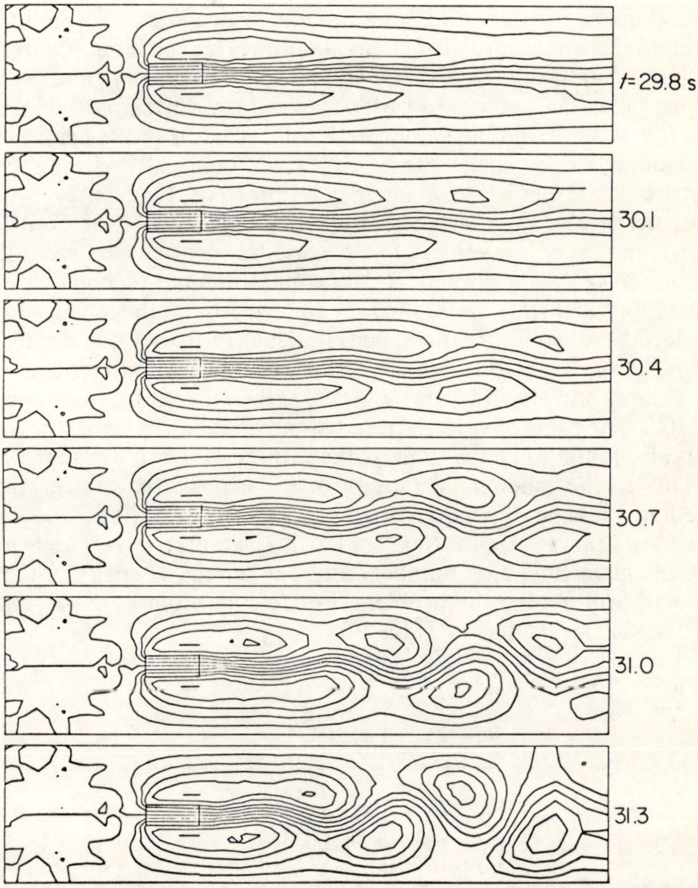

Figure 9.5(e) *Vortex street development for Re = 100 (stationary streamlines)*

sign of decreasing to the value of 0·04. The time step for this case was also found by reducing the original time step in an inversely proportional ratio to the Reynolds number. This criterion was found to produce an overreduction for larger Reynolds numbers.

Instabilities, inherent in numerical solution of the momentum equations at this Reynolds number and propagated by computer round-off error, eventually increased until vortex shedding occurred. Figure 9.5(e) (from $t = 29{\cdot}5$ to $31{\cdot}3$ s) shows this phenomenon, each frame being 0·3 s apart. In order to show the vortex more clearly the free flow streamline values were subtracted from the computed streamline values giving the so called stationary streamlines. The vortices are rather weak, which may be due to the long nature of the obstruction and the coarseness of the mesh. Because of the obstruction length, very low values of velocities and vorticity were recorded at the downstream corners. The Strouhal number (defined as the shedding frequency multiplied by the width of the obstruction divided by the free stream velocity) calculated from these results is approximately $S = 0{\cdot}12$, which agrees favourably with that obtained by Fromm and Harlow,[11] ($S = 0{\cdot}119$ for the same Reynolds number of 100). The ratio of vortex speed to wall speed (0·79 for $Re = 100$) was also found to be the same as the ratio reported in Reference 11.

Due to the inherent instability of the explicit time integration method, the time step used was very small. For $Re = 20$, 50 and 100, the step was $t = 0{\cdot}01$, 0·005 and 0·0025 respectively. These are intermediate Reynolds numbers and neither the viscosity nor the velocity will govern the problem. For reference purposes we summarise the parameters in Table 9.2.

Table 9.2

	Reynolds number		
	20 ($\Delta t = 0{\cdot}01$ s)	50 ($\Delta t = 0{\cdot}005$ s)	100 ($\Delta t = 0{\cdot}0025$ s)
$\dfrac{v\,\Delta t}{(\Delta x)^2}$	0·0666	0·0132	0·0033
$\dfrac{v\,\Delta t}{\Delta x}$	0·40	0·20	0·10
Steady state	reached	reached	not reached

Δx (minimum value) is 0·05 m.
$v = 2$ m/s (free stream velocity).

9.4 PRESSURE AND VELOCITIES APPROACH

Let us consider again the variational statement of equation (9.7), i.e.

$$\iiint \left[\{-p\,\delta e_v - e_v\,\delta p + \tau_{ij}\,\delta e_{ij}\} + \rho\left\{-b_i + \frac{\mathrm{D}v_i}{\mathrm{D}t}\right\}\delta v_i\right]\mathrm{d}V$$

$$= \iint_{S_p} \{\bar{p}_n\,\delta v_n + \bar{p}_s\,\delta v_s\}\,\mathrm{d}S \tag{9.21}$$

One can now substitute for τ_{ij} from equation (9.3) to obtain

$$\iiint \left[\left\{-p\,\delta e_v - e_v\,\delta p + 2\mu e_{ij}\,\delta e_{ij} - \tfrac{2}{3}\mu e_v\,\delta e_v - \rho b_i\,\delta v_i\right.\right.$$

$$\left.\left. + \rho\left\{v_j v_{i,j} + \frac{\partial v_i}{\partial t}\right\}\delta v_i\right]\mathrm{d}V - \iint \{\bar{p}_n\,\delta v_n + \bar{p}_s\,\delta v_s\}\,\mathrm{d}S = 0 \tag{9.22}$$

for arbitrary $\delta p, \delta v_i$. Note that the pressure acts as a Lagrangian multiplier in order to impose the continuity condition.

Consider now that

$$p = \boldsymbol{\phi}^T \mathbf{p}^n, \qquad \mathbf{v}_i = \boldsymbol{\phi}^T \mathbf{v}_i^n \quad (i = 1, 2, 3) \tag{9.23}$$

where $\mathbf{p}^n$ is the element nodal pressure matrix and $\mathbf{v}_i^n$ the nodal velocity matrix. We are assuming that the same functions apply for pressures and velocities. Substituting (9.23) in (9.21) we obtain for the two-dimensional case the following expression for arbitrary variations of δv_1:

$$-\mathbf{C}'^{,T}\mathbf{p}^n + (\mathbf{K}' + \mathbf{A})\mathbf{v}_1^n + \mathbf{K}''^{,T}\mathbf{v}_2^n + \mathbf{M}\dot{\mathbf{v}}_1^n = \mathbf{F}_1 \tag{9.24}$$

for variations of δv_2

$$-\mathbf{C}''^{,T}\mathbf{p}^n + (\mathbf{K}' + \mathbf{A})\mathbf{v}_2^n + \mathbf{K}''^{,T}\mathbf{v}_1^n + \mathbf{M}\dot{\mathbf{v}}_2^n = \mathbf{F}_2 \tag{9.25}$$

plus the continuity equation for arbitrary variations of δp

$$-\{\mathbf{C}'\mathbf{v}_1^n + \mathbf{C}''\mathbf{v}_2^n\} = \mathbf{0} \tag{9.26}$$

where

$$\mathbf{C}' = \iint \boldsymbol{\phi}\boldsymbol{\phi}_{,1}^T\,\mathrm{d}x_1\,\mathrm{d}x_2, \qquad \mathbf{C}'' = \iint \boldsymbol{\phi}\boldsymbol{\phi}_{,2}^T\,\mathrm{d}x_1\,\mathrm{d}x_2$$

$$\mathbf{K}' = \iint \mu\{2\boldsymbol{\phi}_{,1}\boldsymbol{\phi}_{,1}^T + 2\boldsymbol{\phi}_{,2}\boldsymbol{\phi}_{,2}^T - \tfrac{2}{3}\boldsymbol{\phi}_{,1}\boldsymbol{\phi}_{,2}^T\}\,\mathrm{d}x_1\,\mathrm{d}x_2$$

$$\mathbf{K}'' = \iint \mu\{\boldsymbol{\phi}_{,2}\boldsymbol{\phi}_{,1}^T - \tfrac{2}{3}\boldsymbol{\phi}_{,1}\boldsymbol{\phi}_{,2}^T\}\,\mathrm{d}x_1\,\mathrm{d}x_2$$

$$\mathbf{A} = \iint \rho\boldsymbol{\phi}\{v_1\boldsymbol{\phi}_{,1}^T + v_2\boldsymbol{\phi}_{,2}^T\}\,dx_1\,dx_2$$

$$\mathbf{M} = \iint \rho\boldsymbol{\phi}\boldsymbol{\phi}^T\,dx_1\,dx_2$$

$$\mathbf{F}_i = \iint \rho\boldsymbol{\phi}b_i\,dx_1\,dx_2 + \int \boldsymbol{\phi}p_i\,dS \qquad (i = 1, 2)$$

The equation for the whole system can be written after assembling

$$-\mathscr{C}^T\mathscr{Q}_p + (\mathscr{K} + \mathscr{A})\mathscr{Q}_v + \mathscr{M}\dot{\mathscr{Q}}_v = \mathscr{F}_p \tag{9.27}$$

$$\mathscr{C}\mathscr{Q}_v = \mathbf{0} \tag{9.28}$$

$\mathscr{Q}_p$ are the pressure unknowns for the system and $\mathscr{Q}_v$ the velocity unknowns (which include the v_1 and v_2 components).

After applying the velocity boundary conditions one generally has a right-hand side for the continuity equation, i.e. (9.28) becomes

$$\mathscr{C}\mathscr{Q}_v = \mathscr{F}_v \tag{9.29}$$

where $\mathscr{C}$ is now the modified matrix but we write it with the same symbol to avoid proliferation of notation. Equally some of the matrices in the momentum equations (9.27) are changed as a result of applying boundary conditions.

Once the boundary conditions have been applied we can solve (9.27) in terms of acceleration, as follows

$$\dot{\mathscr{Q}}_v = \mathscr{M}^{-1}\{\mathscr{F}_p + \mathscr{C}^T\mathscr{Q}_p - (\mathscr{K} + \mathscr{A})\mathscr{Q}_v\} \tag{9.30}$$

We can now differentiate (9.29) with respect to time

$$\mathscr{C}\dot{\mathscr{Q}}_v = \dot{\mathscr{F}}_v \tag{9.31}$$

and substitute (9.30) into it, which gives

$$(\mathscr{C}\mathscr{M}^{-1}\mathscr{C}^T)\mathscr{Q}_p = \mathscr{F} \tag{9.32}$$

where

$$\mathscr{F} = \mathscr{F}_v + \mathscr{C}\mathscr{M}^{-1}(\mathscr{K} + \mathscr{A})\mathscr{Q}_v - \mathscr{C}\mathscr{M}^{-1}\dot{\mathscr{F}}_p \tag{9.33}$$

Once equation (9.33) is solved we can find the velocities by integrating (9.31). This integration may be carried out using the Runge–Kutta, Euler or some other explicit method. If the flow is slow we can also set $\mathscr{A} = 0$ and obtain the *Stokes* flow. This flow is sometimes used as the starting point for the steady state solution for Reynolds numbers different from 0 (i.e. Stokes flow).

This continuous pressure and velocity formulation has been applied to *steady flow* by Hood and Taylor[12] for Reynolds numbers up to 200. The same authors[13] have more recently proposed the use of linear pressure and quadratic velocity fields in order to obtain more accurate results for pressures.

Kawahara *et al.*[14] have solved steady state flow problems using the same approximations as Hood.[13] They apply an improved Newton–Raphson iteration method with which flows for Reynolds numbers of up to 1200 are investigated.

Oden[15] solved unsteady flow problems using a six node interpolation function for both velocities and pressures. His solution scheme will be reviewed in the following example.

9.5 FREE SURFACE FLOW

The 'natural' way of solving transient flows with free surfaces or interfaces is to use a Lagrangian system of coordinates instead of an Eulerian one.[16,17]

The vorticity–stream function formulations are not very convenient for free surface problems as the boundary conditions are difficult to define. Furthermore this formulation becomes very involved in three-dimensional problems. Thus in what follows we will be only concerned with the velocities–pressure formulation.

The use of a Lagrangian system of coordinates reduces the total derivatives such as Dv/Dt to simply $\partial v/\partial t$. The momentum equations are now

$$-\frac{1}{\rho}\frac{\partial p}{\partial x_i} + b_i + \frac{\mu}{\rho}\nabla^2 v_i = \frac{\partial v_i}{\partial t} \qquad (i = 1, 2, 3) \tag{9.34}$$

and the continuity equation for incompressible flow remains

$$e_v = \frac{\partial v_i}{\partial x_i} = 0 \tag{9.35}$$

Time independent boundary conditions are as shown in formulae (9.5) and (9.6).

The variational statement corresponding to equations (9.34) and (9.35) is similar to formula (9.7) but with the total derivatives substituted by time only derivatives; e.g.

$$\iiint \left[\{-p\,\delta e_v - e_v\,\delta p + \tau_{ij}\,\delta e_{ij}\} + \rho\left\{-b_i + \frac{\partial v_i}{\partial t}\right\}\delta v_i \right] \mathrm{d}V = \iint_{S_p} (\bar{p}_n\,\delta v_n + \bar{p}_s\,\delta v_s)\,\mathrm{d}S \tag{9.36}$$

Applying finite elements to this equation, i.e.

$$v_i = \boldsymbol{\phi}^T \mathbf{v}_i^n \qquad p = \boldsymbol{\phi}^T \mathbf{p}^n$$

we obtain, after assembling,

$$\begin{aligned} -\mathscr{C}^T \mathscr{Q}_p + \mathscr{K}\mathscr{Q}_v + \mathscr{M}\dot{\mathscr{Q}}_v &= \mathscr{F}_p \\ \mathscr{C}\mathscr{Q}_v &= \mathbf{0} \end{aligned} \tag{9.37}$$

Note the absence of a (convective) matrix when comparing this with formulae (9.27–28).

Equation (9.37) can be integrated in time as shown in Section 9.4, using explicit methods (Runge–Kutta, Euler, etc) or, owing to the absence of convective terms, we may efficiently use some implicit method.

The Lagrangian mesh will freely distort, which can be inconvenient if the distortions are large. A redefinition of the grid every now and then may, for these cases, be required. This implies the use of a mixed Lagrangian–Eulerian approach, similar to the one discussed in Chapter 8 (see Section 8.4).

REFERENCES

1. SCHLIGHTING, H., *Boundary Layer Theory*, McGraw-Hill, New York (1968)
2. TONG, P., 'The Finite Element Method for Fluid Flow', *Recent Advances in Matrix Method of Structural Analysis and Design*, R. Gallagher *et al.* (Eds.), University of Alabama Press (1971)
3. OLSON, M. D., 'Formulation of a Variational Principle—Finite Element Method for Viscous Flows', *Variational Methods in Engineering*, Vol. I, C. A. Brebbia and H. Tottenham (Eds.), Southampton University Press (1973)
4. DENNIS, S. C. R. and CHANG, G. Z., 'Numerical Solutions for Steady Flow Past a Circular Cylinder at Reynolds Numbers up to 100', *J. Fluid Mechanics*, **42**, Part 3, 471–489 (1970)
5. LIEBER, P., WEN, K. S. and ATTIA, A. V., 'Finite Element Methods as an Aspect of the Principle of Maximum Uniformity', *Proc. Int. Symp. on Finite Element Methods in Flow Problems*, Swansea (Jan. 1974)
6. BAKER, A. J., 'Finite Element Solution Algorithm for Viscous Incompressible Fluid Dynamics', *Int. J. Numerical Methods Engng.*, **6**, No. 1 (1973)
7. GOSMAN, A. D., PUN, W., RUNCHAL, A., SPALDING, D. and WOLFSHTEIN, M., *Heat and Mass Transfer in Recirculating Flows*, Academic Press, London (1969)
8. BAKER, A. J., 'A Highly Stable Explicit Integration Technique for Computational Continuum Mechanics', in *Numerical Methods in Fluid Dynamics*, C. Brebbia and J. J. Connor (Eds.), Pentech Press (1974)
9. SMITH, S. and BREBBIA, C. A., 'Finite Element Solution for Vortex Stress Development', *J. Computational Physics*, **17**, No. 3 (March 1975)
10. LEE, C. H., 'Finite Element Method for Transient Linear Viscous Flow Problems', *Proc. Int. Conf. Numerical Methods Fluid Dynamics*, Southampton (1973)
11. FROMM, J. E. and HARLOW, F. H., 'Numerical Solution of the Problem of Vortex Stress Development', *Phys. Fluids*, **6**, No. 7 (July 1963)

12. TAYLOR, C. and HOOD, P., 'A Numerical Solution of the Navier–Stokes Equations Using the Finite Element Method', *Comput. Fluids J.*, **1,** 73–100 (1973)
13. HOOD, P. and TAYLOR, C., 'Navier–Stokes Equations using Mixed Interpolation', *Proc. Int. Symp. on Finite Element Methods in Flow Problems*, Swansea (Jan. 1974)
14. KAWAHARA, M., YOSHIMURA, N. and NAKAGAWA, K., 'Analysis of Steady Incompressible Viscous Flow', *Proc. Int. Symp. on Finite Elements in Flow Problems*, Swansea (Jan. 1974)
15. ODEN, J. T. and CARTER WELLFORD, L., 'Analysis of Flow of Viscous Fluids by the Finite Element Method', *J.A.I.A.A.*, **10,** No. 12 (1972)
16. HARLOW, F. G. and WELCH, J. E., 'Numerical Calculations of Time-dependent Viscous Incompressible Flow of Fluid with Free Surface', *Phys. Fluids*, **8,** No. 12 (Dec. 1965)
17. HIRT, C. W., COOK, J. L. and BUTLER, T. D., 'A Lagrangian Method for Calculating the Dynamics of an Incompressible Fluid with Free Surface', *J. Comput. Phys.*, **5,** No. 1 (Feb. 1970)

Appendix: Numerical Integration Formulae

1. ONE-DIMENSIONAL GAUSSIAN QUADRATURE

$$I \equiv \int_{-1}^{+1} f(\xi)\, d\xi = \sum_{i=1}^{n} w_i f(\xi_i) + E_n \qquad \text{(A1)}$$

where

w_i = weighting factor
ξ_i = coordinate of the ith integration point
n = total number of integration points
E_n = error = 0 $(d^{2n}f/d\xi^{2n})$

n	i	ξ_i	w_i
1 (linear)	1	0	2
2 (cubic)	1	$+1/\sqrt{3}$	+1
	2	$-1/\sqrt{3}$	+1
3 (quintic)	1	0	8/9
	2	$+\sqrt{15}/5$	5/9
	3	$-\sqrt{15}/5$	5/9
4 (septimal)	1	+0·861 136 31	0·347 854 85
	2	−0·861 136 31	0·347 854 85
	3	+0·339 981 04	0·652 145 15
	4	−0·339 981 04	0·652 145 15

2. TWO- AND THREE-DIMENSIONAL GAUSSIAN QUADRATURE FOR RECTANGLES AND RECTANGULAR HEXAHEDRA

Two-and three-dimensional formulae are obtained by combining one-dimensional formulae according to

$$\int_{-1}^{+1}\int_{-1}^{+1} f(\xi,\eta)\,\mathrm{d}\xi\,\mathrm{d}\eta = \sum_{j=1}^{n}\sum_{i=1}^{n} w_i w_j f(\xi_i,\eta_j) \tag{A2}$$

$$\int_{-1}^{+1}\int_{-1}^{+1}\int_{-1}^{+1} f(\xi,\eta,\zeta)\,\mathrm{d}\xi\,\mathrm{d}\eta\,\mathrm{d}\zeta = \sum_{k=1}^{n}\sum_{j=1}^{n}\sum_{i=1}^{n} w_i w_j w_k f(\xi_i,\eta_j,\zeta_k) \tag{A3}$$

where the weighting factors and integration point coordinates are listed above. The numbering scheme for the cubic and quintic two-dimensional rules are shown below.

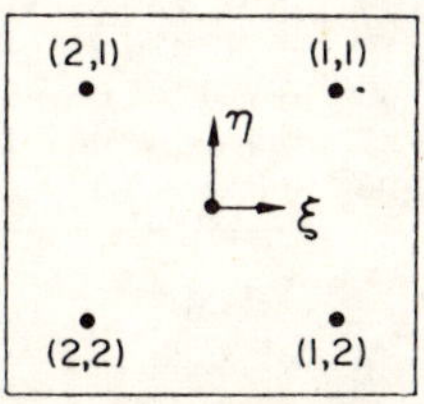

(Cubic)

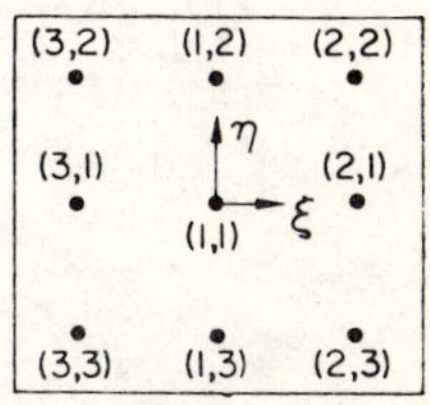

(Quintic)

3. TRIANGULAR DOMAIN

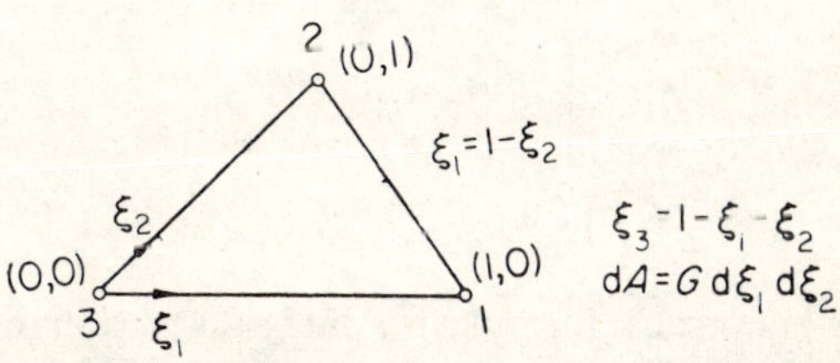

$$I = \int_0^1 \left[\int_0^{1-\xi_2} f(\xi_1,\xi_2,\xi_3)\,\mathrm{d}\xi_1\right]\mathrm{d}\xi_2 = \sum_{i=1}^{n} w_i\, f(\xi_1^i,\xi_2^i,\xi_3^i) \tag{A4}$$

ξ_1, ξ_2, ξ_3 are triangular coordinates (see Section 3.2). The exact integral for a combination of integral powers of the triangular coordinates is

$$I = \int_0^1 \left(\int_0^{1-\xi_2} \xi_1^i \xi_2^j \xi_3^k\,\mathrm{d}\xi_1\right)\mathrm{d}\xi_2 = \frac{i!\,j!\,k!}{(i+j+k+2)!}$$

Quadrature formulae suggested by Hammer *et al.*[1] are listed in the following table.

n	i	ξ_1^i	ξ_2^i	ξ_3^i	w_i
1 (linear)	1	1/3	1/3	1/3	1/2
	1	1/2	1/2	0	1/6
2 (quadratic)	2	0	1/2	1/2	1/6
	3	1/2	0	1/2	1/6
	1	1/3	1/3	1/3	− 9/32
4 (cubic)	2	3/5	1/5	1/5	+25/96
	3	1/5	3/5	1/5	+25/96
	4	1/5	1/5	3/5	+25/96
	1	0·333 333 33	0·333 333 33	0·333 333 33	0·112 500 00
	2	0·797 426 99	0·101 286 51	0·101 286 51	0·062 969 59
	3	0·101 286 51	0·797 426 99	0·101 286 51	0·062 969 59
7 (quintic)	4	0·101 286 51	0·101 286 51	0·797 426 99	0·062 969 59
	5	0·059 715 87	0·470 142 06	0·470 142 06	0·066 197 08
	6	0·470 142 06	0·059 715 87	0·470 142 06	0·066 197 08
	7	0·470 142 06	0·470 142 06	0·059 715 87	0·066 197 08

4. PRISM

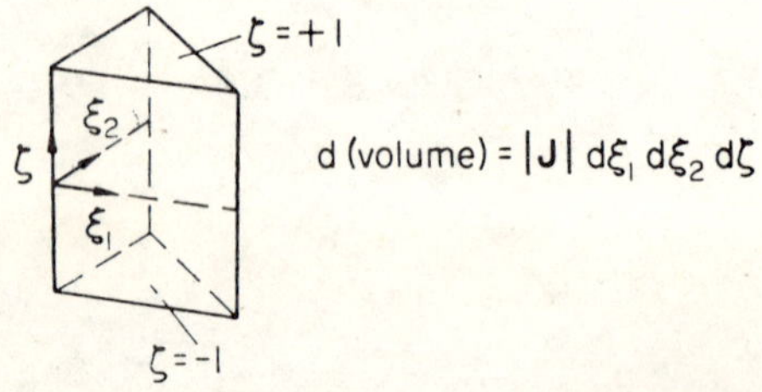

We combine Hammer's formula for integration over the triangular domain with one dimensional gaussian quadrature for the ζ direction.

$$I = \int_{-1}^{+1}\left[\int_0^1\left(\int_0^{1-\xi_2} f\,d\xi_1\right)d\xi_2\right]d\zeta = \sum_{j=1}^{n_\zeta}\sum_{i=1}^{n_t} w_j w_i\, f(\xi_1^i\xi_2^i\xi_3^i\zeta^j) \quad \text{(A5)}$$

For example, the cubic rule corresponds to

$$n_t = 4$$
$$n_\zeta = 2$$

REFERENCES

1. HAMMER, P. C., MARLOVE, O. J., and STROUD, A. H., 'Numerical Integration Over Simplex and Cones'. *Math. Tables and Other Aids to Computation*, Vol. 10 (1956)

BIBLIOGRAPHY

HAMMER, P. C., and STROUD, A. H., 'Numerical Evaluation of Multiple Integrals', *Math. Tables and Other Aids to Computation*, Vol. 12 (1958)

IRONS, B. M., 'Engineering Applications of Numerical Integration in Stiffness Method', *J. IAA*, **4**, 2035–37 (1966)

IRONS, B. M., 'Economical Computer Techniques for Numerically Integrated Finite Elements', *Int. J. Numerical Meth. in Engng*, **1**, 201–203 (1969)

KOPAL, Z., *Numerical Analysis*, 2nd edn, Chapman & Hall (1961)

MILLER, J. C. P., 'Numerical Quadrature Over a Rectangular Domain in Two or More Dimensions', *Mathematics of Computation*, Vol. 14, 13–20 (1960)

Index

RY